THERMAL DESIGN OF HEAT EXCHANGERS

THERMAL DESIGN OF HEAT EXCHANGERS

A NUMERICAL APPROACH:
DIRECT SIZING AND STEPWISE RATING

Eric M. Smith

*Energy Systems,
Dundee, UK*

JOHN WILEY & SONS
Chichester · New York · Weinheim · Brisbane · Singapore · Toronto

Copyright © 1997 by John Wiley & Sons Ltd,
Baffins Lane, Chichester,
West Sussex PO19 1UD, England

National 01243 779777
International (+44) 1243 779777

e-mail (for orders and customer service enquiries): cs-books@wiley.co.uk.
Visit our Home Page on http://www.wiley.co.uk.
or http://www.wiley.com

All Rights Reserved. No part of this publication may be reproduced, stored in a retrieval system, or transmitted, in any form or by any means, electronic, mechanical, photocopying, recording, scanning or otherwise, except under the terms of the Copyright, Designs and Patents Act 1988 or under the terms of a licence issued by the Copyright Licensing Agency, 90 Tottenham Court Road, London, UK W1P 9HE, without the permission in writing of the publisher

Other Wiley Editorial Offices

John Wiley & Sons, Inc., 605 Third Avenue,
New York, NY 10158-0012, USA

VCH Verlagsgesellschaft mbH, Pappelallee 3,
D-69469 Weinheim, Germany

Jacaranda Wiley Ltd, 33 Park Road, Milton,
Queensland 4064, Australia

John Wiley & Sons (Asia) Pte Ltd, 2 Clementi Loop #2-01
Jin Xing Distripark, Singapore 129809

John Wiley & Sons (Canada) Ltd, 22 Worcester Road,
Rexdale, Ontario M9W 1L1, Canada

Library of Congress Cataloging-in-Publication Data

Smith, Eric M.
    Thermal design of heat exchangers : a numerical approach—direct sizing and stepwise rating / Eric M. Smith.
      p. cm.
    Includes bibliographical references and index.
    ISBN o 471 96566 9 (hb : alk. paper)
    1. Heat exchangers—Design and construction. I. Title.
TJ263.S58 1996
621.402'5—dc20   95–54158
                                                                   CIP

British Library Cataloguing in Publication Data

A catalogue record for this book is available from the British Library

ISBN 0 471 96566 9

Typeset in 10/12pt Times by Pure Tech India Limited, Pondicherry
Printed and bound in Great Britain by Bookcraft (Bath) Ltd
This book is printed on acid-free paper responsibly manufactured from sustainable forestation, for which at least two trees are planted for each one used for paper production.

This volume is dedicated to Dorothy my wife
for her unfailing kindness and understanding,
and to my three sons for their consistent support

'If you can build hotter or colder than anyone else,
If you can build higher or faster than anyone else,
If you can build deeper or stronger than anyone else,
If ....
Then, in principle, you can solve all the other problems in between'

(Attributed to Sir Monty Finniston, FRS)

PREFACE

'I would like to extend the way in which you may think about design of heat exchangers...'

(Author, 10th International Heat Transfer Conference, Brighton, 1994)

THE PLACE OF THE VOLUME

Of the many texts available on heat transfer, the majority concentrate on developing correlations for heat transfer and flow friction in a variety of flow situations. This is backed by a fundamental approach, starting from the basic axioms of continuum mechanics, and by a careful examination of experimental results.

Such texts are valuable, but when it comes to consideration of heat exchangers, this topic is often relegated to one chapter with a very sketchy outline of basic theory. The subject of thermal design of heat exchangers deserves fuller treatment, and this text is a contribution to development of the direct-sizing approach in thermal design of the exchanger surface. It does not address mechanical design of the exchanger for which the many International and National Standards should be consulted.

The thrust of this volume is towards numerical methods of performance evaluation, while bringing theory to the point where computer software may be written. Some schematic algorithms are presented in Appendix B to assist in the transition from theory to software. In the course of preparing this volume, approximately four similar-sized volumes containing source listings have been produced, hence the subtitle 'A Numerical Approach'.

The reader will find that the first few chapters on the fundamentals of heat exchanger theory and the concepts of direct sizing provide a gentle development from undergraduate studies. Later chapters, on transient response of heat exchangers and on the related single-blow method of obtaining experimental results, should be of more interest to the practising engineer who may wish to continue his/her engineering development.

HISTORICAL DEVELOPMENT OF THE SUBJECT

Up until the early 1940s virtually all papers employ mean temperature difference as the design parameter; a good collection can be found in the two reference volumes by Jakob (1949, 1957). Around 1942 the method of designing contra/parallel flow heat exchangers was effectively changed by London and Seban (1980) from using LMTD (log mean temperature difference) to using the ϵ-Ntu approach, partly because the LMTD approach did not give explicit results in some elementary cases, and partly because the concept of effectiveness provided a measure of the approach to ultimate performance of the exchanger.

An assessment of design methods prepared by Shah (1982) is explicit in preferring ϵ-Ntu methods to the LMTD approach, but there was no reason to reject LMTD on grounds of non-explicitness in 1942. A simple piece of algebra is sufficient to show that an LMTD-Ntu approach is as explicit as the ϵ-Ntu approach, and both 'rate' and 'energy' equations are seen to exist in the LMTD-Ntu approach.

The consequence has been that since 1942 many important papers have concentrated on expressing results in terms of effectiveness in preference to mean temperature difference, which in this author's view has not been entirely beneficial, particularly in the case of variable thermophysical properties (Soyars 1992), and in the case of cross-flow. Mean temperature difference and effectiveness each have useful roles to play in assessing the performance of heat exchangers and should be used in combination.

With a numerical approach, the case of unmixed/unmixed two-pass cross-flow is examined in some detail in this volume; it has been found that the use of effectiveness alone in design could be misleading. The very comprehensive analytical paper by Baclic (1990), which examines 72 possible configurations for two-pass cross-flow, concentrates on presenting results in terms of effectiveness alone. We have lost understanding by the neglect of mean temperature difference and in not computing temperature sheets; it is time to redress the balance. In numerical work it is natural to compute mean temperature difference and effectiveness together.

DIRECT-SIZING METHODS

Sizing methods have traditionally posed more problems than rating methods. This is because guessing one principal dimension of the exchanger may be necessary before the full performance of the guessed heat exchanger can be evaluated for comparison with design requirements. For the class of heat exchangers in which local geometry of the heat transfer surface is fully representative of the whole geometry, this is no longer necessary. Methods of direct sizing exist which go straight to the optimum size of heat transfer surface, while satisfying all thermal performance constraints.

In direct sizing of the heat exchange surface, the design approach is limited to that class of heat exchangers in which local geometry is fully representative of the complete heat exchange surface. This is not such a severe restriction as might be imagined, for it includes such different designs as compact plate-fin exchangers and the RODbaffle shell-and-tube exchanger. More exotic designs are possible and one of these, the involute-curved tube panel exchanger, has a possible application in future power gas turbines.

As all terminal temperatures may be determined in advance of direct sizing, the necessary input data for complete sizing takes the following form:

- exchanger duty (Q)
- mean temperature difference for heat exchange ($\Delta\theta_m$)
- local geometry on both sides
- mass flow rates of both fluids ($\dot{m}$)
- physical properties for both fluids at mean bulk temperature (Pr, C, η, λ)
- allowable pressure loss data (Δp, p, T_{bulk}, R_{gas})
- physical properties of material of construction (λ, ρ, C)

For the selected geometry, the standard procedure is to evaluate heat transfer performance over the range of valid Reynolds numbers for both sides of the exchanger. This provides a heat transfer curve. Pressure loss performance is similarly evaluated for both sides over the same range of valid Reynolds numbers, providing two separate pressure loss curves. Both pressure loss curves will intersect the heat transfer curve, and the intersection furthest to the right provides the initial design point.

In cases where heat transfer and pressure loss correlations are suitable, a fully algebraic solution may be possible. This approach is adopted for the helical-coil heat exchanger, but the numerical approach is generally preferred because the associated graphics provide a good indication of the possible performance envelope.

LONGITUDINAL CONDUCTION

Techniques for minimising longitudinal conduction effects in both contraflow and cross-flow exchangers are described. Longitudinal conduction reduces exchanger performance and it possible to design out most of the effect in compact heat exchangers by varying local surface geometries.

This is an essential part of the design process for heat exchangers in systems which experience transient temperature disturbances, for longitudinal conduction terms in the set of three simultaneous partial differential equations may then be neglected.

TRANSIENT RESPONSE

The full transient temperature response equations for contraflow and cross-flow are established in some detail, where the concept of *local* Ntu values is found to be an essential part of the analysis. The equations are linear for Newtonian fluids and may be separated into steady-state and perturbance sets. When longitudinal conduction terms are not present, it becomes possible to solve the perturbance set of three partial differential equations using the method of characteristics. The transient temperature response is then added to steady-state values to produce the actual response.

For determining heat transfer and flow friction performance of heat exchanger surfaces the contraflow transient equations can be simplified to the point where they become the single-blow transient test equations. This method of testing matrices is now well established as a reliable method of determining heat exchanger surface performance, providing some simple rules are followed, which may be understood by examining the mathematical analysis and constructing the test rig so as to match closely the mathematical model.

STEPWISE RATING

When changes in thermophysical properties become significant, it is necessary to design by stepwise rating. To start this process, an initial cross-section of the exchanger is required, and direct sizing can be helpful in providing this information. An additional problem exists with multi-stream exchangers, caused by fluids streaming in the same direction having different temperature profiles along the length of the exchanger. This leads to the additional problem of cross-conduction which must be taken into account.

VARIABLE FLUID PROPERTIES

To this point, only single-phase exchangers with constant thermophysical properties have been examined, but fluids which experience changes in thermal capacity, e.g. some phase-change applications, can sometimes be handled by splitting an exchanger into two or more sections in which single-phase behaviour, or indeed different stages of two-phase behaviour, can be assumed to exist. The two final chapters outline some considerations in both stepwise rating and variable fluid properties.

NOTATION

International Standards for nomenclature are generally adopted, with the exception when no definition is available. Hewitt *et al.* (1993) provide good arguments for accepting new notation in a preface to their handbook on process heat transfer. Full listings for each chapter are provided at the start of the appendices.

Much of the difficulty which arises in reading the literature on heat exchangers stems from the way in which temperatures are labelled at the ends of the exchanger. For a heat exchanger under steady-state conditions two possibilities exist:

- Label one end with subscripts 1 and the other end with subscripts 2.
- Label the exchanger to give all inlet temperatures subscript 1 and all outlet temperatures subscript 2.

The second option leads to confusion as one is always referring back to ascertain if the analysis has been correctly assembled. The first option is to be preferred and is used in this text.

APPLICATIONS

The possible applications for exchangers suitable for direct sizing are quite wide, including aerospace, marine propulsion systems, land-based power plant and chemical engineering plant. It is possible to go directly to the optimum exchanger core and minimise the choice of core volume, core mass, core frontal area, etc. Multistream exchangers for cryogenic duty must usually be designed by stepwise rating.

REFERENCES

The reader will find that some references listed at the end of each chapter are not directly mentioned in the text. These are papers which may indicate possible further directions of development, and when reading a chapter it is worth scanning the references for interesting titles. References for this preface are listed below.

MECHANICAL DESIGN

Although it was stated earlier that stressing of exchanger shells and components is specifically excluded from this volume, it can be said in passing that the author also found the correct solution to the problem of creep in thick tubes using numerical methods (Smith 1963, 1965). This is a natural extension of the work of Lamé and Clapeyron (1831) on elastic cylinders, of Duhamel (1838) on thermoelastic cylinders, and of Hill *et al.* (1947) on combined elastic-plastic deformation of cylinders, although in the last paper time-independent plasticity is strictly a thermodynamic impossibility.

ACKNOWLEDGEMENTS

Encouragement has been provided over a number of years by Professor Alan Jeffrey, head of Engineering Mathematics at the University of Newcastle upon Tyne, and it is a pleasure to acknowledge his support.

Some of the material in this volume is developed from papers published by the author, with the permission of the Institution of Chemical Engineers, the Institution of Mechanical Engineers, the American Society of Mechanical Engineers, Rolls-Royce plc, and Elsevier Science Ltd.

Eric M. Smith
August 1995

REFERENCES

Baclic, B. S. (1990) ε-Ntu analysis of complicated flow arrangements, In: *Compact Heat Exchangers – A Festschrift for A. L. London*, Eds. R. K. Shah, A. D. Kraus and D. Metzger, Washington: Hemisphere, pp. 31–90.

Duhamel, J. M. C. (1838) Memoire sur le calcul des actions moleculaires developpées par les changements de température dans les corps solides, *Memoires presentes pars divers savants a l'Academie Royale des Sciences de l'Institut de France*, **5**, 440–498.

Hewitt, G. F., Shires, G. L. and Bott, T. R. (1993) *Process Heat Transfer*, Boca Raton, FL: CRC Press.

Hill, R., Lee, E. H. and Tupper, S. J. (1947) The theory of combined plastic and elastic deformation with particular reference to a thick tube under internal pressure, *Proceedings of the Royal Society, London*, **191 A**, 278–303.

Jakob, M. (1949 and 1957) *Heat Transfer*, Chichester, UK: John Wiley, Vol. I (1949) and Vol. II (1957).

Lamé, G. and Clapeyron, B. P. E. (1833) Memoire sur l'equilibre interieure des corps solides homogenes, *Memoires presentes pars divers savants a l'Academie Royale des Sciences de l'Institut de France*, **4**, 463–562, (esp. 516–525).

London, A. L. and Seban, R. A. (1980) A generalisation of the methods of heat exchanger analysis, *International Journal of Heat and Mass Transfer*, **23**, 5–16. (Release of unpublished 1942 paper.)

Shah, R. K. (1982) Heat exchanger basic design methods, In: *Low Reynolds Number Flow Heat Exchangers*, Eds. S. Kacac, R. K. Shah and A. E. Bergles, Washington: Hemisphere, pp. 21–72.

Smith, E. M. (1963) Primary creep behaviour of thick tubes, *Proceedings of the Institution of Mechanical Engineers, Conference on Thermal Loading and Creep, London 1964*, Vol. 178 Pt. 3L, paper 4, pp. 27–33.

Smith, E. M. (1965) Analysis of creep in cylinders, spheres and thin discs, *Journal of Mechanical Engineering Science*, **7**, 82–92.

Soyars, W. M. (1992) The applicability of constant property analyses in cryogenic helium heat exchangers, *Advances in Cryogenic Engineering*, **37A**, 217–223.

CONTENTS

Preface vii

1 Classification 1

 1.1 Class definition 1
 1.2 Helical-tube multi-start coil 1
 1.3 Involute-curved serpentine-tube panel 2
 1.4 Plate-fin 4
 1.5 RODbaffle 5
 1.6 Helically twisted flattened-tube 6
 1.7 Spirally wire-wrapped 6
 1.8 Bayonet-tube 7
 1.9 Wire-woven heat exchangers 8
 1.10 Porous matrix heat exchangers 8
 1.11 Cryogenic heat exchangers 9
 1.12 Some possible applications 9
 1.13 Exclusions and extensions 13
 References 14

2 Fundamentals 17

 2.1 Simple temperature distributions 17
 2.2 Log mean temperature difference 19
 2.3 LMTD-Ntu rating problem 21
 2.4 LMTD-Ntu sizing problem 23
 2.5 Link between Ntu values and LMTD 24
 2.6 The theta methods 24
 2.7 Effectiveness and number of transfer units 25
 2.8 ε-Ntu rating problem 29
 2.9 ε-Ntu sizing problem 29
 2.10 Comparison of LMTD-Ntu and ε-Ntu approaches 30
 2.11 Sizing when Q is not specified 31

2.12	Most efficient temperature difference in contraflow	32
2.13	Required values of Ntu in cryogenics	36
2.14	To dig deeper	38
2.15	Dimensionless groups	40
	References	50

3 Steady-State Temperature Profiles — 51

3.1	Linear temperature profiles in contraflow	51
3.2	General cases of contraflow and parallel flow	53
3.3	Condensation and evaporation	58
3.4	Longitudinal conduction in contraflow	59
3.5	Mean temperature difference in unmixed cross-flow	65
3.6	Extension to two-pass unmixed cross-flow	70
3.7	Involute-curved plate-fin exchangers	73
3.8	Longitudinal conduction in one-pass unmixed cross-flow	74
3.9	Determined and undetermined cross-flow	83
3.10	Possible optimisation criteria	84
3.11	Cautionary remark about core pressure loss	86
3.12	Mean temperature difference in complex arrangements	87
3.13	Exergy destruction	88
	References	88

4 Direct Sizing of Plate-Fin Exchangers — 91

4.1	Exchanger layup	91
4.2	Flow friction and heat transfer correlations	93
4.3	Plate-fin surface geometries	95
4.4	Direct sizing of an unmixed cross-flow exchanger	97
4.5	Direct sizing of a contraflow exchanger	98
4.6	Fine tuning of rectangular offset strip fins	104
4.7	Optimisation of a contraflow exchanger	106
4.8	Distribution headers	110
4.9	Multi-stream design (cryogenics)	111
4.10	Buffer-zone or leakage plate 'sandwich'	111
4.11	Consistency in design methods	113
4.12	Conclusions	115
	References	117

5 Direct Sizing of Helical-Tube Exchangers — 120

5.1	Design framework	120
5.2	Basic geometry	122
5.3	Simplified geometry	129
5.4	Thermal design	131
5.5	Completion of the design	138

5.6	Thermal design for $t/d=1.346$	141
5.7	Fine tuning	141
5.8	Design for curved tubes	146
5.9	Discussion	151
5.10	Part-load operation with bypass control	153
5.11	Conclusions	153
	References	154

6 Direct Sizing of Bayonet-Tube Exchangers — 156

6.1	Isothermal shell-side conditions	156
6.2	Evaporation	157
6.3	Condensation	168
6.4	Design illustration	169
6.5	Non-isothermal shell-side conditions	171
6.6	Explicit solution	173
6.7	Complete solution	175
6.8	Non-explicit solutions	179
6.9	Pressure loss	181
6.10	Conclusions	184
	References	185

7 Direct Sizing of RODbaffle Exchangers — 187

7.1	Design framework	187
7.2	Configuration of the RODbaffle exchanger	188
7.3	Approach to direct sizing	188
7.4	Characteristic dimensions	189
7.5	Flow areas	190
7.6	Design correlations	191
7.7	Reynolds numbers	191
7.8	Heat transfer	192
7.9	Pressure loss tube-side	193
7.10	Pressure loss shell-side	194
7.11	Direct sizing	196
7.12	Tube-bundle diameter	197
7.13	Practical design	197
7.14	Generalised correlations	201
7.15	Recommendations	203
7.16	Other shell-and-tube designs	203
7.17	Conclusions	204
	References	204

8 Transients in Contraflow Exchangers — 207

8.1	Solution methods	207

	8.2	Fundamental equations	209
	8.3	Analytical considerations	211
	8.4	Method of characteristics	214
	8.5	Laplace transforms with numerical inversion	225
	8.6	Direct solution by finite differences	228
	8.7	Engineering applications	229
	8.8	Conclusions	229
		References	230

9 Transients in Cross-Flow Exchangers — 235

	9.1	Solution methods	235
	9.2	Fundamental equations	235
	9.3	Method of characteristics	238
	9.4	Review of existing solutions	240
	9.5	Engineering applications	241
	9.6	Conclusions	242
		References	242

10 Single-Blow Testing and Regenerators — 244

	10.1	Analytical and experimental background	244
	10.2	Physical assumptions	245
	10.3	Theory	246
	10.4	Relative accuracy of using mathematical outlet-response curves in experimentation	251
	10.5	Choice of test method	255
	10.6	Practical considerations	255
	10.7	Equations with longitudinal conduction	256
	10.8	Regenerators	257
		References	259

11 Cryogenic Heat Exchangers and Stepwise Rating — 263

	11.1	Background	263
	11.2	Liquefaction concepts and components	265
	11.3	Liquefaction of nitrogen	273
	11.4	Hydrogen liquefaction plant	279
	11.5	Preliminary direct sizing of multi-stream heat exchangers	281
	11.6	Stepwise rating of multi-stream heat exchangers	283
	11.7	Future commercial applications	287
	11.8	Conclusions	288
		References	288

12 Variable Heat Transfer Coefficients — 290

	12.1	With and without phase change	290

	12.2	Two-phase flow regimes	291
	12.3	Two-phase pressure loss	292
	12.4	Two-phase heat transfer correlations	296
	12.5	Two-phase design of a double-tube exchanger	298
	12.6	Discussion	302
	12.7	Flow maldistribution	305
	12.8	Some further problems	306
	12.9	Rate processes	306
		References	307

A Transient Equations with Longitudinal Conduction and Wall Thermal Storage — 313

A.1	Temperature transients in contraflow	313
A.2	Temperature transients in unmixed cross-flow	318
A.3	Mass flow and temperature transients in contra-flow	323

B Algorithms and Schematic Source Listings — 326

B.1	Algorithms for mean temperature distribution in one-pass unmixed cross-flow	326
B.2	Schematic source listing for direct sizing of compact cross-flow exchanger	330
B.3	Schematic source listing for direct sizing of compact contraflow exchanger	331
B.4	Parameters for rectangular offset strip-fins	333
B.5	Typical input/output for direct sizing of compact contraflow exchanger	334
B.6	Algorithm for transient response of contraflow exchanger	343
B.7	Splinefitting of data	344

C Optimisation of Rectangular Offset-Strip Plate-Fin Surfaces — 347

C.1	Trend curves	347
C.2	Block volume	348
C.3	Block mass (excluding fluids)	349
C.4	Block length	350
C.5	Frontal area	351
C.6	Total surface area	352

D Performance Data for RODbaffle Exchangers — 353

D.1	Further heat transfer and flow friction data	353

E Evaluation of Single-Blow Outlet Response — 357

E.1	Laplace transforms	357
E.2	Numerical evaluation of outlet response	358

1

CLASSIFICATION

1.1 CLASS DEFINITION

Direct-sizing is concerned with the class of heat exchangers that has consistent geometry throughout the exchanger core, such that local geometry is fully representative of the whole surface. The following designs are included in that class.

1.2 HELICAL-TUBE MULTI-START COIL

This design shown in Fig. 1.1 has no internal baffle leakage problems, it permits uninterrupted cross-flow through the tube bank for high heat transfer coefficients, and it provides advantageous counterflow terminal temperature distribution in the whole exchanger. Some modification to LMTD is necessary when the number of tube-turns is less than about 10 and this analysis has been provided by Hausen (1950, 1983) in both his German and his English texts.

Although exchangers of this type had been in use since the first patents by Hampson (1895) and L'Air Liquide (1934), consistent geometry in the coiled tube-bundle does not seem to have been known before Smith (1960). Since then programmes of work on helical-coil tube-bundles have appeared – Gilli (1965), Smith and Coombs (1972), Smith and King (1978), Gilli (1983) – and a method of direct sizing has been obtained (Smith 1986), which is further reported in this text.

Cryogenic heat exchangers to this design have been built by Linde AG and are illustrated in both editions of Hausen (1950, 1983); further examples can be found in the papers by Bourguet (1972), Abadzic and Scholz (1976) and Weimar and Hartzog (1976). High temperature, nuclear heat exchangers have

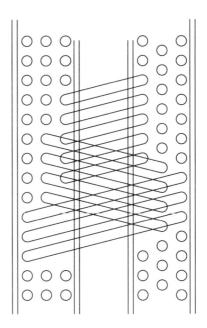

Fig. 1.1 Helical-tube multi-start coil exchanger

been constructed in very large multiple units by Babcock Power Ltd for two AGR reactors (Perrin 1976), and by Sulzer and others for several HTGR reactors (Kalin 1969, Profos 1970, Bachmann 1975, Chen 1978, Anon 1979). A single unit may exceed 18 metres in length and 25 tonnes in mass with a rating of 125 MWt.

The PWR nuclear ship *Otto Hahn* was provided with a helical-coil integral boiler built by Deutsche Babcock (Ulken 1971). For LNG applications, Weimer and Hartzog (1976) report that coiled heat exchangers are preferred for reduced sensitivity to flow maldistribution. Not all of the above heat exchangers have consistent geometry within the tube-bundle.

1.3 INVOLUTE-CURVED SERPENTINE-TUBE PANEL

Related to the helical-tube multi-start coil heat exchanger is the involute-curved serpentine-tube panel design used by Pratt & Whitney in one of their experimental engines powered by liquid hydrogen (Mulready 1979). The conventional tube panel is a single serpentine tube arranged such that the shell-side fluid may flow transversely over the tube as in cross-flow over rows of tubes (Fig. 1.2a).

When each panel is given an involute curve and is placed with others in an annular pattern as in Fig. 1.2b, the pitch between adjacent panels is constant with radius and the shell-side fluid sees the same geometry everywhere.

1.3 INVOLUTE-CURVED SERPENTINE-TUBE PANEL

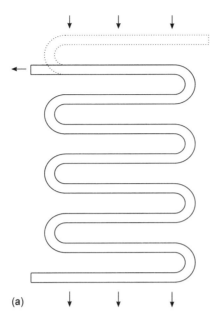

Fig. 1.2(a) Serpentine tube panel

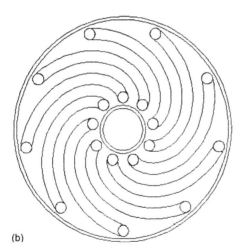

Fig. 1.2(b) involute-curved pattern

Thermal performance of flat serpentine panels has been discussed by Hausen (1950), and will also apply to involute-curved serpentine panels. The only difference is a secondary effect due to involute curvature, which may hardly affect tube-side heat transfer and pressure loss.

4 1 CLASSIFICATION

1.4 PLATE-FIN

The compact plate-fin exchanger is now well known due to the work of Kays and London (1964), London and Shah (1968) and many others. Manufactured in several countries, its principal use has been in cryogenics and aerospace, where high performance with low mass and volume are important. Constructional materials include aluminium alloys, nickel, stainless steel and titanium. The layup is a stack of plates and finned surfaces which are either brazed or diffusion bonded together. Flat plates separate the two fluids, to which the finned surfaces are attached. The finned surfaces are generally made from folded and cut sheet and serve both as spacers separating adjacent plates, and as providers of channels in which the fluids may flow (Fig. 1.3a).

Many types of finned surface have been tested (see e.g. Kays and London 1964); Fig. 1.3b shows an example of a rectangular offset strip-fin surface which is one of the best-performing geometries. The objective is to obtain high heat transfer coefficients without correspondingly increased pressure loss

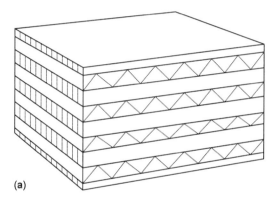

Fig. 1.3(a) Compact plate-fin heat exchanger

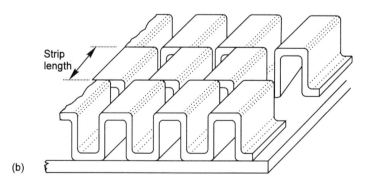

Fig. 1.3(b) Rectangular offset strip-fin surface

penalties. As the strip-fins act as flat plates in the flowing fluid, each new edge starts a new boundary layer which is very thin, thus high heat transfer coefficients are obtained.

1.5 RODBAFFLE

The RODbaffle exchanger is essentially a shell-and-tube exchanger with conventional plate-baffles (segmental or disc-and-doughnut) replaced by grids of rods. Unlike plate-baffles, RODbaffle-sections extend over the full transverse cross-section of the exchanger.

Originally the design was produced to eliminate tube failure due to transverse vortex-shedding induced vibration of unsupported tubes in cross-flow (Eilers and Small 1973) but the new configuration also provided enhanced performance and has been developed further by Gentry (1990) and others.

Only square pitching of the tube-bundle is practicable with rod-baffles, and circular rods are placed between alternate tubes to maintain spacing. To minimise blockage, one set of vertical rods in a baffle-section is placed between every second row of tubes. At the next baffle-section the vertical rods are placed in the alternate gaps between tubes not previously filled at the first baffle-section. The next two baffle-sections have horizontal rod spacers, similarly arranged. Thus each tube in the bank receives support along its length.

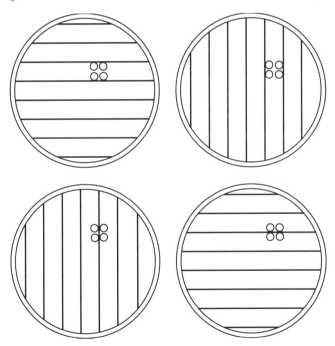

Fig. 1.4 RODbaffle set of four baffles

It might be argued that the RODbaffle geometry is not completely consistent throughout its shell side, and that it should not therefore be included in this study. However, the spacing rods in the shell-side fluid were found to shed von Kármán vortex streets *longitudinally*, vortex streets which persist up to the next baffle rod. Thus as far as the shell-side fluid is concerned, there is consistent geometry in the exchanger, even though the RODbaffles themselves are placed 150 mm apart.

Tube counts are possible for square pitching using the Phadke (1984) approach. Figure 1.4 illustrates the arrangement of baffles in the RODbaffle design.

1.6 HELICALLY-TWISTED FLATTENED-TUBE

This compact shell-and-tube design was developed by Dzyubenko *et al.* (1990) for aerospace use, and it complies with the requirement of consistent local geometry in every respect when triangular pitching is used. The outside of the tube-bundle requires a shield to ensure correct shell-side flow geometry, and the space between the exchanger pressure shell and the shield can be filled with internal insulating material. The performance of this design, illustrated in Fig. 1.5, is discussed thoroughly in the recent textbook by Dzyubenko *et al.* (1990), although its title is somewhat misleading. Tube counts on triangular pitching are possible using the Phadke (1984) approach.

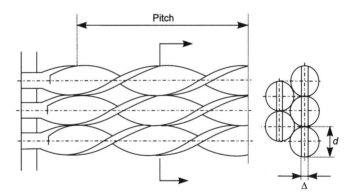

Fig. 1.5 Helically-twisted flattened-tube

1.7 SPIRALLY WIRE-WRAPPED

A further shell-and-tube concept is based on providing spiral wire-wraps to plain tubes – a concept used with nuclear fuel rods. With triangular pitching it is possible to arrange a mixture of right-hand (R), plain (O) and left-hand (L) wire-wraps so as to reinforce mixing in the shell-side fluid. This concept has

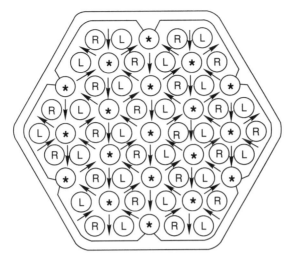

Fig. 1.6 Cross-section of R-O-L spirally wire-wrapped layout

not been tested for heat exchangers, and it does not quite fulfil the requirements of consistent local geometry, as the plain tubes lack the finning effect of the wire-wrap. The cross-section of a tube-bundle is shown in Fig. 1.6, and the wire-wraps approximately extend over the central 90% of the tube length.

The most common spiral wire-wrap configuration is to have *all* nuclear fuel rods with spirals of the same handedness. This leads to opposing streams at the point of closest approach of the rods, and swirling in the truncated triangularly cusped flow principal flow channels. The spiral wrap is slow, about 12–18° to the longitudinal axis of the rod. In several nuclear fuel rod geometries the arrangement of rods does not follow a regular triangular pattern, and correlations need to be assessed accordingly.

The (R-O-L) configuration provides even shell-side fluid distribution and mixing. The Phadke (1984) tube-count method will apply to triangular pitching.

1.8 BAYONET-TUBE

Both bayonet-tube and double-pipe heat exchangers satisfy the concept of consistent shell-side and tube-side geometry; both have been discussed in other works, e.g. Martin (1992). Hurd (1946) appears to be the first to have analysed the performance of the bayonet-tube heat exchanger, but his analysis was not complete and further results are reported in this book. The upper diagram in Fig. 1.7 show a typical exchanger. Practical uses include heating of batch processing tanks, sometimes having vertical bayonet-tubes for condensation of steam in the annuli (Holger 1992), freezing of ground, cooling of cryogenic storage tanks, and high temperature recuperators using silicon carbide tubes.

8 1 CLASSIFICATION

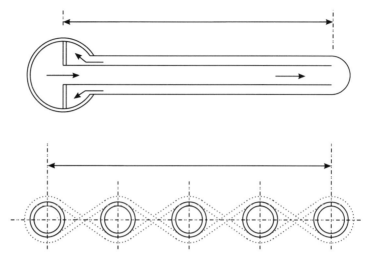

Fig. 1.7 Bayonet-tube exchanger (top) and wire-woven tubes

Residence time of the fluid in the annulus may be extended by adding a spiral wire-wrap to the outside of the inner tube, thus forcing fluid in the annulus to follow a helical path. When this is combined with insulating the inner tube, improved external heat transfer will result.

1.9 WIRE-WOVEN HEAT EXCHANGERS

The concept of fine tubes woven with wire threads into a flat sheet is a recent proposal by Echigo *et al.* (1992). Given the right layout this arrangement could easily qualify for direct sizing. The lower diagram in Fig. 1.7 shows the arrangement.

1.10 POROUS MATRIX HEAT EXCHANGERS

The surface of the porous matrix heat exchanger described by Hesselgreaves (1995) is built-up from flattened sections of perforated plate, or flattened expanded mesh metal, stacked so that each section is offset half a pitch from its immediate neighbours (Fig. 1.8). The fluid flows in and out of the plane of the fins in its passage through the exchanger, coupled with diverging and converging flow, thus creating a three-dimensional flow field in the matrix. Individual plate thicknesses are much thinner than with conventional plate-fin geometries, presently ranging from 0.137 mm to 0.38 mm.

The new geometry offers an increased number of 'flat plate' edges to the flow stream, plus greater cross-sectional area for heat to flow towards the channel separating plates. The layout is thus better configured for heat transfer than conventional plate-fin geometries. An infinite number of geometries are possi-

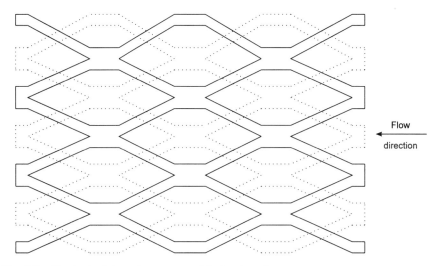

Fig. 1.8 Stacked plates of porous matrix heat exchanger

ble, with the possibility of changing mesh size along the length of the exchanger. Presently only preliminary test results are available, but there is every indication that the pressure loss will be lower and the heat transfer higher than for plate-fin designs.

So far, the flattened expanded mesh plates have been diffusion bonded together in packs of between 6 and 15 layers, with separating plates between streams; this forms a very strong exchanger. As such a construction seems amenable to forming plate-packs with involute curvature, as illustrated in Figs. 1.9 and 1.10, it offers the prospect of constructing a completely bonded two-pass annular flow exchanger. This arrangement could prove suitable for the clean air side of the vehicular gas-turbine application shown in Fig. 1.11.

Sufficient examples of exchangers with a recognisable local geometry have now been given to allow the reader to recognise new types of exchanger which conform to requirements for direct sizing.

1.11 CRYOGENIC EXCHANGERS

The sizing of cryogenic heat exchangers will be discussed in Chapter 11, as they tend to be designed on mean temperature difference, not LMTD. This is because a different approach is required, especially near to the critical point; once that technique has been developed, it is straightforward to apply similar methods to cryogenic designs in which the LMTD concept would be perfectly viable.

1.12 SOME POSSIBLE APPLICATIONS

At this stage it is only possible to indicate a few applications for the heat exchanger configurations described earlier. Not all of this technology is yet in

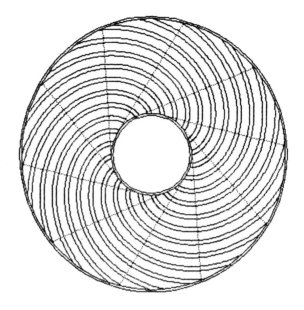

Fig. 1.9 Cross-section of involute-curved plate-fin heat exchanger; plate spacing on the high pressure, cold air side is narrow whereas plate spacing on the low pressure, hot gas side is wide

service, or even constructed, and the reader is simply asked to appreciate some of the possible applications.

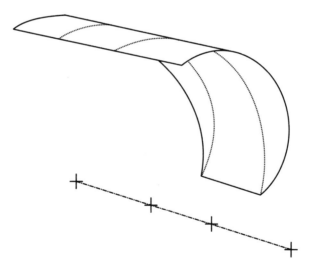

Fig. 1.10 Single curved plate of involute plate-fin exchanger in contraflow with triangular extensions forming inlet and outlet ducts

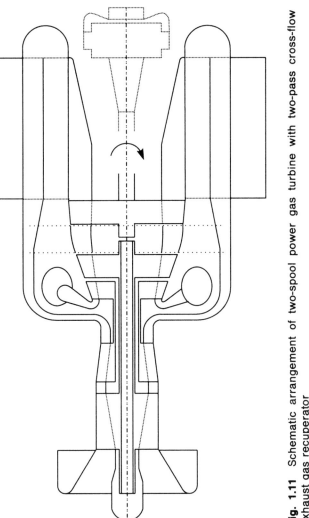

Fig. 1.11 Schematic arrangement of two-spool power gas turbine with two-pass cross-flow exhaust gas recuperator

Propulsion systems

Intercooled and recuperated gas turbine cycles for marine propulsion are presently being developed for the considerable fuel savings that are possible (Cownie 1993, Crisalli and Parker 1993). A contraflow arrangement is always the first choice because it provides more energy recovery than multipass cross-flow, but the practicalities of inlet and outlet ducting have also to be considered.

For the *recuperator* of a smaller gas turbine we might consider a two-pass cross-flow design in which the separating plates are curved to involute form (Fig. 1.9). In the compact vehicle propulsion system of Fig. 1.11, after Collinge (1994), the power turbine exhaust flows outwards through the exchanger core while high pressure combustion air flows axially through the exchanger in two passes. Swirling exhaust gases can be directed by an outlet scroll before entering the exhaust stack. Some development work would be required to realise the involute-curved plate-fin exchanger. Thermal sizing is identical to that for the compact flat-plate design.

However, one problem with the involute exchanger is the difficulty of clearing involute curved channels. Wilson (1995) believes that a rotating ceramic regenerator should be preferred, as it could be more easily cleaned, but it introduces the problem of sliding seals.

For the *recuperator* of a larger gas turbine, a plate-and-frame design with U-type headering has been developed for marine propulsion (Valenti 1995). An *intercooler* is also fitted between LP and HP compressors (Crisalli and Parker 1993, Bannister et al. 1994). This has to be a contraflow plate-fin design for compactness, with freshwater/glycol supply and return from external annular header pipes. The exchanger can be segmented for ease of maintenance. With this arrangement the gas turbine would not become exposed to seawater leaking from a damaged intercooler. Pressure in the closed-loop freshwater/glycol system can be adjusted to suit the desired operating conditions.

Cryogenic storage tank

One problem which has troubled cryogenic and petrochemical industries is roll-over in cryogenic storage tanks. The liquid cryogen is at a higher pressure at the bottom of the tank compared with the surface, due to liquid density. The saturation pressure is thus higher at the bottom of the tank. If there is external heat leak into the tank, it is possible for the liquid at the bottom of the tank to be at a temperature higher than liquid at the top. If liquid is convected from bottom to top, the massive evaporation which ensues at the lower pressure can be sufficient to rupture the tank.

The bayonet-tube heat exchanger requires only single penetration of a pressure vessel. When fitted to the top of a cryogenic storage tank, cryogen in the exchanger cools the contents of the tank and may set up controlled circulation. Colder fluid at the centre of the tank falls to the bottom and warmer fluid at the side walls rises to the free surface (Fig. 1.12). The possible

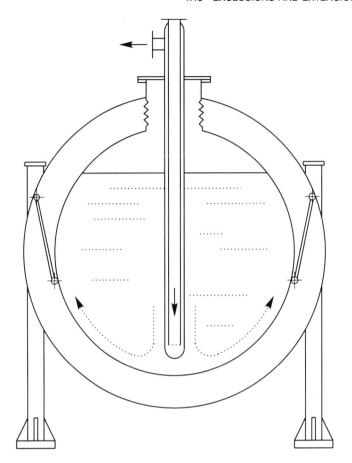

Fig. 1.12 Cryogenic storage tank with bayonet-tube exchanger (Schematic)

effectiveness of the bayonet-tube exchanger in inhibiting roll-over seems worthy of investigation.

It is hoped that the above long-range concepts may stimulate the reader to consider other arrangements for heat exchangers that can be directly sized.

1.13 EXCLUSIONS AND EXTENSIONS

Exclusions

Not every heat exchanger design is considered in this volume, for the objective is to proceed via steady-state direct sizing, through optimisation, to the study of transients.

Segmentally baffled and disc-and-doughnut baffled shell-and-tube designs are not included because the exchanger core does not have sufficiently regular flow geometry. However, there have been some attempts to develop a direct-sizing approach for these exchangers, and they are covered in Chapter 7.

Also excluded is a study of the spiral heat exchanger, because it does not seem suitable for direct sizing. Papers analysing performance of this exchanger design have been published by Bes and Roetzel (1991, 1992, 1993).

The exclusion of these designs is not a criticism of their usefulness; in the appropriate application they may be more economic, or more suitable for corrosive or fouling service.

Extensions

The plate-and-frame heat exchanger is not specifically considered because steady-state design follows standard contraflow or parallel-flow procedures. It is only necessary to source sets of heat transfer and flow friction correlations before proceeding.

Plate-and-frame designs can be similar in flow arrangement to plate-fin designs, but there is restriction on the headering geometry. Optimisation may proceed in a similar way as for compact plate-fin heat exchangers but is likely to be less comprehensive until universal correlations for the best plate-panel corrugations become available. The text by Hewitt *et al.* (1994) provides an introduction to steady-state design using plates with standard corrugations, and it contains further references. The paper by Focke (1985) considers asymmetrically corrugated plates.

Inlet and return headering for plate-and-frame designs, and the same arrangement for plate-fin designs, may add a phase shift to the outlet transient response following an inlet disturbance. Effects of this headering arrangement have been considered by Das and Roetzel (1995). Discussion of the work of Roetzel and co-workers transients in heat exchangers comes towards the end of Chapters 8 and 9.

Faster response is obtained with U-type headering than with Z-type headering, and the choice of U-type headering is evident in the paper by Crisalli and Parker (1993) describing a recuperated gas-turbine plant using plate-fin heat exchangers.

REFERENCES

Abadzic, E. E. and Scholz, H. W. (1972) Coiled tubular heat exchangers, *Proceedings of the 1972 Cryogenic Engineering Conference, Advances in Cryogenic Engineering* **18**, Paper B-1, 42–51.

Adderley, C. and Hallgren, L. H. (1994) The design and manufacture of diffusion bonded plate-fin heat exchangers, *The Industrial Sessions Papers, 10th International Heat Transfer Conference, Brighton, UK, 14–18 August 1994*, Institution of Chemical Engineers, UK.

Anon (1979) The high temperature reactor and process applications, *Proceedings of an International Conference, British Nuclear Energy Society, London, 26–28 November 1979*. (See also IAEA Specialist meeting, Julich, 1979 in *Nuclear Engineering International* March 1980, p. 24.)

Bachmann, U. (1975) Steam generators for the 300-MWe power station with a thorium high-temperature reactor, *Sulzer Technical Review*, **57**(4), 189–194.

Bannister, R. L., Cheruvu, N. S., Little, D. A. and McQuiggan, G. (1994) Turbines for the turn of the century, *Mechanical Engineering* **116**(6), 68–75.

Bes, Th. and Roetzel, W. (1991) Approximate theory of the spiral heat exchanger, In: *Design and Operation of Heat Exchangers*, Eds. W Roetzel, P. J. Heggs, and D. Butterworth, *Proceedings of the EUROTHERM Seminar no. 18, Hamburg, 27 February–1 March 1991*, Berlin: Springer Verlag, pp. 223–232.

Bes, Th. and Roetzel, W. (1992) Distribution of heat flux density in spiral heat exchangers, *International Journal of Heat and Mass Transfer*, **35**(6), 1331–1347.

Bes, Th. and Roetzel, W. (1993) Thermal theory of the spiral heat exchanger, *International Journal of Heat and Mass Transfer*, **36**(3), 765–773.

Bourguet, J. M. (1972) Cryogenics technology and scaleup problems of very large LNG plants, *Proceedings of the 1972 Cryogenic Engineering, Conference, Advances in Cryogenic Engineering*, **18**, Paper A-2, 9–26.

Chen, Y. N. (1978) General behaviour of flow induced vibrations in helical tube bundle heat exchangers, *Sulzer Technical Review, Special Number 'NUCLEX 78'*, 59–68.

Collinge, K. (1994) Lycoming AGT1500 powers the M1 Abrams, *IGTI Global Gas Turbine News*, May, 4–7.

Cownie, J. (1993) Aerospace 90 years on, *Professional Engineering*, **6**(11), 17–19.

Crisalli, A. J. and Parker, M. L. (1993) Overview of the WR-21 intercooled recuperated gas turbine engine system. A modern engine for a modern fleet, *International Gas Turbine and Aeroengine Congress and Exposition, Cincinnati, 24–27 May 1993. ASME Paper 93-GT-231*.

Das, S. K. and Roetzel, W. (1995) Dynamic analysis of plate heat exchangers with dispersion in both fluids (plate-and-frame exchanger), *International Journal of Heat and Mass Transfer*, **38**(6), 1127–1140.

Dzyubenko, B. V., Dreitser, G. A. and Ashmantas, L-V. A. (1990) *Unsteady Heat and Mass Transfer in Helical Tube Bundles* Hemisphere.

Echigo, R., Yoshida, H., Hanamura, K. and Mori, H. (1992) Fine-tube heat exchanger woven with threads, *International Journal of Heat and Mass Transfer* **35**(3), 711–717.

Eilers, J. F. and Small, W. M. (1973) Tube vibration in a thermosiphon reboiler, *Chemical Engineering Progress*, **69**(7), 57–61.

Focke, W. W. (1985) Asymmetrically corrugated plate heat exchanger plates (plate-and-frame exchanger), *International Communications in Heat and Mass Transfer*, **12**(1), 67–77.

Gentry, C. C. (1990) RODbaffle heat exchanger technology, *Chemical Engineering Progress*, **86**(7), 48–57.

Gill, G. M., Harrison, G. S. and Walker, M. A. (1983) Full scale modelling of a helical boiler tube, *International Conference on Physical Modelling of Multi-Phase Flow*, BHRA Fluid Engineering Conference, April, Paper K4, pp. 481–500.

Gilli, P. V. (1965) Heat transfer and pressure drop for crossflow through banks of multistart helical tubes with uniform inclinations and uniform longitudinal pitches, *Nuclear Science and Engineering*, **22**, 298–314.

Hausen, H. (1950) Kreuzstrom in Verbindung mit Parallelstrom im Kreuzgegenstromer, *Wärmeübertragung im Gegenstrom, Gleichstrom und Kreuzstrom*, Berlin: Springer Verlag, pp. 213–228.

Hausen, H. (1983) Crossflow combined with parallel flow in the cross-counterflow heat exchanger, *Heat Transfer in Counter Flow, Parallel Flow and Cross Flow*, New York: McGraw-Hill, pp. 232–248.

Hesselgreaves, J. (1995) Concept proving of a novel compact heat exchanger surface, *Proceedings of the 4th National Conference on Heat Transfer, Manchester, 26–27 September 1995*. IMechE Paper C5109/082/95, pp. 479–486.

Hewitt, G. F., Shires, G. L., and Bott, T. R. (1994) *Process Heat Transfer*, CRC Press.

Holger, M. (1992) *Heat Exchangers*, Washington: Hemisphere.

Hurd, N. L. (1946) Mean temperature difference in the field or bayonet tube, *Industrial & Engineering Chemistry*, **38**(12), 1266–1271.

Kalin, W. (1969) The steam generators of Fort St. Vrain nuclear power plant, USA, *Sulzer Technical Review*, **52**(3), 167–174, Special number 'NUCLEX 1969', 12–34.

Kays, W., and London, A. L. (1964) Compact Heat Exchangers, New York: McGraw-Hill, 2nd edn.

L'Air Liquide (1934) Improvements relating to the progressive refrigeration of gases, *British Patent* 416 096.

London, W. L. and Shah, R. K. (1982) Offset rectangular plate-fin surfaces – heat transfer and flow-friction characteristics *ASME Journal of Engineering for Power* **90**, 218–228.

MacDonald, C. F. (1996) Compact buffer zone plate-fin IHX—the key component for high-temperature nuclear process heat realisation with advanced MHR, *Applied Thermal Engineering*, **16**(1), 3–32.

Martin, H. (1992) *Heat Exchangers*, Washington: Hemisphere.

Mulready, R. C. (1979) Liquid hydrogen engines, *DFVLR International Symposium: Hydrogen in Air Transportation, Stuttgart, 11–14 September 1979*.

Perrin, A. J. (1976) Hartlepool and Heysham pod boilers, *Nuclear Engineering International*, **21**(239), 48–51.

Phadke, P. S. (1984) Determining tube counts for shell-and-tube exchangers, *Chemical Engineering*, September, 65–68.

Profos, O. (1970) The new steam generating system for the French nuclear power station EL-4, *Sulzer Technical Review*, **53**(2), 69–83.

Smith, E. M. (1960) The geometry of multi-start helical coil heat exchangers, *Nuclear Research Centre*, London: C. A. Parsons & Co. Ltd. *Internal Report: NRC 60–121*, December, 1–17.

Smith, E. M. (1986) Design of helical-tube multi-start coil heat exchangers, *Advances in Heat Exchanger Design, ASME Winter Annual Meeting, Anaheim CA, 7–12 December 1986, ASME Publication HTD* **66**, 95–104.

Smith, E. M. and Coombs, B. P. (1972) Thermal performance of cross-inclined tube bundles measured by a transient technique, *Journal of Mechanical Engineering Science*, **14**(3), 205–220.

Smith, E. M. and King, J. L. (1978) Thermal performance of further cross-inclined in-line and staggered tube banks, *6th International Heat Transfer Conference, Toronto*, Paper HX-14, pp. 276–272.

Ulken, D. (1971) N. S. Otto Hahn, *Transactions of the Institute of Marine Engineers*, **83**, 65–83.

Valenti, M. (1995) A turbine for tomorrow's navy (recuperated WR-21), *Mechanical Engineering*, **117**(9), 70–73.

Weimer, R. F. and Hartzog, D. G. (1972) Effects of maldistribution on the performance of multistream multi-passage heat exchangers, *Advances in Cryogenic Engineering*, **18**(B-2), 52–64.

Wilson, D. G. (1995) Automotive gas turbines: government funding and the way ahead, *IGTI Global Gas Turbine News*, **35**(4), 17–21.

2

FUNDAMENTALS

2.1 SIMPLE TEMPERATURE DISTRIBUTIONS

In considering overall heat transfer coefficients, the temperature difference is assumed to remain constant across an elementary length of surface. In general, temperatures of the fluids change over the length of the surface, and a principal method of classifying heat exchangers is according to the directions of fluid flow on each 'side' and the effect upon the temperatures in the system (Figs. 2.1 to 2.4).

The first design problem is to determine the mean temperature difference for heat transfer between the two fluids, so as to be able to use the design equation

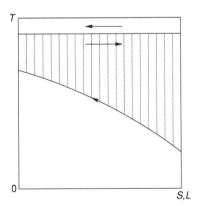

Fig. 2.1 Condenser profile: fluid at constant temperature gives up heat to a colder fluid whose temperature increases

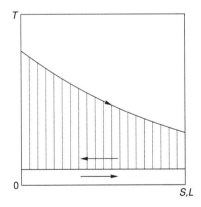

Fig. 2.2 Evaporator profile: fluid at constant temperature receives heat from a hotter fluid whose temperature decreases

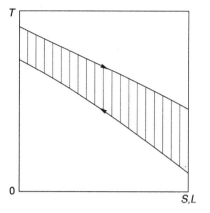

Fig. 2.3 Contraflow profile: fluids flow in opposite directions; more efficient use of temperature difference allows colder fluid to exit at higher temperature than the hotter fluid

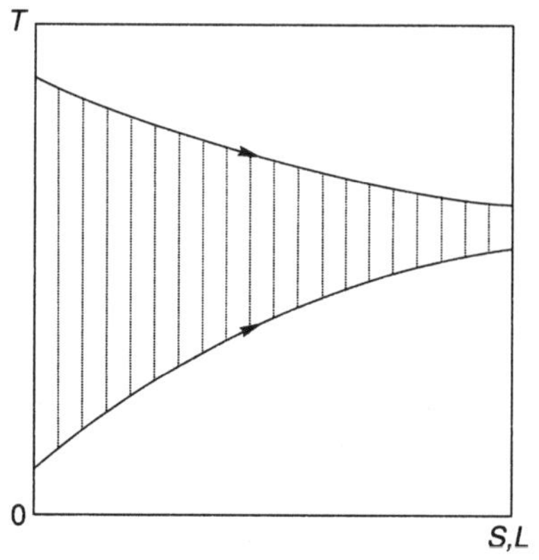

Fig. 2.4 Parallel flow profile: both fluids flow in the same direction, one increasing in temperature, the other decreasing in temperature

$$Q = US\Delta\theta_m \tag{2.1}$$

where $\Delta\theta_m$ is the mean temperature between fluids.

It is essential to watch out for *phase changes*, or situations in which *enthalpy changes* of single-phase fluids do not change linearly with temperature (as in some cryogenic heat exchangers). In the desuperheating *feed heater* (Fig. 2.5) terminal temperature differences may be positive, but temperature crossover might be present if each section is not designed separately (Fig. 2.6).

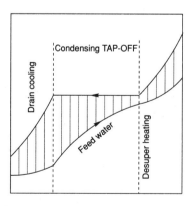

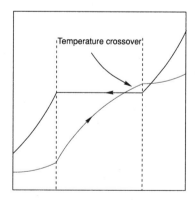

Fig. 2.5 Desuperheating feed heater with phase changes

Fig. 2.6 Temperature crossover in feed heater

Cryogenic recuperators can exhibit similar behaviour when one fluid at a pressure just above its critical value is cooled around its critical temperature, against the flow of a colder fluid at lower pressure well away from its critical point.

The contraflow feed heater may be properly designed as three separate exchangers. The cryogenic exchanger may require incremental design along its length. For the simplest exchangers, the problem reduces either to *rating* an existing design, or *sizing* a new exchanger.

RATING	SIZING
Given: geometry $\dot{m}_h$, C_h, T_{h1}, Δp_h $\dot{m}_c$, C_c, T_{c2}, Δp_c	Given: Q (duty) $\dot{m}_h$, C_h, T_{h1}, Δp_h $\dot{m}_c$, C_c, T_{c2}, Δp_c
Find: Q (duty)	Find: geometry

Note: The allowable pressure losses (Δp_h, Δp_c) have a role to play in design which will be explored in later chapters.

Two design approaches which can be used are the LMTD-Ntu method and the ϵ-Ntu method. Notation is explained in the sections which follow.

2.2 LOG MEAN TEMPERATURE DIFFERENCE

Consider parallel and contraflow heat exchangers (Figs. 2.7 and 2.8) assuming

- fluids with steady mass flow rates ($\dot{m}_h$, $\dot{m}_c$)
- constant overall heat transfer coefficient (U)
- constant specific heats (C_h, C_c)
- negligible heat loss to surroundings

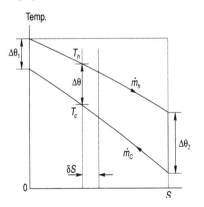

Fig. 2.7 Contraflow

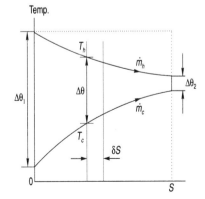

Fig. 2.8 Parallel flow

2 FUNDAMENTALS

For an element of area δS, then

$$\delta Q = U \delta S \Delta \theta_m \tag{2.2}$$

Also, considering each fluid

$$\delta Q = \dot{m}_h \, C_h \, (-\delta T_h) = \dot{m}_c \, C_c \, (\pm \delta T_c) \tag{2.3}$$

Now increment in temperature difference $\Delta \theta$ is

$$\delta(\Delta \theta) = \delta(T_h - T_c) = (\delta T_h - \delta T_c)$$

$$= \left[\frac{-1}{\dot{m}_h \, C_h} - \frac{\pm 1}{\dot{m}_c \, C_c} \right] \delta Q = \left[\frac{-1}{\dot{m}_h \, C_h} - \frac{\pm 1}{\dot{m}_c \, C_c} \right] U \delta S \, \Delta \theta \tag{2.4}$$

Thus

$$\frac{\delta(\Delta \theta)}{\Delta \theta} = \left[\frac{-1}{\dot{m}_h \, C_h} - \frac{\pm 1}{\dot{m}_c \, C_c} \right] U \delta S$$

Integrating between stations 1 and 2

$$\ln \frac{\Delta \theta_2}{\Delta \theta_1} = \left[\frac{-1}{\dot{m}_h \, C_h} - \frac{\pm 1}{\dot{m}_c \, C_c} \right] US \tag{2.5}$$

also integrating equation (2.4) between stations 1 and 2

$$\Delta \theta_2 - \Delta \theta_1 = \left[\frac{-1}{\dot{m}_h \, C_h} - \frac{\pm 1}{\dot{m}_c \, C_c} \right] Q \tag{2.6}$$

Eliminating the square bracket term between equations (2.5) and (2.6)

$$Q = US \, \frac{(\Delta \theta_2 - \Delta \theta_1)}{\ln(\Delta \theta_2 / \Delta \theta_1)}$$

and comparing this expression with equation (2.1), $\Delta \theta_m$ can be written as the *logarithmic mean temperature difference*, $\Delta \theta_{lmtd}$

$$\Delta \theta_{lmtd} = \frac{(\Delta \theta_2 - \Delta \theta_1)}{\ln(\Delta \theta_2 / \Delta \theta_1)} \tag{2.7}$$

In real exchangers the mean temperature difference $\Delta \theta_m$ may not necessarily equal the mathematical LMTD expression based on four terminal temperatures. This is true, for example, when specific heats vary along the length of exchanger. In practice, physical values of mean temperature difference should always be employed.

Equation (2.7) is also correct for condensers and evaporators. In the special case of contraflow with equal water equivalents ($\dot{m}_h C_h = \dot{m}_c C_c$), equation (2.7) is indeterminate, because $\Delta\theta_1 = \Delta\theta_2$, and the temperature profiles are parallel straight lines.

$$\Delta\theta_{lmtd} = \Delta\theta_1 = \Delta\theta_2 \tag{2.8}$$

2.3 LMTD-Ntu RATING PROBLEM

CONTRAFLOW LMTD-Ntu rating	PARALLEL FLOW LMTD-Ntu rating
Given: geometry $\dot{m}_h$, C_h, T_{h1} $\dot{m}_c$, C_c, T_{c2}	Given: geometry $\dot{m}_h$, C_h, T_{h1} $\dot{m}_c$, C_c, T_{c1}
Find: Q (duty)	Find: Q (duty)

$$Q = US\Delta\theta_{lmtd} = \dot{m}_h C_h(T_{h1} - T_{h2})$$
$$= \dot{m}_c C_c(T_{c1} - T_{c2}) \tag{2.9}$$

$$Q = US\Delta\theta_{lmtd} = \dot{m}_h C_h(T_{h1} - T_{h2})$$
$$= \dot{m}_c C_c(T_{c2} - T_{c1}) \tag{2.9}$$

Since the geometry is defined (Figs. 2.9 and 2.10) the surface area S is known. Also hot-side, solid-wall and cold-side heat transfer coefficients α_h, α_c and $\alpha_w = (\lambda_w/t_w)$ may be evaluated from heat transfer correlations and physical properties, giving the overall heat transfer coefficient U.

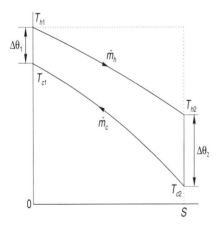

Fig. 2.9 Contraflow profiles

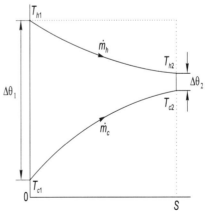

Fig. 2.10 Parallel flow profiles

$$\frac{1}{U} = \frac{1}{\alpha_h} + \frac{1}{\alpha_w} + \frac{1}{\alpha_c}$$

all coefficients are referred to the same reference surface area S.
For duty Q, proceed as follows:

From equation (2.9)

$$\Delta\theta_{lmtd} = \frac{T_{h1} - T_{h2}}{N_h} = \frac{T_{c1} - T_{c2}}{N_c} \quad (2.10)$$

where $N_h = \dfrac{US}{\dot{m}_h C_h}$ and $N_c = \dfrac{US}{\dot{m}_c C_c}$

Then

$$T_{c1} = \frac{N_c}{N_h}(T_{h1} - T_{h2}) + T_{c2} \quad (2.11)$$

$$\Delta\theta_{lmtd} = \frac{\Delta\theta_1 - \Delta\theta_2}{\ln(\Delta\theta_1/\Delta\theta_2)}$$

$$= \frac{(T_{h1} - T_{c1}) - (T_{h2} - T_{c2})}{\ln(\Delta\theta_1/\Delta\theta_2)}$$

The first bracket in the denominator may be written

$$\left[T_{h1} - \frac{N_c}{N_h}(T_{h1} - T_{h2}) - T_{c2}\right]$$

hence

$$\Delta\theta_{lmtd} = \frac{(T_{h1} - T_{h2})(1 - N_c/N_h)}{\ln(\Delta\theta_1/\Delta\theta_2)} \quad (2.12)$$

Equating (2.10) and (2.12)

$$\frac{(T_{h1} - T_{h2})}{N_h} = \frac{(T_{h1} - T_{h2})(1 - N_c/N_h)}{\ln(\Delta\theta_1/\Delta\theta_2)}$$

$$\ln\left(\frac{\Delta\theta_1}{\Delta\theta_2}\right) = (N_h - N_c) \quad (2.13)$$

From equation (2.9)

$$\Delta\theta_{lmtd} = \frac{T_{h1} - T_{h2}}{N_h} = \frac{T_{c2} - T_{c1}}{N_c} \quad (2.10)$$

where $N_h = \dfrac{US}{\dot{m}_h C_h}$ and $N_c = \dfrac{US}{\dot{m}_c C_c}$

Then

$$T_{c2} = \frac{N_c}{N_h}(T_{h1} - T_{h2}) + T_{c1} \quad (2.11)$$

$$\Delta\theta_{lmtd} = \frac{\Delta\theta_1 - \Delta\theta_2}{\ln(\Delta\theta_1/\Delta\theta_2)}$$

$$= \frac{(T_{h1} - T_{c1}) - (T_{h2} - T_{c2})}{\ln(\Delta\theta_1/\Delta\theta_2)}$$

The second bracket in the denominator may be written

$$\left[T_{h2} - \frac{N_c}{N_h}(T_{h1} - T_{h2}) - T_{c1}\right]$$

hence

$$\Delta\theta_{lmtd} = \frac{(T_{h1} - T_{h2})(1 + N_c/N_h)}{\ln(\Delta\theta_1/\Delta\theta_2)} \quad (2.12)$$

Equating (2.10) and (2.12)

$$\frac{(T_{h1} - T_{h2})}{N_h} = \frac{(T_{h1} - T_{h2})(1 + N_c/N_h)}{\ln(\Delta\theta_1/\Delta\theta_2)}$$

$$\ln\left(\frac{\Delta\theta_1}{\Delta\theta_2}\right) = (N_h + N_c) \quad (2.13)$$

From (2.10) and (2.13) the LMTD-Ntu equation pairs for contraflow and parallel flow are

Contraflow		Parallel flow
$\dfrac{\Delta\theta_1}{\Delta\theta_2} = \dfrac{T_{h1} - T_{c1}}{T_{h2} - T_{c2}} = \exp(N_h - N_c)$	'rate'	$\dfrac{\Delta\theta_1}{\Delta\theta_2} = \dfrac{T_{h1} - T_{c1}}{T_{h2} - T_{c2}} = \exp(N_h + N_c)$
$\dfrac{\Delta T_c}{\Delta T_h} = \dfrac{T_{c1} - T_{c2}}{T_{h1} - T_{h2}} = \dfrac{N_c}{N_h}$	'energy'	$\dfrac{\Delta T_c}{\Delta T_h} = \dfrac{T_{c2} - T_{c1}}{T_{h1} - T_{h2}} = \dfrac{N_c}{N_h}$

Given (N_h, T_{h1}) and (N_c, T_{c2}) for contraflow, or (N_h, T_{h1}) and (N_c, T_{c1}) for parallel flow, numerical values may be inserted in the simultaneous LMTD-Ntu equations to find the two unknown temperatures. (In a slightly different form, these equations were anticipated by Clayton (1984), whose paper was not seen until this text was in its final editing stage.)

Explicit algebraic solutions for the unknowns in contraflow can be written

$$T_{h2} = \dfrac{(a-1)T_{h1} - (b-1)T_{c2}}{(a-1) - (b-1)} \qquad T_{c1} = \dfrac{a(b-1)T_{h1} - b(a-1)T_{c2}}{(b-1) - (a-1)}$$

where $a = N_c/N_h$ and $b = \exp(N_h - N_c)$, with similar expressions for parallel flow. Exchanger duty can now be determined from $Q = US\Delta\theta_{lmtd}$.

2.4 LMTD-Ntu SIZING PROBLEM

CONTRAFLOW LMTD-Ntu sizing	PARALLEL FLOW LMTD-Ntu sizing
Given: Q(duty) $\dot{m}_h, C_h, T_{h1}$ $\dot{m}_c, C_c, T_{c2}$	Given: Q(duty) $\dot{m}_h, C_h, T_{h1}$ $\dot{m}_c, C_c, T_{c1}$
Find: geometry	Find: geometry

Writing energy balances

$$Q = US\Delta\theta_{lmtd} = \dot{m}_h C_h (T_{h1} - T_{h2}) \qquad Q = US\Delta\theta_{lmtd} = \dot{m}_h C_h (T_{h1} - T_{h2})$$
$$= \dot{m}_c C_c (T_{c1} - T_{c2}) \qquad\qquad\qquad = \dot{m}_c C_c (T_{c2} - T_{c1})$$

2.5 LINK BETWEEN Ntu VALUES AND LMTD

For the contraflow heat exchanger

$$Q = US\Delta\theta_{lmtd} = \dot{m}_h C_h (T_{h1} - T_{h2}) = \dot{m}_c C_c (T_{c1} - T_{c2})$$

$$\Delta\theta_{lmtd} = \frac{\dot{m}_h C_h}{US}(T_{h1} - T_{h2}) = \frac{\dot{m}_c C_c}{US}(T_{c1} - T_{c2}) \qquad (2.14)$$

$$\Delta\theta_{lmtd} = \frac{T_{h1} - T_{h2}}{N_h} = \frac{T_{c1} - T_{c2}}{N_c}$$

Since $\frac{a}{b} = \frac{c}{d} = \frac{a-c}{b-d}$ $\left(\frac{a+c}{b+d}\right.$ for parallel flow$\left.\right)$, then

$$\Delta\theta_{lmtd} = \frac{(T_{h1} - T_{h2}) - (T_{c1} - T_{c2})}{N_h - N_c} = \frac{(T_{h1} - T_{c1}) - (T_{h2} - T_{c2})}{N_h - N_c} = \frac{\Delta\theta_1 - \Delta\theta_2}{N_h - N_c} \qquad (2.15)$$

but

$$\Delta\theta_{lmtd} = \frac{\Delta\theta_1 - \Delta\theta_2}{\ln(\Delta\theta_1/\Delta\theta_2)} \qquad (2.16)$$

Combining equations (2.15) and (2.16) gives

$$\Delta\theta_{lmtd} = \frac{T_{h1} - T_{h2}}{N_h} = \frac{T_{c1} - T_{c2}}{N_c} = \frac{\Delta\theta_1 - \Delta\theta_2}{N_h - N_c} = \frac{\Delta\theta_1 - \Delta\theta_2}{\ln(\Delta\theta_1/\Delta\theta_2)}$$

which includes the relationship $N_h - N_c = \ln(\Delta\theta_1/\Delta\theta_2)$. Thus explicit expressions for LMTD may be written in the form given by Spalding (1990).

$$\Delta\theta_{lmtd} = \left|\frac{\Delta\theta_1 - \Delta\theta_2}{N_h - N_c}\right| \qquad \Delta\theta_{lmtd} = \left|\frac{\Delta\theta_1 - \Delta\theta_2}{N_h + N_c}\right|$$

contraflow parallel flow

2.6 THE THETA METHODS

Alternative methods of representing the performance of heat exchangers may be found in Hewitt et al. (1994). The basic parameter, theta, was

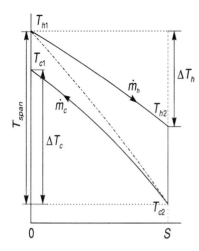

Fig. 2.11 Contraflow

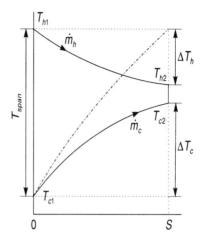

Fig. 2.12 Parallel flow

devised by Taborek (1983), where in the notation of this text (Figs. 2.11 and 2.12)

$$\Theta = \frac{\Delta\theta_m}{T_{span}} \text{ where } \Delta\theta_m \text{ is mean temperature difference}$$

The theta method is related to associated E and P methods by the expressions

$$\Theta = \frac{\Delta\theta_m}{T_{span}} = \frac{P}{\text{Ntu}_{cold}} = \frac{E}{\text{Ntu}_{min}}$$

and the relationships between parameters are often presented in graphical form. However they all depend on finding $\Delta\theta_m$ or $\Delta\theta_{lmtd}$.

2.7 EFFECTIVENESS AND NUMBER OF TRANSFER UNITS

Considering contraflow and parallel-flow exchangers (Figs. 2.11 and 2.12) the assumptions remain the same as in Section 2.2.

Define $\epsilon = \Delta T_c/T_{span}$ or $\epsilon = \Delta T_h/T_{span}$ whichever is the greater, and assume $\dot{m}_c C_c < \dot{m}_h C_h$, then $N_c > N_h$.

$\Delta T_c = T_{c1} - T_{c2}$	$\Delta T_c = T_{c2} - T_{c1}$
$T_{span} = T_{h1} - T_{c2}$	$T_{span} = T_{h1} - T_{c1}$
$\epsilon = \dfrac{\Delta T_c}{T_{span}} = \dfrac{T_{c1} - T_{c2}}{T_{h1} - T_{c2}}$ (2.17)	$\epsilon = \dfrac{\Delta T_c}{T_{span}} = \dfrac{T_{c2} - T_{c1}}{T_{h1} - T_{c1}}$ (2.17)

Equations (2.17) may be written

$$\exp(N_h - N_c)$$
$$= \frac{(T_{h1} - T_{c2}) - (T_{c1} - T_{c2})}{(T_{h1} - T_{c2}) - (T_{h1} - T_{h2})}$$
$$= \frac{1 - \epsilon}{1 - \left(\frac{T_{h1} - T_{h2}}{T_{h1} - T_{c2}}\right)}$$
$$= \frac{1 - \epsilon}{1 - \epsilon(N_h/N_c)}$$

Equations (2.17) may be written

$$\exp(N_h + N_c)$$
$$= \frac{(T_{h1} - T_{c1})}{(T_{h1} - T_{c1}) - (T_{C2} - T_{c1}) - (T_{h1} - T_{h2})}$$
$$= \frac{1}{1 - \epsilon - \left(\frac{T_{h1} - T_{h2}}{T_{h1} - T_{c1}}\right)}$$
$$= \frac{1}{1 - \epsilon(1 + N_h/N_c)}$$

Solving for effectiveness

$$\epsilon = \frac{1 - \exp(N_h - N_c)}{1 - (N_h/N_c)\exp(N_h - N_c)} \quad (2.18)$$

Contraflow

$$\epsilon = \frac{1 - \exp[-(N_h - N_c)]}{1 + (N_h/N_c)} \quad (2.18)$$

Parallel flow

These equations may be expressed in alternative form by writing

$$W = \frac{\dot{m}_c C_c}{\dot{m}_h C_h} = \frac{N_h}{N_c}$$ (it is necessary to have $W < 1$) and writing $N_c = \text{Ntu}$, then

$$\epsilon = \frac{1 - \exp[-\text{Ntu}(1 - W)]}{1 - W\exp[-\text{Ntu}(1 - W)]} \quad (2.19)$$

Contraflow

$$\epsilon = \frac{1 - \exp[-\text{Ntu}(1 + W)]}{1 + W} \quad (2.19)$$

Parallel flow

The exponentials are written with negative exponents so that limiting values for effectiveness may be obtained

As $S \to \infty$
Ntu $\to \infty$ (2.20)
$\epsilon_{\text{lim}} \to 1$

As $S \to \infty$
Ntu $\to \infty$ (2.20)
$\epsilon_{\text{lim}} \to 1/(1 + W)$

2.7 EFFECTIVENESS AND NUMBER OF TRANSFER UNITS

A special case occurs with *Contraflow* when the water equivalents of the two fluids are equal (Fig. 2.13). Then $\dot{m}_c C_c = \dot{m}_h C_h$ and $W = 1$, and the temperature profiles are parallel straight lines (proof in Section 3.1).

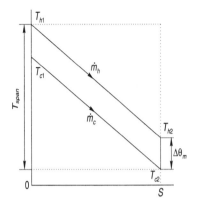

Fig. 2.13 Contraflow

Directly $\Delta\theta_m = T_{h1} - T_{c1}$

and $\epsilon = \dfrac{\Delta T_c}{T_{span}} = \dfrac{T_{c1} - T_{c2}}{T_{h1} - T_{c2}}$

$\text{Ntu} = \dfrac{US}{\dot{m}_c C_c} = \dfrac{T_{c1} - T_{c2}}{\Delta\theta_m} = \dfrac{T_{c1} - T_{c2}}{T_{h1} - T_{c1}}$

$\text{Ntu} + 1 = \dfrac{T_{h1} - T_{c2}}{T_{h1} - T_{c1}}$ thus

$$\boxed{\epsilon = \dfrac{\text{Ntu}}{1 + \text{Ntu}} \quad \text{with} \quad (\text{Ntu} = N_h = N_c)} \quad (2.21)$$

Special case

For *condensation* the condition $\dot{m}_c C_c < \dot{m}_h C_h$ holds for the definition of Ntu, and the effectiveness solution for either contraflow or parallel flow applies (Fig. 2.14).

For *evaporation* the condition $\dot{m}_h C_h < \dot{m}_c C_c$ holds for the definition of Ntu, and the same effectiveness solution for either contraflow or parallel flow applies (Fig. 2.15).

For both cases assume first that a small temperature difference exists in the phase-change fluid. This is not far from the truth because pressure loss always exists in fluid flow. Then

$$\dot{m}_c C_c (T_{c2} - T_{c1}) = \dot{m}_h C_h \delta T_h \qquad \dot{m}_h C_h (T_{h1} - T_{h2}) = \dot{m}_c C_c \delta T_c$$

$$= \dot{m}_h \left(\dfrac{\partial h_h}{\partial T_h}\right)_p \delta T_h \qquad \qquad = \dot{m}_c \left(\dfrac{\partial h_c}{\partial T_c}\right)_p \delta T_c$$

28   2 FUNDAMENTALS

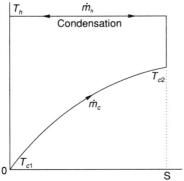

Fig. 2.14 Condensation

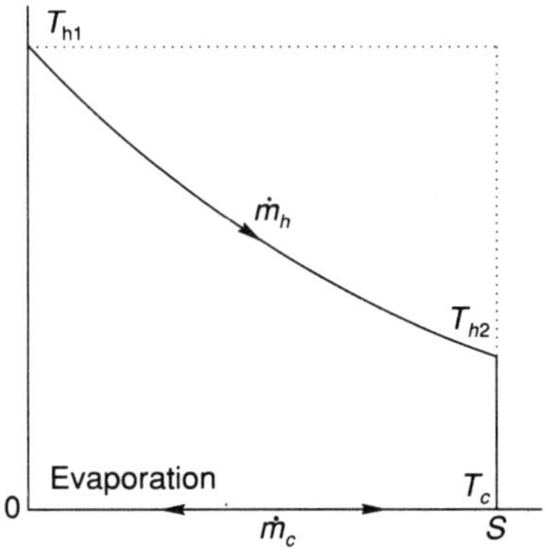

Fig. 2.15 Evaporation

as $\delta T_h \to 0$, $C_h \to \infty$, and
$\dot{m}_c C_c (T_{c2} - T_{c1}) = \dot{m}_h \times h_{fg_h}$

thus $W = \dfrac{\dot{m}_c C_c}{\dot{m}_h C_h} \to 0$ and

$\epsilon = 1 - e^{-\text{Ntu}}$

(2.22)

as $\delta T_c \to 0$, $C_c \to \infty$, and
$\dot{m}_h C_h (T_{c1} - T_{c2}) = \dot{m}_c \times h_{fg_c}$

thus $W = \dfrac{\dot{m}_h C_h}{\dot{m}_c C_c} \to 0$ and

$\epsilon = 1 - e^{-\text{Ntu}}$

(2.22)

Equations (2.22) for effectiveness holds equally for condensation or evaporation.

Results (2.18), (2.19) and (2.22) are presented graphically in Fig. 2.16. To maintain reasonable terminal temperature differences, the value $N_h + N_c$ for

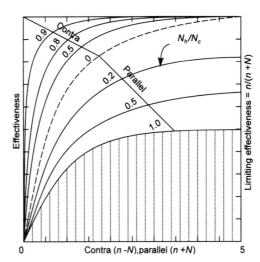

Fig. 2.16 Generalised effectiveness plot with parameter N_h/N_c for contraflow and parallel-flow heat exchangers

parallel flow should not exceed about 4.0, thus for equal values, $N_h = N_c = 2.0$. In a later section it will be shown that for cross-flow with both fluids unmixed, equal values of Ntu should not exceed about $N_h = N_c = 4.0$. For contraflow, comfortable equal values of Ntu lie well above 10.0.

2.8 ε-Ntu RATING PROBLEM

CONTRAFLOW ε-Ntu rating	PARALLEL FLOW ε-Ntu rating
Given : geometry $\dot{m}_h, C_h, T_{h1}$ $\dot{m}_c, C_c, T_{c2}$	Given : geometry $\dot{m}_h, C_h, T_{h1}$ $\dot{m}_c, C_c, T_{c1}$
Find: Q (duty)	Find: Q (duty)

Referring to figures in Section 2.4, for $\dot{m}_c C_c < \dot{m}_h C_h$ the values

$$\text{Ntu} = \frac{US}{\dot{m}_c C_c} \quad \text{and} \quad W = \frac{\dot{m}_c C_c}{\dot{m}_h C_h} \quad \text{apply, thus}$$

$$\epsilon = \frac{1 - e^{-Ntu(1-W)}}{1 - We^{-Ntu(1-W)}} \qquad\qquad \epsilon = \frac{1 - e^{-Ntu(1+W)}}{1+W}$$

$$= \frac{\Delta T_c}{T_{span}} \text{ (for } \dot{m}_c C_c < \dot{m}_h C_h\text{)} \qquad = \frac{\Delta T_c}{T_{span}} \text{ (for } \dot{m}_c C_c < \dot{m}_h C_h\text{)}$$

$$= \frac{T_{c1} - T_{c2}}{T_{h1} - T_{c2}} \qquad\qquad = \frac{T_{c2} - T_{c1}}{T_{h1} - T_{c1}}$$

Hence T_{c1} may be found, then T_{h2} | Hence T_{c2} may be found, then T_{h2}

$$Q = \dot{m}_c C_c (T_{c1} - T_{c2}) \qquad\qquad Q = \dot{m}_c C_c (T_{c2} - T_{c1})$$

2.9 ε-Ntu SIZING PROBLEM

CONTRAFLOW ε-Ntu sizing	PARALLEL FLOW ε-Ntu sizing
Given : Q (duty) $\dot{m}_h, C_h, T_{h1}$ $\dot{m}_c, C_c, T_{c2}$	Given : Q (duty) $\dot{m}_h, C_h, T_{h1}$ $\dot{m}_c, C_c, T_{c1}$
Find: geometry	Find: geometry

For $\dot{m}_c C_c < \dot{m}_h C_h$, $W = \dot{m}_c C_c/(\dot{m}_h C_h)$ and unknown temperatures may be found immediately.

$$Q = \dot{m}_h C_h(T_{h1} - T_{h2})$$

$$= \dot{m}_c C_c(T_{c2} - T_{c1})$$

$$\epsilon = \frac{\Delta T_c}{T_{span}} \quad \text{(for } \dot{m}_c C_c < \dot{m}_h C_h\text{)}$$

$$= \frac{T_{c1} - T_{c2}}{T_{h1} - T_{c2}}$$

Then solve for Ntu from

$$\epsilon = \frac{1 - \exp[-\text{Ntu}(1 - W)]}{1 - W \exp[-\text{Ntu}(1 - W)]}$$

$$Q = \dot{m}_h C_h(T_{h1} - T_{h2})$$

$$= \dot{m}_c C_c(T_{c1} - T_{c2})$$

$$\epsilon = \frac{\Delta T_c}{T_{span}} \quad \text{(for } \dot{m}_c C_c < \dot{m}_h C_h\text{)}$$

$$= \frac{T_{c2} - T_{c1}}{T_{h1} - T_{c1}}$$

Then solve for Ntu from

$$\epsilon = \frac{1 - \exp[-\text{Ntu}(1 + W)]}{1 + W}$$

Since $\dot{m}_c C_c < \dot{m}_h C_h$, the value of Ntu = $US/(\dot{m}_c C_c)$ gives the product US, leaving the geometry to be determined as in Section 2.4.

2.10 COMPARISON OF LMTD-Ntu AND ϵ – Ntu APPROACHES

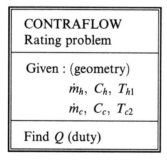
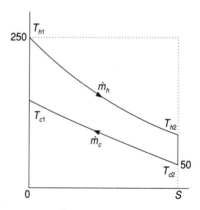

Fig. 2.17 Contraflow profiles

Consider the problem of rating a contraflow exchanger (Fig. 2.17) with surface area $S = 40.0 \text{ m}^2$, and for which the overall heat transfer coefficient $U = 100 \text{ J/(kg K)}$ has been estimated from appropriate correlations by taking physical properties at the assumed mean bulk temperature $(T_{h1} + T_{c2})$, then with

2.11 SIZING WHEN Q IS NOT SPECIFIED

Mass flow rates (kg/s)	Inlet temperatures (°C)	Specific heats (J/kg K)
$\dot{m}_h = 1.60$	$T_{h1} = 250$	$C_h = 1000$
$\dot{m}_c = 1.00$	$T_{c2} = 50$	$C_c = 4000$

LMTD-Ntu approach

$$N_h = \frac{US}{\dot{m}_h C_h} = \frac{100 \times 40}{1.6 \times 1000} = 2.5$$

$$N_c = \frac{US}{\dot{m}_c C_c} = \frac{100 \times 40}{1.0 \times 4000} = 1.0$$

$$\frac{\Delta\theta_1}{\Delta\theta_2} = \frac{T_{h1} - T_{c1}}{T_{h2} - T_{c2}} = \exp(N_h - N_c)$$

$$\frac{250 - T_{c1}}{T_{h2} - 50} = \exp(2.5 - 1.0) = 4.4817$$

$$T_{c1} - 4.4817 \times T_{h2} = 474.08 \quad \text{(a)}$$

$$\frac{\Delta T_c}{\Delta T_h} = \frac{T_{c1} - T_{c2}}{T_{h1} - T_{h2}} = \frac{N_c}{N_h}$$

$$\frac{T_{c1} - 50}{250 - T_{h2}} = \frac{1.0}{2.5} = 0.4$$

$$T_{c1} + 0.4 \times T_{h2} = 150.0 \quad \text{(b)}$$

Subtracting equations (a) from (b)

$$4.0817 \times T_{h2} = 324.084$$

ε-Ntu approach

$$\text{Ntu} = \frac{US}{\dot{m}_h C_h} = \frac{100 \times 40}{1.6 \times 1000} = 2.5$$

$$W = \frac{\dot{m}_h C_h}{\dot{m}_c C_C} = 0.4$$

$$\varepsilon = \frac{1 - \exp[-\text{Ntu}(1 - W)]}{1 - W\exp[-\text{Ntu}(1 - W)]}$$

$$\varepsilon = \frac{1 - \exp[-2.5(1 - 0.4)]}{1 - 0.4\exp[-2.5(1 - 0.4)]}$$

$$\varepsilon = 0.853$$

$$\varepsilon = \frac{\Delta T_h}{T_{span}} = \frac{T_{h1} - T_{h2}}{T_{h1} - T_{c2}}$$

$$0.853 = \frac{250 - T_{h2}}{250 - 50}$$

$$T_{h2} = 79.4\,°C$$

$$\dot{m}_c C_c(T_{c1} - T_{c2}) = \dot{m}_h(T_{h1} - T_{h2})$$

$$4000(T_{c1} - 50) = 1600(250 - 79.4)$$

$$T_{c1} = 118.24\,°C$$
$$T_{h1} = 79.40\,°C$$

2.11 SIZING WHEN Q IS NOT SPECIFIED

Occasionally a sizing problem may arise when Q is not known but 'best practicable recuperation' is desired (Fig. 2.18)

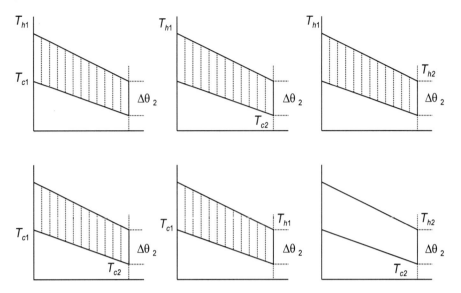

Fig. 2.18 Six possible cases when Q is unknown

The normal procedure is to determine the end of the exchanger at which the temperature pinch point occurs, and give $\Delta\theta$ an appropriate value, say 4 °C. With two known terminal temperatures it is then possible to solve the heat balance equations for the unknown temperatures directly in five out of six possible cases.

The remaining case arises when the pinch point is at the same end as the two known temperatures. When an exchanger of this type is required, there is usually a limiting temperature for one of the unknowns. If this is not a restriction, an alternative approach may be to determine the limiting value of effectiveness, e.g.

$$\epsilon_{\lim} = 1 \text{ for Contraflow} \qquad \epsilon_{\lim} = \frac{1}{1+W} \text{ for Parallel flow}$$

and obtain the actual effectiveness from $\epsilon = f\, \epsilon_{\lim}$ where from practical experience $0.7 < f < 0.9$ approximately. Alternatively, and preferably, it is possible to proceed along lines suggested by Grassman and Kopp, discussed in the next section.

2.12 MOST EFFICIENT TEMPERATURE DIFFERENCE IN CONTRAFLOW

The paper by Grassman and Kopp (1957) presents the following analysis based on minimising exergy loss for temperatures below the dead state T_0. Above the dead state the same relationships apply.

2.12 OPTIMUM TEMPERATURE DIFFERENCE

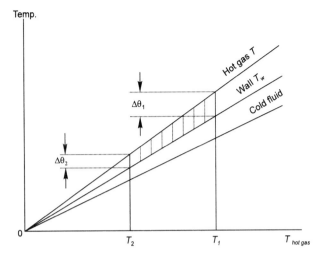

Fig. 2.19 Schematic temperature profiles

Consider the hot fluid side of the heat exchanger depicted in Fig. 2.19. Over a differential length of the exchanger δx (x is not shown above), the energy δq is transferred from the hot fluid to the wall. Using Carnot efficiencies this leads to the following expressions for differential exergy changes.

Below T_0

$$\delta E = \delta q \left(\frac{T - T_0}{T} \right) \quad \text{for fluid}$$

$$\delta E_w = \delta q \left(\frac{T_w - T_0}{T_w} \right) \quad \text{for wall}$$

As $T_w < T$, then $\delta E_w > \delta E$

The net exergy loss is

$$\delta E_{loss} = \delta E_w - \delta E = \delta q T_0 \left(\frac{T - T_w}{T_w T} \right)$$

Above T_0

$$\delta E = \delta q \left(\frac{T_0 - T}{T} \right) \quad \text{for fluid}$$

$$\delta E_w = \delta q \left(\frac{T_0 - T_w}{T_w} \right) \quad \text{for wall}$$

As $T_w < T$, then $\delta E_w < \delta E$

The net exergy loss is

$$\delta E_{loss} = \delta E - \delta E_w = \delta q T_0 \left(\frac{T - T_w}{T_w T} \right)$$

Writing $\Delta \theta = T - T_w$ then $\delta E_{loss} = \delta q T_0 \frac{\Delta \theta}{(T - \Delta \theta) T}$

Over the whole exchanger

$$E_{loss} = T_0 \int_{T_2}^{T_1} \frac{\Delta \theta}{(T - \Delta \theta) T} \, dq \quad (2.23)$$

since $\delta q = C\,\delta T$ then

$$E_{loss} = T_0 \int_{T_2}^{T_1} \frac{C\,\Delta\theta}{(T-\Delta\theta)T}\,dq \tag{2.24}$$

and the problem reduces to the variational problem: find a function $\Delta\theta = \phi(T)$ such that the integral in equation (2.24) is a minimum, subject to the condition that the surface area S has a fixed value.

For specific heat C constant

$$E_{loss} = T_0 \int_{T_2}^{T_1} \frac{\Delta\theta}{(T-\Delta\theta)T}\,dT = T_0\,C\left[\int_{T_2}^{T_1} \frac{dT}{(T-\Delta\theta)} - \int_{T_2}^{T_1} \frac{dT}{T}\right] \tag{2.25}$$

and it becomes evident that $\Delta\theta = \phi(T)$ should be as small as practicable, but not zero which would be for zero heat exchange.

Values of $\Delta\theta$ must also satisfy the energy balance equation

$$\delta Q = \dot{m}\,C\,\delta T = -\alpha\,\delta S(T - T_w) = -\alpha\,\delta S\,\Delta\theta \tag{2.26}$$

where the negative sign is present because T reduces as S increases going from T_1 to T_2 (Fig. 2.19). With heat transfer coefficient α also constant, equation (2.26) may be integrated to the energy equation

$$\frac{\alpha S}{\dot{m}C} = -\int_{T_1}^{T_2} \frac{dT}{\Delta\theta} \tag{2.27}$$

Bearing in mind the desirability of solving equation (2.25) easily, subject to the constraint of equation (2.27), the simplest choices for the temperature difference function are

$$\left.\begin{array}{l}\Delta\theta = a \\ \Delta\theta = bT\end{array}\right\} \text{ where } a \text{ and } b \text{ are constants}$$

Assume that cooling of a perfect gas takes place from 290 K to 20 K. For the heat transfer to be the same in both cases, the energy balance requires

$$\frac{\alpha S}{\dot{m}C} = \int_{20}^{290} \frac{dT}{a} = \int_{20}^{290} \frac{dT}{bT}$$

2.12 OPTIMUM TEMPERATURE DIFFERENCE

$$\frac{1}{a}(290 - 20) = \frac{1}{b}\ln\left(\frac{290}{20}\right)$$

$$\alpha = (100.9667) \times b \qquad (2.28)$$

Choose $a = 2.5$, then $b = 0.02476$.

Evaluating the exergy loss in each case, using equation (2.25), when $\Delta\theta = a = 2.5$

$$E_{loss} = T_0 C \left[\int_{20}^{290} \frac{dT}{T-a} - \int_{20}^{290} \frac{dT}{T} \right] = 0.124\,87 \times T_0 C$$

When $\Delta\theta = bT = 0.02476 \times T$

$$E_{loss} = T_0 C \left[\frac{b}{1-b} \int_{20}^{290} \frac{dT}{T} \right] = T_0\, C\left(\frac{b}{1-b}\right)\ln\left(\frac{290}{20}\right) = 0.076\,89 \times T_0 C$$

and the exergy loss is roughly halved for the same heat transfer.

Grassmann and Kopp (1957) carried out a more sophisticated analysis using variational calculus, and found that the only non-trivial solution was $\Delta\theta = bT$. When they made heat transfer coefficient a function of temperature, too, the influence of temperature was very weak, thus it is reasonable to assume that

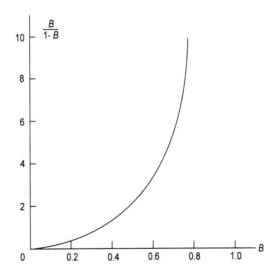

Fig. 2.20 Variation of $B/(1 - B)$ with B

exergy loss is approximately proportional to $b/(1-b)$, and is to be minimised, see Appendix F.

So far $\Delta\theta$ has been taken as the temperature difference between one fluid and the exchanger wall, but it can equally be taken as the temperature difference between the two fluids, say $\Delta\theta = BT$. For the plot of $B/(1-B)$ versus B (Fig. 2.20) a reasonable value for cryogenics is $B = 0.05$ giving

$$\Delta\theta = T/20, \text{ (typically)} \quad (2.29)$$

The value 20 may be increased for exchangers operating at temperatures above T_0.

2.13 REQUIRED VALUES OF Ntu IN CRYOGENICS

The problem of lifting energy at cryogenic temperatures is best illustrated by Fig. 2.21, which makes use of the basic expressions for Carnot efficiency above and below the dead state to define three regions: cryogenics, heat pumps and engines.

The work required to lift energy from cryogenic temperatures and reject it to the dead state places a premium on achieving the best possible heat exchange conditions.

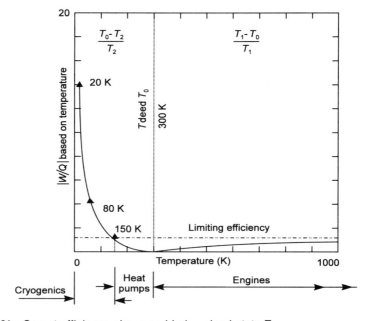

Fig. 2.21 Carnot efficiency above and below dead state T_0

2.13 REQUIRED VALUES OF Ntu IN CRYOGENICS

A thermophysically efficient means of creating a refrigerating stream is to expand high pressure gas in an inward radial-flow cryoturbine. The temperature reduction so caused depends on the isentropic index $\gamma = C_p/C_v$ of the expanded gas. For the cryogenic gases of interest – H_2, He, Ne, N_2, O_2, Ar – γ is around 1.66 for monatomic gases and around 1.4 for diatomic gases (see Chapter 11).

To avoid loss from shock waves, a limiting Mach number of 1 in the inlet nozzles of the cryoturbine produces limiting pressure ratios of around 10:1 and 6:1 for monatomic and diatomic gases, respectively, which correspond to maximum temperature reduction ratios of around 1:2 and 2:3, respectively. In cryogenic practice, desirable pressure ratios are about one-third of the limiting values.

This information, together with the result obtained by Grassman and Kopp that the temperature difference in cryogenic heat exchange should be proportional to absolute temperature (e.g. $\Delta\theta = T/20$), allows evaluation of the levels of Ntu required in cryogenic heat exchange.

Let us assume there is cooling of high pressure nitrogen product stream by a second refrigerating stream of nitrogen first expanded in a cryoturbine so that linearisation of the H–T curve for each fluid is a reasonable assumption, both streams being sufficiently far away from the critical point. The outlet stream from the cryoexpander is at 100 K and is rewarmed to 150 K while cooling the product stream.

Terminal temperature differences are 150/20 = 7.5 and 100/20 = 5.0, providing inlet and outlet temperatures of 157.5 K and 105 K for the product stream being cooled.

Applying the 'rate' equation

$$\frac{\Delta\theta_1}{\Delta\theta_2} = \frac{T_{h1} - T_{c1}}{T_{h2} - T_{c2}} = \exp(N_h - N_c)$$

$$N_h - N_c = 0.405\,465 \tag{2.30}$$

Applying the 'energy' equation

$$\frac{\Delta T_c}{\Delta T_h} = \frac{T_{c1} - T_{c2}}{T_{h1} - h_{h2}} = \frac{N_c}{N_h}$$

$$N_c = 0.952\,381 \times N_h \tag{2.31}$$

Solving simultaneous equations (2.30) and (2.31), $N_h = 8.5148$ and $N_c = 8.1093$; these values must be achieved or exceeded, indicating use of high performance surfaces. With these overall values of Ntu, the resulting temperature profiles are as shown in Fig. 2.22, providing the specific heats of both fluids remain constant.

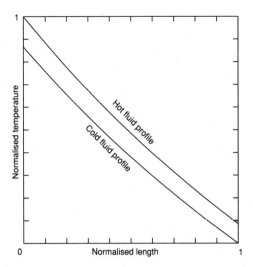

Fig. 2.22 Normalised temperature profiles for contraflow with nearly constant temperature difference

2.14 TO DIG DEEPER

It is useful to think a little more carefully about Figs. 2.1 to 2.4 before proceeding to the rest of the text, as a number of features may be seen by careful observation.

1. With the condenser and the evaporator, it does not matter whether the constant temperature fluid flows to the right or to the left. In a real exchanger, e.g. one of shell-and-tube design, this also allows the condensing or evaporating fluid to flow at right angles to the tubes and to the indicated flow directions shown, without changing the temperature field.
2. By comparing Figs. 2.1 and 2.3 (the condenser and contraflow pair), it becomes obvious that, as the hotter fluid in the contraflow exchanger increases in water equivalent beyond all bounds $\dot{m}C \to \infty$, the condenser is a limiting case of contraflow. The same is true for the parallel-flow exchanger (Fig. 2.4 can be flipped about its vertical axis).

    Similarly, by comparing Figs. 2.2 and 2.4, the evaporator is a limiting case of parallel flow. Thus all four exchanger configurations are closely related, and this observation is expressed formally in the general effectiveness plot (Fig. 2.16). The reader may like to think about where the condenser and evaporator might fit into this diagram.
3. The curved temperature profiles in Figs. 2.1 and 2.2 can be flipped about their vertical axes without changing the concept. Shifting the origin can be helpful in simplifying mathematics cf. Figs. 3.9 and 3.10 on condensation and evaporation). In Chapter 6, on bayonet-tube exchangers, shift-

ing the origin from one end of the exchanger to the other greatly simplifies the mathematics for the isothermal and non-isothermal cases.
4. Anything connected with temperatures and temperature differences involves rate processes, which are usually governed by exponentials. Exponentials should be expected in the solutions to most of the cases examined in this text. For neatness, whenever possible, the final expressions are given as dimensionless ratios.
5. Figures 2.1 to 2.4 can be drawn with the vertical scales corresponding to real temperatures and the horizontal scales corresponding to either exchanger length or surface area (the class of exchanger examined here has this constraint). But these figures can be redrawn so that the maximum dimension in each direction is unity. This 'normalisation' does not change the relation of the curves to each other, but simplifies the mathematics. However, normalised results must be converted back to engineering dimensions before they can be applied.

Engineers may find that full normalisation of the mathematics sometimes takes away too much from the solution. A good example of this is to be found in equation (3.25), where full normalisation would produce the following canonical equation pair (Nusselt equations) at the expense of obscuring the problem.

$$\frac{\partial T_h}{\partial x} = -(T_h - T_c)$$
$$\frac{\partial T_c}{\partial y} = +(T_h - T_c)$$

The effectiveness concept

Effectiveness is a measure of how closely the temperature of the fluid with the least water equivalent approaches the maximum possible temperature rise (*Tspan*) in the exchanger. For the contraflow arrangement, and to some extent for the cross-flow arrangement, this corresponds to seeking the closest temperature approach between fluids. When care is taken to keep the temperature approach as small as practicable, good effectiveness values should be achieved without the need to address the effectiveness issue specifically in design.

Thinking is different for parallel-flow arrangements because parallel-flow applications are usually more concerned with limiting the maximum temperature of the cold fluid being heated, or to controlling the drop in temperature of the hot fluid being cooled, while recovering energy. Here the closest temperature approach in the exchanger is related to temperatures of fluids at the same end, and the actual value of effectiveness achieved can usefully be compared to its limiting value.

Units in differential equations

SI units are used throughout this text. It is perhaps not always realised that ordinary and partial differential equation have units, and checking them is a valuable way of confirming whether the equation has been correctly formulated.

Consider the symbols x, t representing distance (metres) and time (seconds). It is familiar terriroty to recognise velocity and acceleration respectively is

$$\dot{x} = \frac{dx}{dt}, \qquad \ddot{x} = \frac{d^2 x}{dt^2}$$

and it is but a short step to recognise that the units 'go' as the back end of the differential expressions: for velocity and acceleration respectively

$$\dot{x} = \frac{d}{d}\left(\frac{x}{t}\right) \Rightarrow \left(\frac{m}{s}\right), \qquad \ddot{x} = \frac{d^2}{d}\left(\frac{x}{t^2}\right) \rightarrow \left(\frac{m}{s^2}\right)$$

Where differential terms are themselves raised to powers, then units are obtained as

$$\dot{x}^2 = \left(\frac{dt}{dx}\right)^2 \Rightarrow \left[\frac{d}{d}\left(\frac{x}{t}\right)\right]^2 \Rightarrow \left(\frac{m}{s}\right)^2$$

The central partial differential equation of (A.1) in Appendix A has dimensionless parameters n_h, n_c, R_h, R_c and the individual terms must have identical dimensions for the equation to make sense, namely

$$\frac{\partial T_w}{\partial t} - \kappa_x \frac{\partial^2 T_w}{\partial x^2} = +\frac{n_h}{R_h \tau_h}(T_h - T_w) - \frac{n_c}{R_c \tau_c}(T_w - T_c)$$

$$\left(\frac{K}{s}\right) - \left(\frac{m^2}{s}\right)\left(\frac{K}{m^2}\right) = +\left(\frac{1}{s}\right)(K) \quad - \left(\frac{1}{s}\right)(K)$$

2.15 DIMENSIONLESS GROUPS

It would not be proper to proceed further without some discussion of dimensionless groups which arise in both heat transfer and flow-friction correlations used in the design of heat exchangers. A deeper study may require other texts because here it has been simplified as far as possible, without destroying fundamental concepts of dimensional analysis of linear systems.

Rayleigh's method and Buckingham's π-theorem

The reader may come across one or both of these algebraic approaches used in finding dimensionless groups. It is useful to know both methods because there are situations where the form may not be known for the differential equations governing the phenomena under consideration.

Both approaches first require an intelligent guess of the number of independent variables involved in the problem. If too many are guessed, the number of

2.15 DIMENSIONLESS GROUPS

dimensionless groups may become too large. If too few are guessed, valid groups will still be produced but they will be unfamiliar and difficult to apply.

When the governing differential equations are known in advance, the exact number of dimensionless groups can be extracted from them quite naturally. This is the approach adopted below.

Fundamental approach via differential equations

A differential equation is a mathematical model of a whole class of phenomena (Luikov 1966). To obtain one particular solution from the multitude of possible solutions we must provide additional information, the conditions of single-valuedness. These conditions include the following:

(a) geometrical properties of the system
(b) physical properties of the bodies involved in the phenomena under consideration
(c) initial conditions describing the state of the system at the first instant
(d) boundary conditions giving the interation of the system with its suroundings

Two conditions are similar if they are described by one and the same system of differential equations and have similar conditions of single-valuedness.

We recognise the following concepts:

- a class of phenomena: partial differential equations
- a group of phemomena: similarity
- a single phemomenon: partial differential equations plus conditions of single-valuedness

Similarity in transient thermal conduction

Examine the case without internal heat generation. There is no increased difficulty with heat generation, but it introduces another parameter. Consider the Cartesian form of the (energy balance + Fourier constitutive) differential equation with constant physical properties.

$$\rho C \frac{\partial T}{\partial t} = \lambda \left(\frac{\partial^2 T}{\partial x^2} + \frac{\partial^2 T}{\partial y^2} + \frac{\partial^2 T}{\partial z^2} \right)$$

For body 1 this becomes

$$\frac{\partial T_1}{\partial t_1} = \kappa_1 \left(\frac{\partial^2 T_1}{\partial x_1^2} + \frac{\partial^2 T_1}{\partial y_1^2} + \frac{\partial^2 T_1}{\partial z_1^2} \right) \quad \text{where } \kappa_1 = \frac{\lambda_1}{\rho_1 C_1}$$

If the surroundings are at T_0 then

$$\frac{\partial \theta_1}{\partial t_1} = \kappa_1 \left(\frac{\partial^2 \theta_1}{\partial x_1^2} + \frac{\partial^2 \theta_1}{\partial y_1^2} + \frac{\partial^2 \theta_1}{\partial z_1^2} \right) \quad \text{where } \theta_1 = T_1 - T_0 \quad (2.32)$$

For body 2, the corresponding equation is

$$\frac{\partial \theta_2}{\partial t_2} = \kappa_2 \left(\frac{\partial^2 \theta_2}{\partial x_2^2} + \frac{\partial^2 \theta_2}{\partial y_2^2} + \frac{\partial^2 \theta_2}{\partial z_2^2} \right) \qquad (2.33)$$

Let the quantities referring to body 2 be related everywhere and for all times to the corresponding quantities of the first body where the Γ's are constants of proportionality, then

$$x_2 = (\Gamma_x)x_1 \qquad t_2 = (\Gamma_t)t_1$$
$$y_2 = (\Gamma_y)y_1 \qquad \kappa_2 = (\Gamma_\kappa)\kappa_1$$
$$z_2 = (\Gamma_z)z_1 \qquad \theta_2 = (\Gamma_\theta)\theta_1$$

In equation (2.33) we can therefore write

$$\frac{\Gamma_\theta}{\Gamma_t} \cdot \frac{\partial \theta_1}{\partial t_1} = \Gamma_\kappa \kappa_1 \left(\frac{\Gamma_\theta}{\Gamma_x^2} \cdot \frac{\partial^2 \theta_1}{\partial x_1^2} + \frac{\Gamma_\theta}{\Gamma_y^2} \cdot \frac{\partial^2 \theta_1}{\partial y_1^2} + \frac{\Gamma_\theta}{\Gamma_z^2} \cdot \frac{\partial^2 \theta_1}{\partial z_1^2} \right) \qquad (2.34)$$

Equation (2.34) will be identical with equation (2.32), thus the heat flow in the two bodies will be similar, providing

$$\frac{\Gamma_\theta}{\Gamma_t} = 1 \qquad (2.35)$$

$$\frac{\Gamma_\kappa \Gamma_\theta}{\Gamma_x^2} = \frac{\Gamma_\kappa \Gamma_\theta}{\Gamma_y^2} = \frac{\Gamma_\kappa \Gamma_\theta}{\Gamma_z} = 1 \qquad (2.36)$$

First, from (2.36) it follows that

$$\Gamma_x = \Gamma_y = \Gamma_z \, (= \Gamma_\ell, \text{ say}) \qquad (2.37)$$

Thus

$$\frac{x_1}{x_2} = \frac{y_1}{y_2} = \frac{z_1}{z_2} = \frac{\ell_1}{\ell_2} \qquad (2.38)$$

where ℓ_1 and ℓ_2 are characteristic (or reference) lengths, similarly defined in the two bodies. In other words, equation (2.38) implies that the bodies must be geometrically similar.

Second, from (2.35) and (2.36) it follows that

$$\frac{\Gamma_\theta}{\Gamma_t} = \frac{\Gamma_\kappa \Gamma_\theta}{\Gamma_x^2} \quad \text{i.e.} \quad \frac{\Gamma_\kappa \Gamma_t}{\Gamma_x^2} = 1$$

or, where ℓ is any characteristic dimension, it follows that

$$\frac{\kappa_1 t_1}{\ell_1^2} = \frac{\kappa_2 t_2}{\ell_2^2} = \text{Fo (Fourier number)} \qquad (2.39)$$

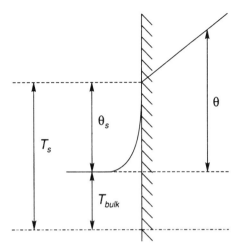

Fig. 2.23 Surface temperature profiles

The Fourier number, which includes the physical constants, is in a sense 'generalised time', and must be dimensionless because the Γ's are dimensionless.

$$\text{Fo} = \frac{\lambda}{\rho C} \cdot \frac{t}{\ell^2} \qquad \frac{J}{(m\, s\, K)} \cdot \frac{m^3}{kg} \cdot \frac{(kg\, K)}{J} \cdot \frac{s}{m^2}$$

If equations (2.38) and (2.39) are satisfied, the temperature distribution in the two bodies will be similar, provided the boundary conditions and the initial conditions are also similar (Fig. 2.23).

$$\left\{ \begin{array}{c} \text{heat transported across} \\ \text{boundary surface} \end{array} \right\} = \left\{ \begin{array}{c} \text{heat flowing in} \\ \text{body at surface} \end{array} \right\}$$

$$\alpha \theta_s = -\lambda \left(\frac{d\theta}{d\ell}\right)_s \qquad (2.40)$$

where α = surface heat transfer coefficient (J/m² s K)
θ_s = temperature excess of surface above reference (K)
ℓ = dimension normal to the surface (m)

Then by same argument as before

$$\alpha_2 = (\Gamma_\alpha)\alpha_1 \qquad \theta_2 = (\Gamma_\theta)\theta_1$$
$$\lambda_2 = (\Gamma_\lambda)\lambda_1 \qquad \ell_2 = (\Gamma_l)\ell_1$$

thus

$$(\Gamma_\alpha \Gamma_\theta)\alpha_1 \theta_{b1} = -\left(\frac{\Gamma_\lambda \Gamma_\theta}{\Gamma_\ell}\right)\lambda_1 \left(\frac{d\theta_1}{d\ell_1}\right)_b$$

and the further condition is required that

$$\Gamma_\alpha = \frac{\Gamma_\lambda}{\Gamma_t}, \quad \text{i.e.} \quad \frac{\Gamma_\alpha \Gamma_t}{\Gamma_\lambda} = 1$$

or where ℓ is a characteristic dimension

$$\frac{\alpha_1 \ell_1}{\lambda_1} = \frac{\alpha_2 \ell_2}{\lambda_2} = \text{Bi (Biot number)} \quad (2.41)$$

The Biot number differs from the Nusselt number in that λ refers to the solid, not to the fluid surrounding the body.

The condition that there is a constant ratio of the temperatures at any point in the bodies λ to their surface temperatures (θ_s) must also apply. This gives the condition of similarity of temperature distribution throughout the bodies at all times, including similarity at the start, i.e. similar initial conditions. Thus the relationships

$$\frac{x}{\ell} = \text{const.}, \quad \text{Fo} = \text{const.}, \quad \text{Bi} = \text{const.}, \quad \frac{\theta}{\theta_1} = \text{const.}$$

define the conditions for similarity of heat conduction in a solid body. Transient heat flow is therefore characterised by relations of the form

$$f\left(\text{Fo, Bi, } \frac{\theta}{\theta_s}, \frac{x}{\ell}\right) = 0 \quad \text{or} \quad \frac{\theta}{\theta_s} = \phi\left(\text{Fo, Bi, } \frac{x}{\ell}\right)$$

Comparison with analytical solution

To illustrate the connection between analytical solutions and conditions of similarity, consider the problem of a wall of finite thickness (ℓ), heated on both sides in such a way that the surface temperatures are suddenly raised and maintained constant at temperature T_s.

The basic (energy balance + Fourier constitutive) differential equation governing this problem is

$$\rho C \frac{\partial T}{\partial t} = \lambda \frac{\partial^2 T}{\partial x^2} \quad (2.42)$$

with initial conditions $T = 0$ at $0 < x < \ell$ and surface conditions $T = T_s$ at $x = 0$ and $x = L$.

The analytical solution to this problem is given in terms of a Fourier series, which converges in about five terms

$$\frac{\theta}{\theta_s} = \frac{2}{\pi} \sum_1^n \left\{ [1 - \cos(n\pi)] \frac{1}{n} \exp\left[-(n\pi)^2 \left(\frac{\kappa t}{\ell^2}\right)\right] \sin\left[n\pi\left(\frac{x}{\ell}\right)\right] \right\} \quad (2.43)$$

where the term $[1 - \cos(n\pi)]$ is either 2 or 0,

$$\text{thus} \quad \frac{\theta}{\theta_s} = f\left[\left(\frac{\kappa t}{\ell^2}\right), \left(\frac{x}{\ell}\right)\right] = f\left[\text{Fo}, \left(\frac{x}{\ell}\right)\right] \quad \text{where} \quad \frac{\theta}{\theta_s} = \frac{(T - T_0)}{(T_s - T_0)}$$

The Biot number does not enter into this solution because surface temperatures were *specified* in fixing the boundary conditions. If surface heat transfer coefficients had also been involved, the relationship would be of the form

$$\frac{\theta}{\theta_s} = f\left[\text{Fo, Bi}, \left(\frac{x}{\ell}\right)\right]$$

Williamson and Adams (1919) developed analytical solutions for the history of centre-line temperatures for a number of shapes whose surface temperatures were suddenly changed to a new value. The shapes considered were

- infinitely wide slab
- cylinder with length equal to diameter
- cylinder with infinite length
- cube
- square bar
- sphere

Of these, the infinitely wide slab is the simplest case as heat flows along one axis only. A graphical plot prepared for this analytical solutions is presented in Fig. 2.24, illustrating the use of dimensional groups as co-ordinates.

Convective heat transfer

Turning to convective heat transfer, the basic differential equations become extremely complex, and analytical solutions exist only for very simple physical situations. To appreciate this, it would be necessary to solve simultaneously

(a) The three Navier-Stokes equations (momentum balance + Newtonian constitutive), which describe the velocity components of a Newtonian fluid at each point in the fluid and at each instant of time. Only the x-direction equation is given below

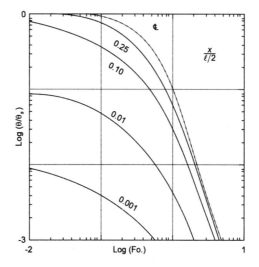

Fig. 2.24 Dimensionless plot of temperature history of an infinite flat plate with step change at the surface

$$\rho\left(u\frac{\partial u}{\partial x} + v\frac{\partial u}{\partial y} + w\frac{\partial u}{\partial z} + \frac{\partial u}{\partial t}\right) = \rho X - \frac{\partial p}{\partial x} + \frac{1}{3}\cdot\eta\frac{\partial}{\partial x}\left(\frac{\partial u}{\partial x} + \frac{\partial v}{\partial y} + \frac{\partial w}{\partial z}\right) + \eta\nabla^2 u$$

(b) The single energy equation (energy balance + Newtonian constitutive), which gives temperature distribution throughout the fluid w.r.t. location of each point in space w.r.t. time

$$\rho C\left(u\frac{\partial T}{\partial x} + v\frac{\partial T}{\partial y} + w\frac{\partial T}{\partial z} + \frac{\partial T}{\partial t}\right) = \left(u\frac{\partial p}{\partial x} + v\frac{\partial p}{\partial y} + w\frac{\partial p}{\partial z} + \frac{\partial p}{\partial t}\right) + \lambda\nabla^2 T + \eta\Phi$$

where Φ is the Rayleigh dissipation function

$$\Phi = 2\left[\left(\frac{\partial u}{\partial x}\right)^2 + \left(\frac{\partial v}{\partial y}\right)^2 + \left(\frac{\partial w}{\partial z}\right)^2\right] +$$

$$\left[\left(\frac{\partial u}{\partial y} + \frac{\partial v}{\partial x}\right)^2 + \left(\frac{\partial v}{\partial z} + \frac{\partial w}{\partial y}\right)^2 + \left(\frac{\partial w}{\partial x} + \frac{\partial u}{\partial z}\right)^2\right] - \frac{2}{3}\left[\frac{\partial u}{\partial x} + \frac{\partial v}{\partial y} + \frac{\partial w}{\partial z}\right]^2$$

These equations must be solved in association with

- boundary conditions (velocity and temperature conditions at the surface)
- initial conditions (velocity and temperature conditions at time zero)

Referring back to the very simple conduction equation (2.42) and its more complex analytical solution (2.43), it is not surprising that general solutions for the simultaneous linear differential equations describing fluid flow have not been found. In forced turbulent convection, simple experimental correlations for fluids and gases flowing through pipes, may be written

$$\text{Nu} = 0.023(\text{Re})^{0.8}(\text{Pr})^{0.4} \tag{2.44}$$

Comparing this with equation (2.43) it is easy to induce that it might be better written as a more complicated series expansion

$$\text{Nu} = a_1(\text{Re})^{b1}(\text{Pr})^{c1} + a_2(\text{Re})^{b2}(\text{Pr})^{c2} + \cdots$$

This comparison suggests simply that empirical correlations are at best an approximation to what is actually happening, that they should be used with caution and that their range of applicability must always be known.

Dimensionless groups in heat transfer and fluid flow

It is straightforward to set about extracting dimensionless groups from the Navier-Stokes and Energy balance equations. This is explained in Schlichting (1960) and in other engineering texts.

The extraction will not be repeated here; suffice it to provide physical interpretation of some dimensionless groups which may be encountered in experimental correlations for heat transfer and fluid flow used in the design of heat exchangers.

From similarity of the velocity fields (Navier-Stokes)

$$\text{Reynolds number} \quad \text{Re} = \frac{\ell G}{\eta} = \frac{\text{inertia force}}{\text{viscous force}}$$

$$\text{Grashof number} \quad \text{Gr} = \frac{\text{Ra}}{\text{Pr}} = \frac{(\beta g)\rho^2 \ell^3 \theta}{\eta^2} = \frac{\text{buoyancy force}}{\text{viscous force}}$$

$$\text{Euler number} \quad \text{Eu} = \frac{\Delta p}{\rho u^2} = \frac{\text{pressure force}}{\text{inertia force}}$$

From similarity of the temperature fields (Newtonian energy balance)

$$\text{Eckert number} \quad \text{Ec} = \frac{u^2}{C^2} = \frac{2 \times \text{temperature increase at stagnation}}{\text{temperature difference between wall and fluid}}$$

$$\text{perhaps to be understood from} \quad \frac{mu^2/2}{mC\theta} = \frac{\text{kinetic energy}}{\text{thermal energy}}$$

Mach number $\mathrm{Ma} = \sqrt{\mathrm{Ec}/(\gamma - 1)}$

$$= \frac{u}{a} = \frac{\text{velocity of fluid flow}}{\text{speed of sound in fluid}} \text{ for perfect gas}$$

Peclet number $\mathrm{Pe} = \mathrm{Pr}.\mathrm{Re} = \dfrac{u\ell}{\kappa} = \dfrac{\text{heat transfer by convection}}{\text{heat transfer by conduction}}$

Prandtl number $\mathrm{Pr} = \dfrac{\mathrm{Pe}}{\mathrm{Re}} = \dfrac{C\eta}{\lambda} = \dfrac{\eta/\rho}{\kappa} = \dfrac{\text{momentum diffusivity}}{\text{thermal diffusivity}}$

From similarity at the boundary

$$\text{Nusselt number } \mathrm{Nu} = \frac{\alpha \ell}{\lambda} = \frac{\text{total heat transfer}}{\text{conductive heat transfer of fluid}}$$

From geometric similarity

One, two or three lengths as appropriate, e.g. d/ℓ as one length ratio.

A general function obtained from governing equations for convective heat transfer may look like

$$\mathrm{Nu} = f\left(\mathrm{Re},\ \mathrm{Pr},\ \mathrm{Gr},\ \mathrm{Ec},\ \frac{d}{\ell}\right)$$

Whether the Eckert number need be present may be determined by the Mach number, which is a measure of whether heating effects caused by compressibility are likely to be important. If $\mathrm{Ma} < 0.33$ the fluid may be regarded as incompressible and the Eckert number can be omitted.

It is not usual to have both Reynolds number and Grashof number present together because Reynolds number applies to forced convection whereas Grashof number applies to natural convection. Only when the magnitudes of the two effects are of similar order will it be necessary to include both numbers. The Peclet number is adequately represented by the (Pr, Re) groups and need not be explicitly present. The Stanton number may be used to replace the Nusselt number in some correlations, namely $\mathrm{St} = \dfrac{\mathrm{Nu}}{\mathrm{Re}.\mathrm{Pr}}$.

Typical heat transfer correlations are

$$\mathrm{Nu} = 0.023(\mathrm{Re})^{0.8}(\mathrm{Pr})^{0.4}\left(\frac{\mu_b}{\mu_w}\right)^{-0.14} \qquad \mathrm{Nu} = \left\{0.6 + \frac{0.387\,\mathrm{Ra}^{1/6}}{[1+(0.559/\mathrm{Pr})^{9/16}]^{8/27}}\right\}^2$$

forced turbulent convection inside a tube

natural convection, air over horizontal pipes

where the Rayleigh number Ra is the product of Grashof and Prandtl numbers. The Euler number provides a pressure loss coefficient for flow Eu = ϕ(Re). At the elementary level used in heat transfer, the friction factor (f) provides the link.

For forced turbulent convection inside a tube

$$f = 0.046(\text{Re})^{0.2} \qquad \Delta p = \frac{4 f G^2}{2\rho}\left(\frac{\ell}{d}\right)$$

Flow drag expressions in natural convection may be more complicated.

Coupling between the equation for heat transfer and the equation for pressure loss is through the Reynolds number, and these effects are seperable because the Eckert number is small, i.e. thermal effects due to friction are small. When the Eckert number is large, thermal effects due to compressibility become significant, e.g. aerodynamic heating in high speed aircraft.

The reader is reserred to Bejan (1995) for an up-to-date treatment of correlations.

Applicability of dimensionless groups

There are many applications where dimensional analysis provides valuable information which would not otherwise be easily seen, e.g. Obot *et al* (1991) and Obot (1993), who extended flow similarity concepts to include transition to turbulent flow for different channel geometries.

Similarity can also be applied to mechanical structures; see e.g. Lessen (1953), Dugundji and Calligeros (1962), Hovanesian and Kowalski (1967) and Jones (1974). However, the principal applications have been in the field of fluid mechanics and heat transfer, illustrated by the papers by Boucher and Alves (1959), Klockzien and Shannon (1969) and Morrison (1969).

The reader may be impressed by the number of dimensionless groups listed by Catchpole and Fulford (1966, 1968).

In using heat transfer and flow-friction correlations it is not essential to have a correlation expressed in mathematical form, e.g.

$$\text{Nu} = 0.023(\text{Re})^{0.8}(\text{Pr})^{0.4}$$

This equation is simply a mathematical 'best' fit to a graph of experimental data, and frequently a better fit can be produced employing an interpolating cubic spline-fit which allows for individual experimental errors at each data point. Some recommendations on splinefitting procedures are given in Appendix B.7.

REFERENCES

Bejan, A. (1995) *Convection Heat Transfer, 2nd Edn*, New York: Wiley Interscience.
Boucher, D. F. and Alves, G. E. (1959) Dimensionless numbers for fluid mechanics, heat transfer, mass transfer and chemical reaction, *Chemical Engineering Progress*, **55**(9), 55–83.
Catchpole, J. P. and Fulford, G. (1966) Dimensionless groups, *Industrial and Engineering Chemistry*, **58**(3), 46–60.
Catchpole, J. P. and Fulford, G. (1968) Dimensionless groups, *Industrial and Engineering Chemistry* 60(3), 71–78.
Clayton, D. G. (1984) Increasing the power of the LMTD method for heat exchangers, *International Journal of Mechanical Engineering Education*, 13(3), 183–190.
Dugundji, J. and Calligeros, J. M. (1962) Similarity laws for aerothermoelastic testing, *Journal of the Aerospace Science*, 29, 935–950.
Grassman, P. and Kopp, J. (1957) Zur gunstigen Wahl der temperaturdifferenz und der Wärmeübergangszahl in Wärmeaustauchern, *Kaltetechnik*, 9(10), 306–308.
Hewitt, G. F., Shires G. L. and Bott, T. R. (1994) *Process Heat Transfer*, CRC Press.
Hovanesian, J. D. and Kowalski, H. C. (1967) Similarity in elasticity, *Experimental Mechanics*, 7, 82–84.
Jones, N. (1974) Similarity principles in structural mechanics, *International Journal of Mechanical Engineering Education*, 2(2), 1–10.
Klockzien, V. G. and Shannon, R. L. (1969) Thermal scale modelling of space- craft, *Society of Automotive Engineers, Paper 690196, 13–17 January 1969*.
Lessen, M. (1953) On similarity in thermal stresses in bodies, *Journal of the Aerospace Sciences*, **20**(10), 716–717.
Luikov, A. V. (1966) *Heat and Mass Transfer in Capillary-porous Bodies*, English translation by P. W. B. Harrison and W. M. Pun, Oxford: Pergamon.
Morrison, F. A. (1969) Generalised dimensional analysis and similarity analyses, *Bulletin of Mechanical Engineering Education*, 8, 289–300.
Obot, N. T. (1993) The frictional low of corresponding states: its origin and applications, *Transactions Institution of Chemical Engineering* 71(A), 3–10.
Obot, N. T., Jendrzejczyk, J. A. and Wambsganss, M. W. (1991) Direct determination of the onset of transition to turbulence in flow passages, *Trans. ASME, Journal of Fluids Engineering*, 113, 602–607.
Schlichting, H. (1960) *Boundary Layer Theory*, New York: McGraw-Hill. 4th edn.
Spalding, D. B. (1990) Section 1.3.1–1: Analytical solutions, In: *Hemisphere Handbook of Heat Exchanger Design* Ed. G. F. Hewitt, New York: Hemisphere.
Taborek, J. (1983) In: *Heat Exchanger Design Handbook*, New York: Hemisphere, Vol. 1, Section 1.5.
Williamson, E. D. and Adams, L. H. (1919) Temperature distribution in solids during heating or cooling, *Physical Review*, 14, 99–114.

3

STEADY-STATE TEMPERATURE PROFILES

3.1 LINEAR TEMPERATURE PROFILES IN CONTRAFLOW

This is a special case of contraflow which is of interest for cryogenic heat exchangers. Proof of linear temperature profiles requires a simple introduction to the development of differential equations which govern temperature distributions (Fig. 3.1). Take differential energy balances.

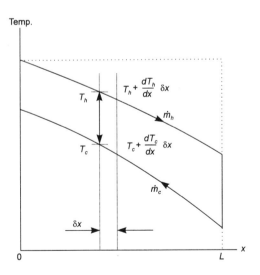

Fig. 3.1 Arbitrary temperature profiles

3 STEADY-STATE TEMPERATURE PROFILES

Hot fluid

$$\left\{\begin{array}{l}\text{energy entering}\\\text{with hot fluid}\end{array}\right\} - \left\{\begin{array}{l}\text{energy leaving}\\\text{with hot fluid}\end{array}\right\} - \left\{\begin{array}{l}\text{heat transferred}\\\text{to cold fluid}\end{array}\right\} = \left\{\begin{array}{l}\text{energy stored}\\\text{in hot fluid}\end{array}\right\}$$

$$\dot{m}_h C_h T_h - \dot{m}_h C_h \left(T_h + \frac{dT_h}{dx}\delta x\right) - U\left(S\frac{\delta x}{L}\right)(T_h - T_c) = 0$$

giving

$$\frac{dT_h}{dx} = -\frac{US}{\dot{m}_h C_h} \cdot \frac{1}{L}(T_h - T_c) \qquad (3.1)$$

Cold fluid

$$\dot{m}_c C_c \left(T_c \frac{dT_c}{dx}\delta x\right) + U\left(S\frac{\delta x}{L}\right)(T_h - T_c) - \dot{m}_c C_c T_c = 0$$

giving

$$\frac{dT_c}{dx} = -\frac{US}{\dot{m}_c C_c} \cdot \frac{1}{L}(T_h - T_c) \qquad (3.2)$$

Writing overall values of Ntu as $N_h = US/(\dot{m}_h C_h)$ and $N_c = US/(\dot{m}_c C_c)$ the coupled equations (3.1) and (3.2) become

$$\boxed{\begin{aligned}\frac{dT_h}{dx} &= -\frac{N_h}{L}(T_h - T_c)\\ \frac{dT_c}{dx} &= -\frac{N_c}{L}(T_h - T_c)\end{aligned}} \qquad (3.3)$$

For equal water equivalents $N_h = N_c$ and it follows that $dT_h/dx = dT_c/dx$, which shows that the gradients are the same at any x. It is a necessary but not sufficient condition for straight and parallel temperature profiles. It only remains to show that one temperature profile is linear.

From equation (3.3)

$$T_h - T_c = \frac{L}{N_h} \cdot \frac{dT_h}{dx}$$

and differentiating

$$\frac{dT_c}{dx} = \frac{dT_h}{dx} - \frac{L}{N_h} \cdot \frac{d^2 T_h}{dx^2}, \quad \text{but} \quad \frac{dT_h}{dx} = \frac{dT_c}{dx}, \quad \text{thus} \quad \frac{d^2 T_h}{dx^2} = 0$$

hence a linear profile exists (Fig. 3.2).

3.2 GENERAL CASES OF CONTRAFLOW AND PARALLEL FLOW

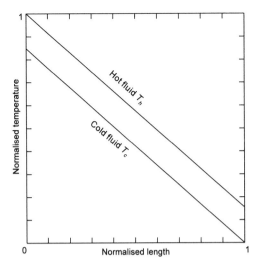

Fig. 3.2 Normalised temperature profiles with $N_h = N_c = 5.0$

3.2 GENERAL CASES OF CONTRAFLOW AND PARALLEL FLOW

In the following treatment there is no longitudinal conduction in the wall, no energy storage in the fluids or the wall (transients), no heat generation in the fluids or the wall, and no external losses.

The analyses for the heat exchanger flow configurations in Figs. 3.3 and 3.4 are practically identical, thus only the contraflow exchanger will be considered. Similarly, the treatments of hot and cold fluids are virtually identical. We examine only the hot fluid in the contraflow case.

Hot fluid

$$\left\{\begin{array}{c} \text{energy entering} \\ \text{with hot fluid} \end{array}\right\} - \left\{\begin{array}{c} \text{energy leaving} \\ \text{with hot fluid} \end{array}\right\} - \left\{\begin{array}{c} \text{heat transferred} \\ \text{to cold fluid} \end{array}\right\} = \left\{\begin{array}{c} \text{energy stored} \\ \text{in hot fluid} \end{array}\right\}$$

$$\dot{m}_h C_h T_h - \dot{m}_h C_h \left(T_h + \frac{dT_h}{dx} \delta x \right) - U \left(S \frac{\delta x}{L} \right) (T_h - T_c) = 0$$

$$\frac{dT_h}{dx} = -\frac{US}{\dot{m}_h C_h} \cdot \frac{1}{L} (T_h - T_c) = -\frac{N_h}{L} (T_h - T_c) \quad (3.4)$$

Cold fluid

$$\frac{dT_c}{dx} = -\frac{US}{\dot{m}_c C_c} \cdot \frac{1}{L} (T_h - T_c) = -\frac{N_c}{L} (T_h - T_c) \quad (3.5)$$

54   3 STEADY-STATE TEMPERATURE PROFILES

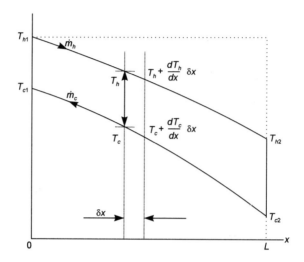

Fig. 3.3   Contraflow

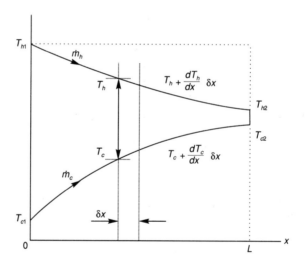

Fig. 3.4   Parallel flow

Scaling of length (x) is possible by writing $\xi = N_h(x/L)$ giving $d\xi/dx = N_h/L$, so that with

$$\frac{dT}{dx} = \frac{dT}{d\xi} \cdot \frac{d\xi}{dx} = \frac{dT}{d\xi} \cdot \frac{N_h}{L}$$

equations (3.4) and (3.5) become for $0 < \xi < 1$

3.2 GENERAL CASES OF CONTRAFLOW AND PARALLEL FLOW

$$\frac{dT_h}{d\xi} = -(T_h - T_c)$$
$$\frac{dT_c}{d\xi} = +\frac{N_c}{N_h}(T_h - T_c)$$

but including N_h in the independent variable takes too much information away from the engineering and it is better to write $\xi = x/L$. Then reverting back to the original notation

$$\frac{dT_h}{dx} = -N_h(T_h - T_c)$$
$$\frac{dT_c}{dx} = +N_c(T_h - T_c)$$

Scaling and normalisation will be useful later when compact notation is helpful, but for the present we can proceed more directly from equations (3.4) and (3.5).

From (3.4)
$$T_c = \frac{L}{N_h} \cdot \frac{dT_h}{dx} + T_h$$

Differentiating
$$\frac{dT_c}{dx} = \frac{L}{N_h} \cdot \frac{d^2 T_h}{dx^2} + \frac{dT_h}{dx}$$

In equation (3.5)

$$\frac{L}{N_c}\left[\frac{L}{N_h} \cdot \frac{d^2 T_h}{dx^2} + \frac{dT_h}{dx}\right] = -T_h + \left[\frac{L}{N_h} \cdot \frac{dT_h}{dx} + T_h\right]$$

$$\frac{d^2 T_h}{dx^2} + \left[\frac{N_h - N_c}{L}\right]\frac{dT_h}{dx} = 0 \qquad (3.6)$$

Similarly for the cold fluid

$$\frac{d^2 T_c}{dx^2} + \left[\frac{N_h - N_c}{L}\right]\frac{dT_c}{dx} = 0 \qquad (3.7)$$

The solution to equation (3.6) is $T_h = Ae^{(N_h - N_c)x/L} + B$ with boundary conditions

$$\left.\begin{array}{c} x = 0 \\ T_h = T_{h1} \end{array}\right\} \qquad \left.\begin{array}{c} x = L \\ T_h = T_{h2} \end{array}\right\}$$

from which the following dimensionless (and normalised) result is obtained, valid for both hot fluid and cold fluid in Contraflow.

$$\frac{T_{h1} - T_h}{T_{h1} - T_{h2}} = \frac{1 - \exp\left[-(N_h - N_c)\frac{x}{L}\right]}{1 - \exp\left[-(N_h - N_c)\right]} = \frac{T_{c1} - T_c}{T_{c1} - T_{c2}} \qquad (3.8)$$

A similar analysis for Parallel flow produces

$$\frac{T_{h1} - T_h}{T_{h1} - T_{h2}} = \frac{1 - \exp\left[-(N_h + N_c)\frac{x}{L}\right]}{1 - \exp\left[-(N_h + N_c)\right]} = \frac{T_{c1} - T_c}{T_{c1} - T_{c2}} \qquad (3.9)$$

Expressions $(N_h - N_c)$ and $(N_h + N_c)$ relate back to both the common effectiveness diagram, and 'rate' and 'energy' equations derived in the (LMTD-Ntu) approach of Section 2.3.

The following dimensionless expressions control the temperature profiles and provide means for assessing the useful length of the exchanger, i.e. the condition when the driving temperature differences in equations (3.8) and (3.9) cease to be effective. They are perhaps more fundamental than the concept of effectiveness itself, in that they come directly from the differential equations.

$$\frac{1 - \exp\left[-(N_h - N_c)\frac{x}{L}\right]}{1 - \exp\left[-(N_h - N_c)\right]} \quad \text{and} \quad \frac{1 - \exp\left[-(N_h + N_c)\frac{x}{L}\right]}{1 - \exp\left[-(N_h + N_c)\right]}$$

Figures 3.5 to 3.8 provide normalised temperature profiles for two cases of contraflow and two cases of parallel flow, respectively. The chain dotted lines

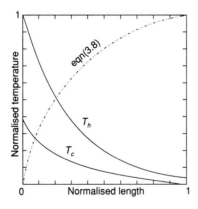

Fig. 3.5 Contraflow: normalised temperature profiles with $N_h=5.0$, $N_c=2.0$

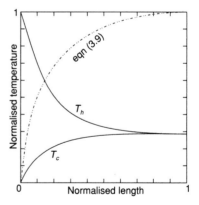

Fig. 3.6 Parallel flow: normalised temperature profiles with $N_h=5.0$, $N_c=2.0$

3.2 GENERAL CASES OF CONTRAFLOW AND PARALLEL FLOW

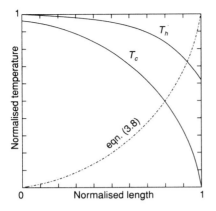

Fig. 3.7 Contraflow: normalised temperature profiles with $N_h=5.0$, $N_c=2.0$

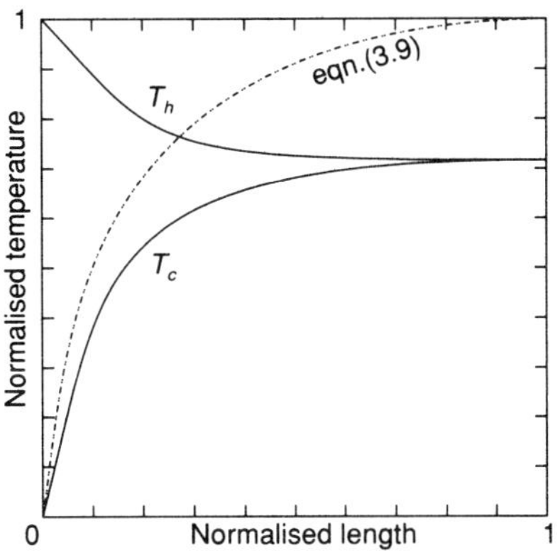

Fig. 3.8 Parallel flow: normalised temperature profiles with $N_h=2.0$, $N_c=5.0$

generated from the above dimensionless exponential ratios allow practical assesment of useful length in the exchanger for effective heat transfer.

Intermediate wall temperature

In steady-state heat exchange it is possible to obtain approximate expressions for wall temperature without formally introducing a differential equation for the wall. In such circumstances the wall has no stored energy (no transients) and there is no longitudinal conduction along the wall, so that wall temperature gradients at the ends are not correctly represented. But the representation is still quite good.

At any intermediate point in the exchanger, differential heat flow and energy expressions may be written

$$dQ = \dot{m}\, C_h dT_h = UdS(T_h - T_c) = \dot{m}_c C_c dT_c$$
$$= \alpha_h dS(T_h - T_w) \qquad\qquad = \alpha_c dS(T_w - T_c) \qquad (3.10)$$

To confirm the consistency of these equations, write

$$T_h - T_w = \frac{dT_h}{dS} \cdot \frac{\dot{m}_h C_h}{\alpha_h} = \frac{U(T_h - T_c)}{\dot{m}_h C_h} \cdot \frac{\dot{m}_h C_h}{\alpha_h}$$

$$T_w - T_c = \frac{dT_c}{dS} \cdot \frac{\dot{m}_c C_c}{\alpha_c} = \frac{U(T_h - T_c)}{\dot{m}_c C_c} \cdot \frac{\dot{m}_c C_c}{\alpha_c}$$

Adding

$$(T_h - T_c) = (T_h - T_c) U \left[\frac{1}{\alpha_h} + \frac{1}{\alpha_c} \right]$$

3 STEADY-STATE TEMPERATURE PROFILES

which is only true if $1/U = 1/\alpha_h + 1/\alpha_c$, seen to be correct for the case when no wall equation is present.

Again from the above derived equation

$$U(T_h - T_c) = \alpha_h(T_h - T_w) \text{ and } U(T_h - T_c) = \alpha_c(T_w - T_c)$$

giving

$$\frac{T_h - T_w}{T_h - T_c} = \frac{U}{\alpha_h} \quad \text{and} \quad \frac{T_w - T_c}{T_h - T_c} = \frac{U}{\alpha_c} \quad (3.11)$$

In practical cases, α_h and α_c will have been evaluated. It is not possible to obtain T_w from knowledge of N_h and N_c alone.

Results (3.11) will apply *locally* to the case of simple cross-flow with both fluids unmixed and, by inspection, may apply to other cases as well.

3.3 CONDENSATION AND EVAPORATION

Two special cases in which the temperature of one fluid remains constant require separate consideration. The analysis can be simplified by considering normalised temperatures from the start.

Although these cases are simple, their solutions also provides the inlet temperature distributions for the case of unmixed cross-flow with constant inlet conditions, and are therefore worthy of consideration.

Consider the case of condensation (Fig. 3.9).

Cold fluid

$$\left\{\begin{array}{c}\text{energy entering}\\\text{with cold fluid}\end{array}\right\} - \left\{\begin{array}{c}\text{energy leaving}\\\text{with cold fluid}\end{array}\right\} - \left\{\begin{array}{c}\text{heat transferred}\\\text{to hot fluid}\end{array}\right\} = \left\{\begin{array}{c}\text{energy stored}\\\text{in cold fluid}\end{array}\right\}$$

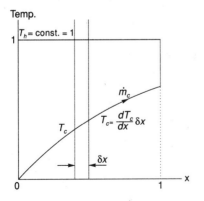

Fig. 3.9 Condensation

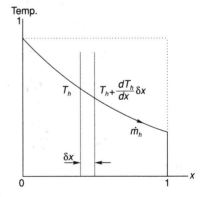

Fig. 3.10 Evaporation

$$\left[\dot{m}_C C T_c + U\left(\frac{S\delta x}{L}\right)(1 - T_c)\right] - \dot{m}_c C_c\left(T_c + \frac{dT_c}{dx}\delta x\right) = 0$$

$$\frac{US}{L}(1 - T_c) = \dot{m}_c C_c \frac{dT_c}{dx}$$

$$\frac{dT_c}{dx} + \frac{N_c}{L}T_c - \frac{N_c}{L} = 0$$

which provides the solution

CONDENSATION
$$T_c = 1 - e^{-N_c(x/L)} \qquad (3.12)$$

Similarly for the case of evaporation (Fig. 3.10)

EVAPORATION
$$T_h = e^{-N_h(x/L)} \qquad (3.13)$$

3.4 LONGITUDINAL CONDUCTION IN CONTRAFLOW

Longitudinal conduction in the direction of falling temperature is a problem in high effectiveness, contraflow heat exchangers, particularly in the design of cryogenic plant. Conduction will occur in each fluid, in the wall separating the two fluids and in the shell of the exchanger. For gases, conduction effects in the fluid can usually be neglected,* and with plate-fin heat exchangers shell conduction losses are also small, leaving only longitudinal conduction in the wall as a significant effect.

The following analysis is based on the paper by Kroeger (1967) which includes a closed-form solution for the case of equal water equivalents – identical to the case of most severe deterioration in heat exchanger performance. Given the terminal temperatures of a contraflow heat exchanger, the problem is to find the corrected LMTD to use in design when longitudinal conduction effects are present.

---

*Thermal conductivity in the supercritical liquid region starts to become important.

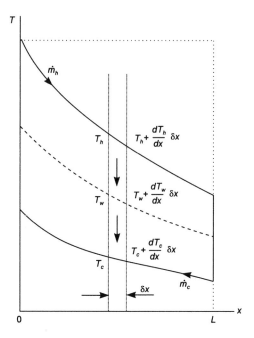

Fig. 3.11 Schematic temperature profiles including the wall

In sizing a heat exchanger, a first design pass is necessary to determine the approximate size of the exchanger and to obtain the wall cross-section involved in longitudinal conduction. The reduced LMTD is then obtained using Kroeger's method, and applied in a second design pass to obtain the final size of the exchanger.

This obviates the need to solve three simultaneous partial differential equations for each individual case. The approach is approximate, but as maximum deterioration in performance has been incorporated, the final design should be conservative.

We strike the usual energy balances (Fig. 3.11).

Hot fluid

$$\left\{\begin{array}{c}\text{energy entering}\\ \text{with hot fluid}\end{array}\right\} - \left\{\begin{array}{c}\text{energy leaving}\\ \text{with hot fluid}\end{array}\right\} - \left\{\begin{array}{c}\text{heat transferred}\\ \text{to solid fluid}\end{array}\right\} = \left\{\begin{array}{c}\text{energy stored}\\ \text{in hot fluid}\end{array}\right\}$$

$$\dot{m}_h C_h T_h - \dot{m}_h C_h T_h\left(T_h + \frac{dT_h}{dx}\delta x\right) = \alpha_h\left(S\frac{\delta x}{L}\right)(T_h - T_w) = 0$$

Solid wall

$$-\lambda A \frac{dT_w}{dx} + \lambda A \frac{d}{dx}\left(T_w + \frac{dT_w}{dx}\delta x\right) + \alpha_h\left(S\frac{\delta x}{L}\right)(T_h - T_w) - \alpha_c\left(S\frac{\delta x}{L}\right)(T_w - T_c) = 0$$

3.4 LONGITUDINAL CONDUCTION IN CONTRAFLOW

Cold fluid

$$\dot{m}_c C_c \left(T_c + \frac{dT_c}{dx} \delta x \right) + \alpha_c \left(S \frac{\delta x}{L} \right)(T_w - T_c) - \dot{m}_c C_c T_c = 0$$

Thus

$$+\frac{dT_h}{dx} = -\frac{\alpha_h S}{\dot{m}_h C_h} \cdot \frac{1}{L}(T_h - T_w)$$

$$-\lambda A \frac{d^2 T_w}{dx^2} = +\alpha_h S \cdot \frac{1}{L}(T_h - T_w) - \alpha_c S \cdot \frac{1}{L}(T_w - T_c)$$

$$-\frac{dT_c}{dx} = +\frac{\alpha_c S}{\dot{m}_c C_c} \cdot \frac{1}{L}(T_w - T_c)$$

The central equation may be written as

$$+\frac{\lambda A L}{M_w C_w} \cdot \frac{d^2 T_w}{dx^2} = -\frac{\alpha_h S}{M_w C_w}(T_h - T_w) + \frac{\alpha_c S}{M_w C_w}(T_w - T_c)$$

and substituting

$$\kappa = \frac{\lambda}{\rho_w C_w}, \quad n_h = \frac{\alpha_h S}{\dot{m}_h C_h}, \quad n_c = \frac{\alpha_c S}{\dot{m}_c C_c},$$

$$\tilde{m}_h = \dot{m}_h \tau_h, \quad \tilde{m}_c = \dot{m}_c \tau_c$$

$$R_h = \frac{M_w C_w}{\tilde{m}_h C_h}, \quad R_c = \frac{M_w C_w}{\tilde{m}_c C_c}$$

the three coupled differential equations governing steady-state heat transfer with longitudinal wall conduction become

$$+\frac{dT_h}{dx} = -\frac{n_h}{L}(T_h - T_w)$$

$$-\kappa \frac{d^2 T_w}{dx^2} = +\frac{n_h}{R_h \tau_h}(T_h - T_w) - \frac{n_c}{R_c \tau_c}(T_w - T_c) \quad (3.14)$$

$$-\frac{dT_c}{dx} = +\frac{n_c}{L}(T_w - T_c)$$

Ntu values are redefined as they are now *local* values designated by the subscripted lower case n. As an aside, if wall conductance is neglected, the relationship between *overall* values of Ntu (N_h, N_c) and *local* values of Ntu (n_h, n_c) can be expressed as

$$\frac{N_h}{n_h} + \frac{N_c}{n_c} = 1$$

Using the first and last of equations (3.14) the central equation may be written as

$$\kappa \frac{d^2 T_w}{dx^2} = +\frac{L}{R_h \tau_h} \cdot \frac{dT_h}{dx} - \frac{L}{R_c \tau_c} \cdot \frac{dT_c}{dx}$$

which may be written

$$\left(\frac{\lambda_w}{\rho_w C_w}\right) \frac{d^2 T_w}{dx^2} = +\left(\frac{\dot{m}_h C_h L}{M_w C_w}\right) \frac{dT_h}{dx} - \left(\frac{\dot{m}_c C_c L}{M_w C_w}\right) \frac{dT_c}{dx}$$

and for the special case of equal water equivalents ($\dot{m}_h C_h = \dot{m}_c C_c = \dot{m}C$) we can write

$$\left(\frac{\lambda A}{\dot{m}C}\right) \frac{d^2 T_w}{dx^2} = +\frac{dT_h}{dx^2} - \frac{dT_c}{dx} \quad (3.15)$$

The set of equations (3.14) now become

$$\left(\frac{d}{dx} + \frac{n_h}{L}\right) T_h \quad\quad\quad -\left(\frac{n_h}{L}\right) T_w = 0$$

$$-\frac{dT_h}{dx} \quad +\frac{dT_c}{dx} + \left(\frac{\lambda A}{\dot{m}C}\right)\frac{d^2 T_w}{dx^2} = 0 \quad (3.16)$$

$$+\left(\frac{d}{dx} - \frac{n_c}{L}\right) T_c \quad +\left(\frac{n_c}{L}\right) T_w = 0$$

Write

$$\beta_1 = \frac{n_h}{L} \quad \beta_2 = \frac{n_c}{L} \quad \sigma = \frac{\lambda A}{\dot{m}C} \quad \mu = 1$$

where μ is the ratio of equal thermal capacity rates. Then normalise the length of the exchanger by putting $\hat{x} = x/L$, then

3.4 LONGITUDINAL CONDUCTION IN CONTRAFLOW

$$\left(\frac{d}{d\hat{x}} + \beta_1\right)T_h \qquad\qquad - \beta_1 T_w = 0$$

$$-\frac{dT_h}{d\hat{x}} + \frac{dT_c}{d\hat{x}} + \left(\frac{\sigma}{\mu}\right)\cdot\frac{d^2 T_w}{d\hat{x}^2} = 0$$

$$+\left(\frac{d}{d\hat{x}} - \beta_2\right)T_c \qquad + \beta_2 T_w = 0$$

corresponding with Kroeger's equations, which have the solution

$$T_h = A_0 - \frac{A_1}{\beta_1} + A_1\hat{x} + \left(\frac{\beta_1}{\beta_1 - r_2}\right)e^{-r_2\hat{x}}A_2 + \left(\frac{\beta_1}{\beta_1 - r_3}\right)e^{-r_3\hat{x}}A_3$$

$$T_w = A_0 \qquad\quad + A_1\hat{x} \qquad\qquad + e^{-r_2\hat{x}}A_2 \qquad\qquad + e^{-r_3\hat{x}}A_3 \qquad (3.17)$$

$$T_c = A_0 + \frac{A_1}{A_3} + A_1\hat{x} + \left(\frac{\beta_2}{\beta_2 + r_2}\right)e^{-r_2\hat{x}}A_2 + \left(\frac{\beta_2}{\beta_2 + r_3}\right)e^{-r_3\hat{x}}A_3$$

where

$$r_2 = (p+q) \quad r_3 = (p-q) \quad p = \frac{(\beta_1 - \beta_2)}{2} \quad q = \sqrt{\frac{(\beta_1 - \beta_2)^2}{4} + \frac{1}{\sigma}(\beta_1 + \beta_2)}$$

The boundary conditions are

$$\hat{x} = 0,\ T_h = 1,\ dT_w/d\hat{x} = 0$$
$$\hat{x} = 0,\ T_c = 0,\ dT_w/d\hat{x} = 0$$

Substituting in the first and last of equations (3.17)

$$1 = A_0 + \left(\frac{-1}{\beta_1}\right)A_1 + \left(\frac{\beta_1}{\beta_1 - r_2}\right)A_2 + \left(\frac{\beta_1}{\beta_1 - r_3}\right) \qquad (3.18)$$

$$0 = A_0 + \left(\frac{1}{\beta_2} + 1\right)A_1 + \left[\left(\frac{\beta_2}{\beta_2 + r_2}\right)e^{-r_2}\right]A_2 + \left[\left(\frac{\beta_2}{\beta_2 + r_3}\right)e^{-r_3}\right]A_3 \qquad (3.19)$$

Differentiating the central equation of (3.17)

$$\frac{dT_w}{d\hat{x}} = A_1 + \left[-r_2 e^{-r_2\hat{x}}\right]A_2 + \left[-r_3 e^{-r_3\hat{x}}\right]A_3 \qquad (3.20)$$

3 STEADY-STATE TEMPERATURE PROFILES

and substituting from the two boundary conditions given above

$$0 = A_1 + (-r_2)A_2 + (-r_3)A_3 \qquad (3.21)$$

$$0 = A_1 + (-r_2 e^{-r_2})A_2 + (-r_3 e^{-r_3})A_3 \qquad (3.22)$$

Subtracting (3.18) from (3.19)

$$-1 = \left[\frac{1}{\beta_2} + 1 + \frac{1}{\beta_1}\right] A_1 + \left[\left(\frac{\beta_2}{\beta_1 - r_2}\right)e^{-r_2} - \left(\frac{\beta_1}{\beta_1 - r_2}\right)\right] A_2$$
$$+ \left[\left(\frac{\beta_2}{\beta_2 + r_3}\right)e^{-r_3} - \left(\frac{\beta_1}{\beta_1 - r_3}\right) A_3\right] \qquad (3.23)$$

Simultaneous equations (3.21), (3.22) and (3.23) may be written

$$-1 = a_{11}A_1 + a_{12}A_2 + a_{13}A_3$$
$$0 = a_{21}A_1 + a_{22}A_2 + a_{23}A_3$$
$$0 = a_{31}A_1 + a_{32}A_2 + a_{33}A_3$$

and solved for A_1, A_2 and A_3. Equation (3.18) may then be solved for A_0 providing the complete solution for temperature profiles (T_h, T_w, T_c) in equations (3.17).

Schematic temperature profiles

The data in Table 3.1 have been selected to illustrate the effect of longitudinal conduction on performance of the contraflow exchanger. The results of computation are presented in Fig. 3.12 which shows both fluid and solid wall-temperatures. Deviation from expected linear temperature profiles for the two fluids is quite clear.

Table 3.1 Data for evaluating longitudinal conduction profiles

Parameter	Hot gas	Cold fluid	Solid wall
Mass flow rate (kg/s)	1.00	1.50	–
Specific heat (J/kg K)	1200.0	800.0	–
Local transfer units (Ntu)	5.00	4.00	–
Density (kg/s)	–	–	1000.0
Cross-section (m^2)	–	–	0.10
Exchanger length (m)	–	–	0.50
Thermal conductivity (J/m s K)	–	–	500.0

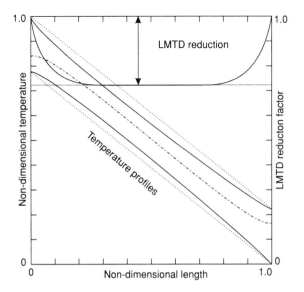

Fig. 3.12 Temperature profiles with longitudinal conduction (equal water equivalents)

It also becomes possible to evaluate the actual temperature difference along the exchanger, and to relate this to the original value assumed with no longitudinal conduction.

We will use this LMTD reduction factor later in design. Note that the solution for equal water equivalents gives the maximum reduction in LMTD, and thus will be conservative in direct sizing when the water equivalents are not equal. In such cases the LMTD calculated for unequal water equivalents can be multiplied by the LMTD reduction factor for equal water equivalents to produce a corrected mean temperature difference to be used in design.

Kroeger's explicit solution is for equal water equivalents only. It will be left for the reader to discover the appropriate values to insert in Table 3.1 when the original water equivalents are not equal. Kroeger's paper provides numerical solutions and graphs for the unequal cases, but does not discuss how to evaluate the reduction in mean temperature difference.

3.5 MEAN TEMPERATURE DIFFERENCE IN UNMIXED CROSS-FLOW

One-pass cross-flow offers several possible arrangements; the fundamental configuration has both fluids unmixed (Fig. 3.13). A simple case is for one fluid mixed. When two or more passes are involved, the number of possible arrangements increases because heat exchange in each pass can be both unmixed or with one fluid mixed, and between passes each fluid can be mixed or unmixed.

66 3 STEADY-STATE TEMPERATURE PROFILES

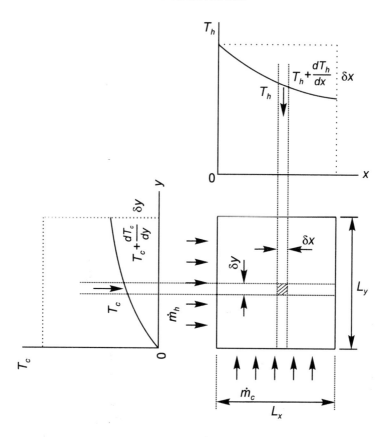

Fig. 3.13 Differential temperatures for one-pass unmixed cross-flow

Here we shall consider the fundamental problem of both fluids unmixed in a one-pass exchanger and neglect effects of longitudinal conduction. A solution to the problem was first obtained by Nusselt (1911) in the form of analytical series expansions. Since then many other workers have sought improved solutions, but it was left to Baclic and Heggs (1985) to show that all of them were mathematically equivalent. All of the expressions obtained require numerical evaluation to be of use. We shall not travel down that path.

Instead the fundamental equations will be obtained in canonical form followed by their direct numerical solution using finite-differences. This prepares the reader for later chapters in which transients are considered, where a numerical approach rather like solving the unmixed cross-flow problem is required using the method of characteristics.

The mass flow in the x-direction entering the element of side δy is $\dot{m}_h \, \delta y / L_y$ Strike energy balances for hot fluid and cold fluid together.

3.5 MEAN TEMPERATURE DIFFERENCE IN UNMIXED CROSS-FLOW

Hot fluid

$$\left\{\begin{array}{l}\text{energy entering}\\ \text{with hot fluid}\end{array}\right\} - \left\{\begin{array}{l}\text{energy leaving}\\ \text{with hot fluid}\end{array}\right\} - \left\{\begin{array}{l}\text{heat transferred}\\ \text{to cold fluid}\end{array}\right\} = \left\{\begin{array}{l}\text{energy stored}\\ \text{in hot fluid}\end{array}\right\}$$

$$\left(\dot{m}_h \frac{\delta y}{L_y}\right) C_h T_h - \left(\dot{m}_h \frac{\delta y}{L_y}\right) C_h \left(T_h + \frac{\partial T_h}{\partial x}\delta x\right) - U(\delta x \delta y)(T_h - T_c) = 0$$

Cold fuid

$$\left(\dot{m}_c \frac{\delta x}{L_x}\right) C_c T_c - \left(\dot{m}_c \frac{\delta x}{L_x}\right) C_c \left(T_c + \frac{\partial T_c}{\partial y}\delta y\right) + U(\delta x \delta y)(T_h - T_c) = 0$$

Thus

$$\frac{\partial T_h}{\partial x} = -\frac{UL_y}{\dot{m}_h C_h}(T_h - T_c) = -\frac{US}{\dot{m}_h C_h} \cdot \frac{1}{L_x}(T_h - T_c)$$

$$\frac{\partial T_c}{\partial y} = +\frac{UL_x}{\dot{m}_c C_c}(T_h - T_c) = +\frac{US}{\dot{m}_c C_c} \cdot \frac{1}{L_y}(T_h - T_c)$$

giving

$$\boxed{\begin{aligned}\frac{\partial T_h}{\partial x} &= -\frac{N_h}{L_x}(T_h - T_c)\\ \frac{\partial T_c}{\partial y} &= +\frac{N_c}{L_y}(T_h - T_c)\end{aligned}} \quad (3.24)$$

Normalising the sides (L_x, L_y) of the exchanger and the temperatures on both sides with linear scalings

$$\hat{x} = \frac{x}{L_x} \qquad \hat{y} = \frac{y}{L_y} \qquad \hat{T} = \frac{T - T_{c1}}{T_{h1} - T_{c1}}$$

resulting in the equations

$$\boxed{\begin{aligned}\frac{\partial \hat{T}_h}{\partial \hat{x}} &= -N_h(\hat{T}_h - \hat{T}_c)\\ \frac{\partial \hat{T}_c}{\partial \hat{y}} &= +N_c(\hat{T}_h - \hat{T}_c)\end{aligned}}$$

To keep the Ntu values visible, refrain from making the complete normalisation

3 STEADY-STATE TEMPERATURE PROFILES

$$\hat{x} = \frac{x}{L_x} N_h \qquad \hat{y} = \frac{y}{L_y} N_c$$

which would produce the canonical pair of Nusselt equations (see Section 2.14). The present choice retains the values of Ntu to help with physical interpretation of the finite-difference solution, and from this point onwards drop the ^ notation for simplicity.

$$\frac{\partial T_h}{\partial x} = -N_h(T_h - T_c)$$
$$\frac{\partial T_c}{\partial y} = +N_c(T_h - T_c)$$
(3.25)

The first analytical solution of the coupled partial differential equation pair (3.25) was obtained by Nusselt (1911). Many further analytical solutions in the form of infinite series, were subsequently published, until Baclic and Heggs (1985) showed that all the solutions were mathematically equivalent.

For temperature distributions in cross-flow, symbolic logic software evaluation of one of the many equivalent mathematical series solutions obtained by Baclic and Heggs (1985) may be less comprehensive than proceeding directly from the finite-difference numerical solution, which provides:

- mean temperature difference
- mean outlet temperatures
- outlet temperature profiles
- temperature sheets
- temperature difference sheet
- effectiveness
- self-checking heat balances

We notice that the independent variables occur in different equations. This allows a straightforward numerical solution similar in style to the method of characteristics.

It is straightforward to set up a finite-difference solution starting with (3.25) and using a unit block (x, y, T) as a means of representing the temperature field. If both T_h and T_c are uniform at inlet, the initial states $(T_h = 1.0, T_c = 0.0)$ can be used to obtain the temperature responses for T_h at $x = 0$ and T_h at $y = 0$, which turn out to be the temperature distributions for condensation and evaporation, respectively.

CONDENSATION   EVAPORATION

$$T_c = 1 - e^{-N_c(y/L_y)} \qquad T_h = e^{-N_h(x/L_x)}$$

3.5 MEAN TEMPERATURE DIFFERENCE IN UNMIXED CROSS-FLOW

Now consider the (x, y) face of the block as a (50×50) square grid, and from the Nusselt slopes at $x=0$ and $y=0$ obtain

$$T_h[1, y] \text{ over } (1 \cdots y \cdots 50) \qquad T_c[x, 1] \text{ over } (1 \cdots x \cdots 50)$$

Since both $T_h[1, 1]$ and $T_c[1, 1]$ are now known, using the modified Euler-Cauchy method it becomes possible to generate values for

$$T_h[x, 1] \text{ over } (2 \cdots x \cdots 50) \qquad T_c[1, y] \text{ over } (2 \cdots y \cdots 50)$$

The process is repeated until the temperature sheets are complete. All essential design parameters may now be found, and results for $(N_h = N_c = 5.0)$ are shown in Figs. 3.14 and 3.15. The above approach is explicit finite difference, and accuracy is affected by error propagation, particularly from the steepest parts of the temperature field. For equally spaced intervals at least a (50×50) mesh should be used, and desirably correct analytical values used as input data.

With $N_h = N_c = 10.0$, the computed temperature difference (TD) sheet $(T_h - T_c)$ is near zero at the corners $(0, 1, T)$ and $(1, 0, T)$, providing no driving force for heat transfer. Clearly the cross-flow exchanger is not an appropriate design selection when Ntu values close to 10.0 are involved. A better choice is for maximum values of Ntu around 4.0 for one-pass unmixed cross-flow.

Schematic algorithms

Codings that allow construction of finite-difference algorithms in any source code are presented in Appendix B. The graphical results of Figs. 3.14 and 3.15 were produced with these listings.

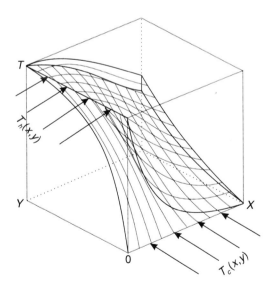

Fig. 3.14 Temperature sheets (T_h, T_c) and $N_h = N_c = 5.0$

70   3 STEADY-STATE TEMPERATURE PROFILES

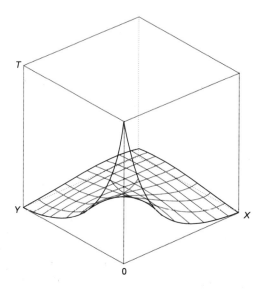

Fig. 3.15 Temperature difference sheet (T_h, T_c) and $N_h = N_c = 5.0$

3.6 EXTENSION TO TWO-PASS UNMIXED CROSS-FLOW

Figure 3.16 illustrates a flow arrangement for a two-pass cross-flow exchanger. Of the possible arrangements suggested by Stevens *et al.* (1957) and Baclic (1990), this is possibly the most practicable configuration.

The exchanger can be considered as two single-pass exchangers of equal surface area; temperature and temperature-difference (TD) distributions may be obtained by using the algorithm for the case of simple unmixed cross-flow. Values of Ntu employed in the solution now refer to half the total surface area, and the solution proceeds by solving the simple cases A and B successively until there is no change in the intermediate temperature distributions for T_h and T_c (Fig. 3.17).

It is convenient to take the hot fluid inlet temperature as $T_h = 1$, and the cold fluid inlet temperature as $T_c = 0$. A solution to case A is first obtained by assuming that the intermediate cold inlet temperature distribution is $T_c = 0.5$ everywhere. An approximate intermediate warm outlet temperature distribution for T_h is calculated, which then becomes input for case B. This produces a better estimate for the intermediate temperature distribution T_c, and the process continues until no significant changes in intermediate temperature distributions for T_h and T_c are obtained.

In this coupled solution, negative temperature differences may appear because nothing has been built in to satisfy the second law of thermodynamics. As soon as crossover temperature appears, an adjusted mean of the two temperatures may be substituted according to the following scheme:

3.6 EXTENSION TO TWO-PASS UNMIXED CROSS-FLOW 71

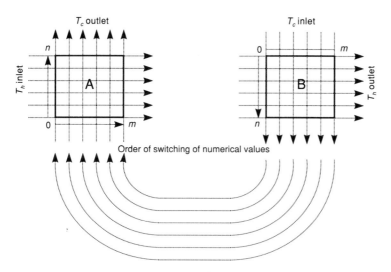

Fig. 3.16 Computation of mean TD for two-pass cross-flow

$$T_h = 1.010 \times (T_h + T_c)/2$$
$$T_c = 0.990 \times (T_h + T_c)/2$$

Mean outlet values for T_h and T_c can then be evaluated, together with all other information concerning temperature distributions.

In preparing the algorithm it is important to note the order of switching of numerical values for intermediate temperature distributions between cases A and B; they are best studied on Fig. 3.16. Appropriate selection of order switching leads to solutions for different exchanger configurations.

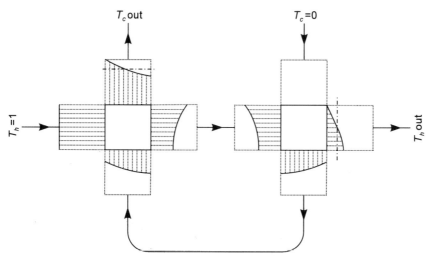

Fig. 3.17 Terminal temperature profiles for $N_h=7.0$, $N_c=7.0$

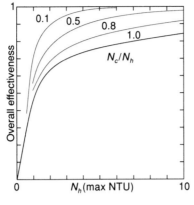

Fig. 3.18 ε-Ntu for one-pass unmixed cross-flow

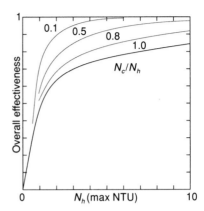

Fig. 3.19 ε-Ntu for two-pass unmixed cross-flow

Little difference in performance is evident from the ε-Ntu curves based on mixed outlet temperature value (Figs 3.18 and 3.19). However, it is important also to examine outlet temperature profiles in Figs. 3.20 and 3.21.

Final mixing involves external thermodynamic irreversibilities, and it is relevant to note the values of Ntu above which any local temperature differences in the exchanger become negligible. Surface areas of such regions are inefficient in heat transfer. Figures 3.22 and 3.23 show mean outlet temperatures together with their associated temperature bands. Band limits should not get too close to the limiting values 1 and 0, indicating that one-pass unmixed cross-flow is limited to Ntu values below 4.0, whereas two-pass unmixed cross-flow might be used at Ntu values around 7.0 whenever exchanger mass is important, e.g. aerospace applications.

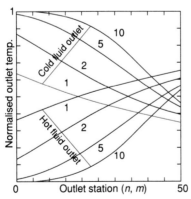

Fig. 3.20 Outlet temperature profiles, for one-pass unmixed cross-flow. Profiles for $N_h = N_c$ ($N_h = 1, 2, 5, 10$)

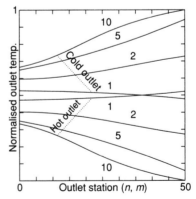

Fig. 3.21 Outlet temperature profiles for two-pass unmixed cross-flow. Profiles for $N_h = N_c$ ($N_h = 1, 2, 5, 10$)

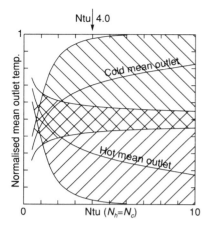
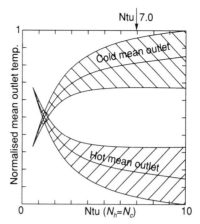

Fig. 3.22 Mean outlet temperatures and temperature bands for one-pass unmixed cross-flow

Fig. 3.23 Mean outlet temperatures and temperature bands for two-pass unmixed cross-flow

There is no restriction in going to higher Ntu values – it just means that some of the surface is doing nothing while the remainder carries more duty – but the designer should be aware of the situation and decide when another pass is to be added. In the limit, the performance of a four-pass cross-flow arrangement approaches that of a contraflow design.

For sizing A and B, there is an alternative design philosophy to having equal surface areas; it involves iterating until effectiveness values are equal for A and B. This results in different paired values (N_h, N_c) for each section, and the surface areas involved will thus be different.

3.7 INVOLUTE-CURVED PLATE-FIN EXCHANGERS

These may be designed as (unmixed/unmixed) single-pass contraflow exchangers using the same theory as developed for flat-plate compact contraflow exchangers in Chapter 4. It is only necessary to relate the curved length of an involute plate s to the inner a and outer b radii of the exchanger core.

Parametric representation of the involute curve depends on angle t measured on the base circle

$$x = a\cos(t) + at\sin(t)$$
$$y = a\sin(t) - at\cos(t) \qquad (3.26)$$

Differential length of arc is given by

$$\frac{ds}{dx} = \sqrt{1 + \left(\frac{dy}{dx}\right)^2}$$

giving $ds = \sqrt{(dx)^2 + (dy)^2}$

and $\left(\dfrac{ds}{dt}\right)^2 = \left(\dfrac{dx}{dt}\right)^2 + \left(\dfrac{dy}{dt}\right)^2$ (3.27)

Using equations (3.26) in (3.27)

$$ds = at\, dt$$
$$s = \int ds = \dfrac{at^2}{2}$$ (3.28)

Outer radius b is given by

$$b^2 = x^2 + y^2$$
$$b = a\sqrt{1 + t^2}$$ (3.29)

One-and two-pass cross-flow arrangements may also be analysed using the same theory developed for flat-plate compact cross-flow exchangers in Chapter 4.

The unmixed one-pass cross-flow arrangement has one fluid flowing axially and the other fluid flowing radially.

Unmixed two-pass cross-flow may have two arrangements. The first arrangement has one radial pass and two axial passes. The second arrangement has two radial passes and one axial pass. Insulation strips would be required in both cases.

Arrangements with higher numbers of passes are also possible but at the expense of greater complexity in manufacture.

3.8 LONGITUDINAL CONDUCTION IN ONE-PASS UNMIXED CROSS-FLOW

There is no clean analytical solution to this problem, or a numerical solution which can be coded without some effort; see Chiou (1978).

Chiou investigated performance deterioration for Ntu values in the range 1 to 100. Elsewhere in this text it is shown that surface area in the (0, 1) and (1, 0) corners of the unit block for one-pass unmixed cross-flow tend to become ineffective at Ntu values greater than about 5.0. Thus Chiou's results are of practical interest perhaps only for Ntu's less than 5.0 for single-phase fluids.

The greatest deterioration occured with equal water equivalents, as was the case with pure contraflow.

A simpler approach based on that of Chiou may be employed. The starting point is the set of three partial differential equations for steady-state cross-flow

3.8 LONGITUDINAL CONDUCTION IN ONE-PASS UNMIXED CROSS-FLOW

$$+u_h \frac{\partial T_h}{\partial x} = -\frac{n_h}{\tau_h}(T_h - T_w)$$

$$-\hat{\kappa}_x \frac{\partial^2 T_w}{\partial x^2} - \hat{\kappa}_y \frac{\partial^2 T_w}{\partial y^2} = +\frac{n_h}{R_h \tau_h}(T_h - T_w) - \frac{n_c}{R_c \tau_c}(T_w - T_c) \quad (3.30)$$

$$+u_c \frac{\partial T_c}{\partial y} = +\frac{n_c}{\tau_c}(T_w - T_c)$$

In considering how we might solve these coupled equations, it is useful first to consider whether solutions can be found for the two flow-inlet faces of the cross-flow exchanger. If these 'boundary' conditions become available, the chances of finding a solution to the complete problem are much improved.

Equation pairs for the inlet faces

The governing equations for both hot fluid and cold fluid conditions may be extracted from the above set, and when normalised to unit length and unit temperature span they provide two separate sets of equations.

Hot inlet face

$$\frac{dT_h}{dX} = -n_h(T_h - T_w)$$

$$-\frac{\hat{\kappa}_x}{L_x^2} \frac{d^2 T_w}{dX^2} = +\frac{n_h}{R_h \tau_h}(T_h - T_w) - \frac{n_c}{R_c \tau_c}(T_w - T_c) \quad (3.31)$$

Cold inlet face

$$\frac{dT_c}{dY} = +n_c(T_w - T_c)$$

$$-\frac{\hat{\kappa}_y}{L_y^2} \cdot \frac{d^2 T_w}{dY^2} = +\frac{n_h}{R_h \tau_h}(T_h - T_w) - \frac{n_c}{R_c \tau_c}(T_w - T_c) \quad (3.32)$$

where $\hat{\kappa}_x = \left(\frac{A_x L_x}{V}\right) \kappa_x$ and $\hat{\kappa}_y = \left(\frac{A_y L_y}{V}\right) \kappa_y$

Solution methods

Analytical approach
Attempts were made first to solve these equations analytically because the presence of T_w alone in the first equation of both sets means that explicit equations for T_w and its derivatives can be found for substitution in the second equation of each pair.

Boundary conditions are $T_h = 1$ for $x = 0$ and for all y, and $T_c = 0$ for $y = 0$ and for all x. Wall temperature at the origin might be obtained from knowledge of the local heat transfer coefficients on each side of the exchanger, $T_w = \alpha_h/(\alpha_h + \alpha_c)$, as there is no longitudinal conduction at this point. But as will be seen later, this may be a questionable assumption when longitudinal conduction is present.

The resulting third-order ordinary differential equation in T_h or T_c may be solved by assuming the usual exponential solution. A cubic equation, which can be solved numerically, permits an analytical solution for both temperature profiles on each face, and the necessary boundary conditions are available for a complete solution.

However, when numerical values for a real exchanger were inserted, although the cubic could be solved, the roots so produced became arguments in exponentials and some of them could not be evaluated on the computer. It then became apparent that the equations were very 'stiff', and some other method of solution would be required.

Explicit finite-difference approach
When an explicit finite-difference method was applied, starting at the origin, again the solution 'blew up'. This may be because the requirement of zero wall temperature gradient at the origin is very swiftly followed by a point of inflexion in the wall profile, and also because the assumption of $T_w = \alpha_h/(\alpha_h + \alpha_c)$ at the origin is required. Another approach thus became necessary.

Implicit modified finite-difference approach
The method of Chiou (1978, 1980) provides a solution for the temperature field in cross-flow when longitudinal conduction is present, but the equations presented by Chiou are quite complicated, as the author freely admitted in his paper.

Examining this solution, it became apparent that it was entirely equivalent to writing simultaneous finite-difference expressions then loading them in a matrix for direct solution. For equations (3.31) the trick is to avoid writing wall equations for $x = 0$ and $x = L_x$; we are only interested in the difference between hot and cold fluid temperatures to determine degradation in exchanger performance, so this seems quite acceptable.

The same remarks apply for equations (3.32) and the y-direction.

Formulation of a matrix

Matrices are most easily generated by writing out fully the equations for five finite-difference intervals then generalising this to m intervals. To simplify notation, temperatures are replaced by their subscripts in upper case (Fig. 3.24).

3.8 LONGITUDINAL CONDUCTION IN ONE-PASS UNMIXED CROSS-FLOW

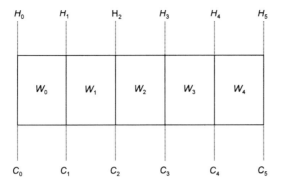

Fig. 3.24 Notation for five finite-difference intervals

For five intervals on the hot face (H), the first equation of (3.31) above may be written

$$j = 0 \qquad \left[1 + \frac{n_h \Delta X}{2}\right] H_1 \qquad - [n_h \Delta X] W_0 = +\left[1 - \frac{n_h \Delta X}{2}\right]$$

$$j = 1, 2, 3 \qquad \left[1 + \frac{n_h \Delta X}{2}\right] H_{j+1} - \left[1 - \frac{n_h \Delta X}{2}\right] H_j - [n_h \Delta X] W_j = 0$$

$$j = 4 \qquad \left[1 + \frac{n_h \Delta X}{2}\right] H_5 \quad - \left[1 - \frac{n_h \Delta X}{2}\right] H_4 - [n_h \Delta X] W_4 = 0$$

The second equation of (3.31) may be written

$j = 0$

$$-\left[\frac{\hat{\kappa}_x}{L_x^2 \Delta X^2}\right](W_1) \quad + \left[\frac{\hat{\kappa}_x}{L_x^2 \Delta X^2} + \frac{n_h}{R_h \tau_h} + \frac{n_c}{R_c \tau_c}\right] W_0 - \left[\frac{n_h}{2 R_h \tau_h}\right] H_1 \quad = \left[\frac{n_h}{2 R_h \tau_h}\right] H_0$$

$j = 1, 2, 3$

$$-\left[\frac{\hat{\kappa}_x}{L_x^2 \Delta X^2}\right](W_{j+1} + W_{j-1}) + \left[\frac{2 \hat{\kappa}_x}{L_x^2 \Delta X^2} + \frac{n_h}{R_h \tau_h} + \frac{n_c}{R_c \tau_c}\right] W_j - \left[\frac{n_h}{2 R_h \tau_h}\right](H_{j+1} + H_j) = 0$$

$j = 4$

$$-\left[\frac{\hat{\kappa}_x}{L_x^2 \Delta X^2}\right](W_3) \quad + \left[\frac{\hat{\kappa}}{L_x^2 \Delta X^2} + \frac{n_h}{R_h \tau_h} + \frac{n_c}{R_c \tau_c}\right] W_4 - \left[\frac{n_h}{2 R_h \tau_h}\right](H_5 + H_4) = 0$$

Writing the coefficients as follows:

$$P = +\left[1 + \frac{n_h \Delta X}{2}\right] \qquad S = -\left[\frac{n_h}{2 R_h \tau_h}\right]$$

3 STEADY-STATE TEMPERATURE PROFILES

$$Q = -\left[1 - \frac{n_h \Delta X}{2}\right] \qquad T = -\left[\frac{\hat{\kappa}_x}{L_x^2 \Delta X^2}\right]$$

$$R = -[n_h \Delta X] \qquad U = +\left[\frac{2\hat{\kappa}_x}{L_x^2 \Delta X^2} + \frac{n_h}{R_h \tau_h} + \frac{n_c}{R_c \tau_c}\right]$$

the 10×11 matrix to be solved is given in Table 3.2.

Inversion of this matrix gives the values of T_h at $X = j\Delta X$, ($j = 1, 2, 3, 4, 5$) and T_w at $X = (j+0.5)\Delta X$, ($j = 0, 1, 2, 3, 4$). The value of T_h at $X = 0$ is 1.0 and the value of T_w at $X = 0$ may be taken as $\alpha_h/(\alpha_h + \alpha_c)$.

Table 3.2   Matrix of coefficients

	1	2	3	4	5	6	7	8	9	10	11
	H_1	H_2	H_3	H_4	H_5	W_0	W_1	W_2	W_3	W_4	rhs
1	P					R					$-Q$
2	Q	P					R				
3		Q	P					R			
4			Q	P					R		
5				Q	P					R	
6	S					$U+T$	T				$-S$
7	S	S				T	U	T			
8		S	S				T	U	T		
9			S	S				T	U	T	
10				S	S				T	$U+T$	

Making use of this matrix as a model, the general PASCAL algorithms for a $2m$, $2m + 1$ matrix may easily be constructed as follows:

```
              FOR j := 1 TO m DO
      BEGIN k := j;
              A[j, k] := P
      END;
              FOR j := 2 TO m DO
      BEGIN K := j - 1;
              A[j, k] := Q
      END;
```

and so on. A similar matrix exists for equations (3.32).

Solution of the inlet problem

Data for the cross-flow exchanger is due to Shah (1981) and is given at the end of this section.

3.8 LONGITUDINAL CONDUCTION IN ONE-PASS UNMIXED CROSS-FLOW

In equations (3.31) and (3.32) coefficients are evaluated in terms of $\hat{\kappa} = \left(\frac{AL}{V}\right)\kappa$, where A is the cross-section for axial conduction, L is the length of the exchanger in the direction of conduction, V is the volume of solid material in the heat exchange surfaces and κ is the thermal diffusivity of the construction material. Evaluation of the cross-section for plain plate-fin surface 19.86 is eased if the centre-line of the geometry is first obtained. In this case it is approximately a sine curve (Fig. 3.25) and the arc length may be determined using standard methods, e.g. Wylie (1953).

Note that we are interested in only half the fin height associated with one plate in the exchanger. The half-height cross-section for rectangular offset strip-fin surface 104(S) designated 1/8-15.61 is more easily determined. This provides values of surface cross-sections useful in determining V, which must also include the plate.

Cross-sections for longitudinal conduction may be obtained at the same time. In the case of plain plate-fin surface 19.86 the conduction section is the same as for determining the volume of material. In the case of rectangular offset strip-fin surface 104(s) the conduction path is effectively reduced to material attached directly to the plate, and this is only half the width of the plate.

A little more thought is required to evaluate the contribution of both surfaces in conduction at right angles to their flow lengths, which contributes to deterioration in performance on the other side of the plate. For plain surface 19.86 this was taken as the cross-section of the material. For the rectangular offset strip-fin surface, again, half the cross-section of the material was assumed effective.

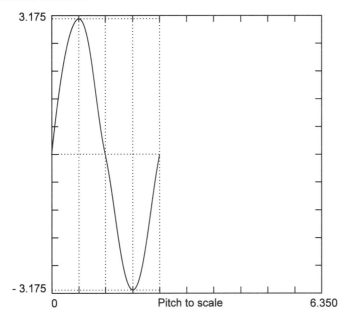

Fig. 3.25 Centre-line of plain plate-fin surface 19.86 (approximately to scale)

The reader may rightly enquire whether such detailed evaluation of the surfaces for longitudinal conduction is necessary; we could more simply use κ instead of $\hat{\kappa}$ and still obtain a conservative design. We still have to find V to calculate the mass of material for evaluating R_h and R_c, but this simpler procedure is altogether less traumatic.

Normalised inlet conditions for both sides of the exchanger were computed using equations (3.31) and (3.32) and are presented graphically in Fig. 3.26. It was found possible to use 50 divisions along each face; the ends of the computed curve for wall temperature are marked by 'plus' symbols to show how distant they were from the end values.

Shown dotted are the steady-state values computed for the case without the wall equation; normalised wall temperature is evaluated by using the appropriate ratios of local heat transfer coefficients, $T_w = \alpha_h/(\alpha_h + \alpha_c)$.

For the exchanger examined, longitudinal conduction in the wall does affect performance. Reduction in mean temperature difference on the 'condenser' face was evaluated as 0.992 58, and reduction in mean temperature on the 'evaporator' face was found to be 0.992 51.

Figure 3.15 shows the temperature ridge starting at (x = 1, y = 1) and finishing at (x = 0, y = 0); as any hill climber knows, the easiest route is along the ridge. It is likely that greatest longitudinal conduction effects are to be found at right angles to this ridge and near to the point (x = 0, y = 0). The two inlet faces provide an immediate approximation to the steepest slopes, and temperature profiles on these faces will be used to assess the amount of longitudinal conduction.

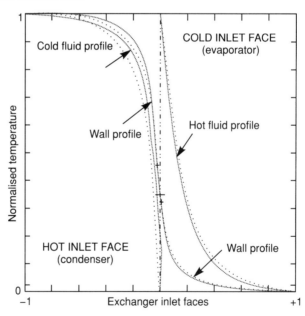

Fig. 3.26 Schematic temperature profiles on the inlet faces of a cross-flow exchanger

3.8 LONGITUDINAL CONDUCTION IN ONE-PASS UNMIXED CROSS-FLOW 81

If we use the maximum reduction factor we have found, dividing the exchanger duty by this factor, then a second run of the program will produce a design which allows for the effect. Of course, it would be better if the mean temperature difference reduction factor for the whole exchanger were computed; the necessary boundary conditions have now been determined.

A power density rating of 1.046 MW/m^3 was obtained for this cross-flow exchanger. Although the exchanger was not optimised, this value may be compared with 4.592 MW/m^3 obtained in Chapter 4 for a nearly optimised contraflow exchanger.

If the objective is to reduce longitudinal conduction to a minimum, so as to be able to calculate transients, it will suffice to use the reduction factor obtained from temperature profiles on the inlet faces. This involves systematically changing individual surface geometries as was done for the contraflow exchanger (see Chapter 4). It is beyond the scope of this book because the heat transfer and flow friction correlations for the plain sinusoidal gas-side surface geometry have not yet been reduced to universal correlations, but the approach should be clear.

The one reservation associated with Chiou's method is that it cannot properly evaluate $dT_w/dx = 0$ at $x = 0$ unless the increments Δx are extremely small. In this mathematical respect, we still do not have an exact solution to the problem (Fig. 3.27).

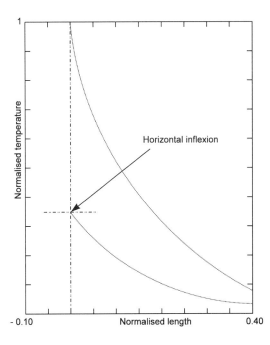

Fig. 3.27 Temperature profiles near the origin for the 'evaporator'

3 STEADY-STATE TEMPERATURE PROFILES

Input data

Performance
exchanger duty, kW $\qquad Q = 163.39$

Flow stream parameter	Hot fluid side	Cold fluid side
mass flow rate, kg/s	$m_h = 0.8962$	$\dot{m}_c = 0.8296$
inlet temperature, K	$T_{h1} = 513.15$	$\Delta P_c = 1670.0$
allowable (core) pressure loss, N/m²	$\Delta P_h = 320.0$	$T_{c1} = 277.15$
absolute inlet pressure, bar	$P_{h1} = 1.100$	$P_{c1} = 1.100$
Prandtl number	$Pr_h = 0.688$	$Pr_c = 0.692$
specific heat, J/(kg K)	$C_h = 1021.0$	$C_c = 1014.0$
absolute viscosity, kg/(m s)	$\eta_h = 0.00023\,82$	$\eta_c = 0.000\,021\,75$
thermal conductivity, J/(m s K)	$\lambda_h = 0.035\,349$	$\lambda_c = 0.031\,87$
mean absolute pressure, bar	$P_{hm} = 1.0984$	$R_c = 287.04$
gas constant, J/(kg K)	$R_h = 287.04$	$T_{cm} = 363.33$
mean bulk absolute temperature, K	$T_{hm} = 433.92$	$p_{cm} = 1.0917$
mean density, kg/m³	$\rho_{hm} = 0.881\,88$	$\rho_{cm} = 1.046\,79$

Material

plate thickness, mm	$tp_w = 0.400$
thermal conductivity, J/(m s K)	$\lambda_w = 190.0$
specific heat, J/(kg K)	$C_w = 2707.0$
density, kg/m³	$\rho_w = 1146.00$
warm gas surface	plain plate-fin 19.86
cool air surface	rectangular offset strip-fin 1/8-15.61

Data highlighted in bold was used to begin assembly of the input information. Using the exchanger duty, mass flow rate and inlet temperature, an inlet value of specific heat can be found and used to obtain a first estimate of the outlet temperature on both sides, from $Q = \dot{m}C(T_1 - T_2)$. This permits an estimate of mean bulk temperatures; similarly, the allowable pressure losses can be used with inlet pressure to obtain an estimate of mean pressures on both sides. Physical properties can now be evaluated for a first design pass. Results from the first pass can then be used to find the reduction factor in mean temperature difference caused by longitudinal conduction.

Once the results of the first pass are available, improved values for physical properties can be recalculated for use in a second design pass. In this final pass, the reduction factor found for mean temperature difference is used as the divisor for the exchanger duty. This means that the exchanger is now being designed for a fictitiously increased duty, and the greater size of core now obtained allows for the deleterious effects of longitudinal conduction.

A final core was obtained of dimensions 0.580 m × 0.277 m × 0.979 m. The inlet face for warm gas is 0.580 m × 0.979 m and the inlet face for cold air is 0.1573 m³, with an exchanger duty of 163.39 kW. The volume of the exchanger is 0.1573 m³, giving a specific performance of 1.046 MW/m³.

Output results

Performance
adjustment to mean temp. difference applied   adjLMTD = 0.99251

Hot fluid side
outlet warm gas temperature, K $T_{h2} = 334.9$
overall value of Ntu (plate surface) $N_h = 6.7643$
local value of Ntu (plate surface) $n_h = 10.4786$
heat transfer coefficient (plate surface), J/(m² s K) $\alpha_h = 411.412$
Reynolds number $Re_h = 327.7$
Stanton number $St_h = 0.016\,792$
friction factor $f_h = 0.055\,020$

Cold fluid side
outlet cool air temperature, K $T_{c2} = 471.01$
overall value of Ntu (plate surface) $N_c = 7.3578$
local value of Ntu (plate surface) $n_c = 21.1328$
heat transfer coefficient (plate surface), J/(m² s K) $\alpha_c = 762.792$
Reynolds number $Re_c = 825.3$
Stanton number $St_c = 0.021\,013$
friction factor $f_c = 0.063\,353$

Overall
overall heat transfer coefficient, J/(m² s K) $U = 267.113$
effectiveness $\epsilon = 0.8214$
number of plates $Z = 145$
aspect ratio of plate   aspect $= 2.0893$
total surface area of plates, m² $S = 23.3055$

3.9 DETERMINED AND UNDETERMINED CROSS-FLOW

A principal difficulty in sizing cross-flow heat exchangers arises when the configuration does not exactly match the mathematical requirements for mixed or unmixed cross-flow, making the mean temperature difference uncertain. Permutation pairs of three flow arrangements – mixed, unmixed and undetermined – are possible.

84 3 STEADY-STATE TEMPERATURE PROFILES

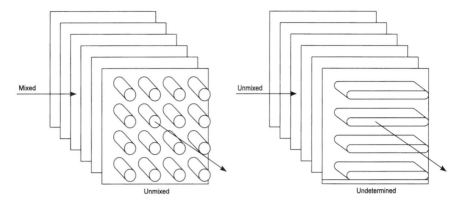

Fig. 3.28 Fin-and-tube exchanger: mixed/unmixed cross-flow

Fig. 3.29 Fin-and-duct exchanger: unmixed/ undetermined cross-flow

Figure 3.28 is an example of mixed/unmixed cross-flow; it is repeated in simpler form in Fig. 3.30a. However, triangular tube pitching is better suited to estimating fin performance of the flat plates, as the performance ratio of circular disc fins may be used instead of the hexagonal shapes. Figure 3.30b is also mixed/unmixed.

Figure 3.29 is an example of unmixed/undetermined cross-flow in which flat ducts are used in place of tubes. Once internal finning is introduced (Fig. 3.30c) the configuration becomes unmixed/unmixed. The performance of the fin plates may be handled as for straight rectangular fins, but the external finning lacks recreation of the boundary layers to obtain higher heat transfer coefficients. To improve heat transfer, the flat plates may be rippled transversely to the flow, but the overall design will still suffer from longitudinal conduction in the flat-plate fins. Slotted or louvred fins may then become necessary both to recreate boundary layers and to control longitudual conduction.

Figure 3.30d breaks up the single long flat ducts into many small staggered flat ducts or flattened tubes, approaching the unmixed/unmixed configuration without quite achieving it. With rippled plates this configuration begins to approximate to the rectangular offset-strip plate-fin surfaces used in compact heat exchangers, but their fins are so small and close together that they produce a good approximation to unmixed/unmixed flow.

When the mean temperature difference cannot easily be determined then recourse to testing may be necessary, even with single-phase cross-flow.

3.10 POSSIBLE OPTIMISATION CRITERIA

It is possible to optimise the performance of an exchanger by progressively changing the local geometry on one or both sides. Optimisation may include minimisation of any one of several criteria, including

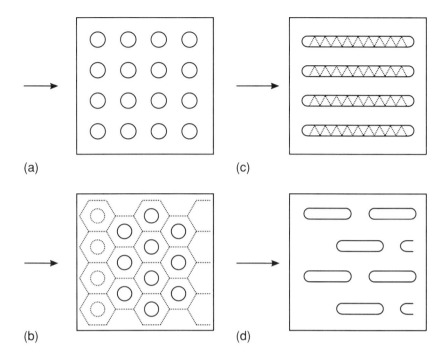

Fig. 3.30 Some possible cross-flow configurations: (a) plate-fin and tube with square pitching (mixed/unmixed cross-flow); (b) plate-fin and tube with triangular pitching (mixed/unmixed cross-flow); (c) plate-fin and flat ducts with internal finning (unmixed/unmixed cross-flow); (d) plate-fin and flattened tubes (approach to unmixed/unmixed cross-flow)

- core volume
- core mass
- frontal area

We should not be surprised if the final results differ from case to case – only one optimum at a time. However, a little thought reveals that when an exchanger is optimised so as to utilise all the allowable pressure loss on both sides, then this also corresponds to minimisation of surface area (see Figs. 5.8 and 5.9).

In cases where the constraint is the *sum* of pressure losses, it is straightforward to write a statement for the allowable pressure loss on one side when the pressure loss on the other side is known. Then proceed with the normal optimisation procedure.

Pinch technology

In a process plant with many heat exchangers, optimisation of the whole system has been made thermodynamically simple by process integration tech-

86 3 STEADY-STATE TEMPERATURE PROFILES

nology developed by Linnhoff (1978) and others. The technique introduces a distinction between 'avoidable' and 'unavoidable' inefficiencies. By first constructing separate composite *T-H* curves for hot streams and cold streams, and by sliding the cold curve horizontally towards the hot curve on the *T-H* diagram, a minimum temperature difference or *pinch point* is found somewhere in the operational field. This represents the best energy recovery configuration for the system that can be achieved with the chosen minimum temperature difference, as long as no heat is transferred across the pinch point. The second stage is to consider the heat exchangers making up the network.

Process integration is best approached by first consulting the user guide prepared by Linhoff *et al.* (1982) and the recent reference text by Hewitt *et al.* (1994).

3.11 CAUTIONARY REMARK ABOUT CORE PRESSURE LOSS

To evaluate core pressure loss, a simple direct expression may be used

$$\Delta p = \frac{4fG^2}{2\rho_m}\left(\frac{L}{d}\right) \tag{3.33}$$

This is entirely equivalent to the core friction loss term given by Kays and London (1964) in their comprehensive expression for complete losses in an exchanger core

$$\Delta p = \frac{G^2 v_1}{2}\left[(Kc + 1 - \sigma^2) + 2\left(\frac{v_2}{v_1} - 1\right) + f \cdot \frac{S_{total}}{A_{flow}} \cdot \frac{v_m}{v_1} - (1 - \sigma^2 - Ke)\frac{v_2}{v_1}\right]$$

$$\{\text{inlet loss}\} \quad \left\{\begin{array}{c}\text{flow}\\\text{acceleration}\end{array}\right\} \quad \left\{\begin{array}{c}\text{core}\\\text{friction}\end{array}\right\} \quad \{\text{outlet loss}\}$$

$$\tag{3.34}$$

Much grief can be avoided by using a consistent method for evaluating mean density (ρ_m) in equation (3.33) and mean specific volume (v_m) in equation (3.34). The mean bulk condition for gases may be approximated by first evaluating the mean pressure and temperature levels, *and only then* evaluating the mean density and mean specific volume. With this method it will be found that the reciprocal of the mean density is equal to the mean specific volume.

This will not happen when the mean specific volume is taken as the mean of specific volumes at inlet and outlet conditions, and the mean density is similarly taken as the mean of densities at inlet and outlet conditions.

In both direct-sizing and rating design methods it is essential to be consistent in the treatment of variables, otherwise significant discrepancies will arise between the two solutions of the same problem.

3.12 MEAN TEMPERATURE DIFFERENCE IN COMPLEX ARRANGEMENTS

The reader will notice that some arrangements of heat exchanger are not considered in this volume; some of them are just more complex arrangements of those which are considered, and some are well analysed in other mathematical publications.

For completeness, it was decided to include the following list of solutions for mean temperature difference in other arrangements, together with the name of the author(s) who first published the solutions. Not all of the arrangements listed are appropriate for direct sizing.

Mean temperature difference in multipass parallel flow

- one pass shell-side, two passes tube-side   Underwood, A. J. V. (1934)
- one pass shell-side, four passes tube-side   Underwood, A. J. V. (1934)
- one pass shell-side, three passes tube-side   Fischer, K. F. (1938)
- one pass shell-side, infinite number of
  passes tube-side   Smith, D. M. (1934)
- two passes shell-side, four passes tube-side   Underwood, A. J. V. (1934)

Mean temperature difference in multiple cross-flow

- single passes, both fluids unmixed   Nusselt, W. (1911)
- single passes, both fluids mixed   Smith, D. M. (1934)
- single passes, first fluid mixed, second fluid
  unmixed   Smith, D. M. (1934)
- first fluid single-pass mixed, second fluid
  two passes unmixed, except between passes   Smith, D. M. (1934)

All the above mathematical analyses (and references) are included in the textbook by Jakob (1957).

Further analyses of single-two-and three-pass cross-flow arrangements are considered by Stevens *et al.* (1957). Four-pass arrangements tend to approach true cross-flow. These results are suitable for use in LMTD-Ntu type solutions, whereas some of the equivalent ε-Ntu solutions are given by Shah (1982).

A more recent paper by Spang and Roetzel (1995) provides an approximate equation with three or four empirical parameters for the calculation of LMTD-correction factors for about 50 different configurations. Comprehensive ε-Ntu type results for ***36 two-pass cross-parallel flow*** arrangements and ***36 two-pass cross-counterflow*** arrangements have been presented by Baclic (1990), which greatly extend the work of Stevens *et al.* (1957). Graphs are presented to help select the best flow arrangements which minimise irreversibilities due to mixing at exit and between passes. Explicit solutions were earlier provided by Baclic and Gvozdenac (1981).

Numerical solution for mean temperature difference for all of these configurations is straightforward, and the possibility of temperature crossover should be examined as discussed in Section 3.6.

3.13 EXERGY DESTRUCTION

By this stage the reader should be aware that the traditional heat exchanger design is based on the *balance of energy axiom*, plus a degenerate form of the *balance of momentum axiom* for pressure loss. Only briefly, in Section 2.12, was there an appeal to the *growth of entropy inequality* (second law, or destruction of exergy). To achieve a balanced understanding of the role of thermodynamics in the design of heat exchangers (plus power and cryogenic cycles) the reader should study the excellent text by Bejan (1988) *Advanced Engineering Thermodynamics*, which presents thorough and readable analyses of many engineering systems of interest.

Exergy destruction occurs by 'heat transfer across a finite temperature difference' and by 'flow with friction'. Bejan's analysis shows that there exists a trade-off in performance between mean temperature difference and pressure loss in heat exchangers, indicating that minimisation of irreversibility is possible. For heat exchangers in which the water equivalents of the two fluids are unequal there also exists a flow-imbalance irreversibility, which can only be designed out by reconfiguring plant to use balanced exchangers. See also Sekulic (1986, 1990) who emphasises the importance of including irreversibility due to pressure loss. Das and Roetzel (1995) apply the approach to plate-fin exchangers.

Herbein and Rohsenow (1988) compare Bejan's entropy generation method of optimisation with conventional methods of optimising a recuperator in a gas turbine cycle; they conclude that the entropy generation method does not necessarily find the best optimum. The value of the entropy generation (or exergy destruction) approach is that it clearly indicates where improvements are both possible and worthwhile. Questions of exergy destruction should be settled at the plant layout stage, well before detailed exchanger design commences.

A direct approach to optimisation is covered in Chapters 4 and 5 of this book. In practice the designer has first to decide what is to be minimised (see Section 3.10 and Appendix C) before seeking a viable approach to optimisation. In the direct-sizing approach it is possible to include all the small deviations and nuances in the splinefitted correlations, and to explore the performance envelope fully. Fluid pressure losses in the exchanger are fully taken into account in the optimisation.

REFERENCES

Baclic, B. S. (1990) ϵ-Ntu analysis of complicated flow arrangements, In: *Compact Heat Exchangers – A Festschrift for A. L. London*, Eds. R. K. Shah, A. D. Kraus and D. Metzger, Washington: Hemisphere, pp. 31–90.

Baclic, B. S. and Gvozdenac, D. D. (1981) Exact explicit equations for some two and three pass cross-flow heat exchangers, In: *Heat Exchangers – Thermal Hydraulic Fundamentals and Design*, Eds. S. Kakac, A. E. Bergles and F. Mayinger, Washington: Hemisphere, pp. 481–494.

Baclic, B. S. and Heggs, P. J. (1985) On the search for new solutions of the single-pass crossflow heat exchanger problem, *International Journal of Heat and Mass Transfer*, **28**(10), 1965–1976.

Bejan, A. (1988) *Advanced Engineering Thermodynamics*, New York: Wiley.

Chiou, J. P. (1978) The effect of longitudinal heat conduction on crossflow exchanger, *Transactions ASME, Journal of Heat Transfer*, **100**, 346–351.

Chiou, J. P. (1980) The advancement of compact heat exchanger theory considering the effects of longitudinal heat conduction and flow nonuniformity, In: *Symposium on Compact Heat Exchangers – History, Technological Advancement and Mechanical Design Problems*, Eds. R. K. Shah, C. F. McDonald and C. P. Howard, New York: ASME, book no. G00183.

Das, S. K. and Roetzel, W. (1995) Exergetic analysis of plate heat exchanger in presence of axial dispersion in fluid, *Cryogenics*, **35**(1), 3–8.

Fischer, K. F. (1938) Mean temperature difference correction in multipass exchanges, *Industrial and Engineering Chemistry*, **30**, 377–383.

Hausen, H. (1950) *Wärmeübertragung im Gegenstrom, Gleichstrom und Kreuzstrom*, Berlin: Springer Verlag.

Herbein, D. S. and Rohsenow, W. M. (1988) Comparison of entropy generation and conventional method of optimising a gas turbine regenerator, *International Journal of Heat and Mass Transfer*, **31**(2), 241–244. (Rohsenow uses the word 'regenerator' in the sense of heat recovery in a thermodynamic cycle; it is actually a recuperator.)

Hewitt, G. F., Shires, G. L. and Bott, T. R. (1994) *Process Heat Transfer*, Boca Raton, FL: CRC Press.

Jakob, M. (1949 and 1957) *Heat Transfer*, Chichester, UK: John Wiley, Vol. I (1949), and Vol. II (1957).

Kays, W. M. and London, A. L. (1964) *Compact Heat Exchangers*, New York: McGraw-Hill, pp. 250–257.

Kroeger, P. G. (1966) Performance deterioration in high effectiveness heat exchangers due to axial conduction effects, *Proceedings of the 1966 Cryogenic Engineering Conference, Advances in Cryogenic Engineering*, **12**, Paper E-5, 363–372.

Linnhoff, B. (1983) New concepts in thermodynamics for better chemical process design, *Proceedings of the Royal Society, Series A*, **386**(1790), 1–33; *Chemical Engineering Research and Design*, **61**(4), 207–223.

Linnhoff, B. and Flower, J. R. (1978) Synthesis of heat exchanger networks: 1. Systematic generation of energy optimal networks, 2. Evolutionary generation of networks with various criteria of optimality, *American Institute of Chemical Engineers Journal*, **24**, 633–642, 642–654.

Linnhoff, B. and Hindmarsh, E. (1983) The pinch design method for heat exchanger networks, *Chemical Engineering Science*, **38**(5), 745–763.

Linnhoff, B., Townsend, D. W., Boland, D., Hewitt, G. F., Thomas, B. E. A., Guy, A. R. and Marsland, R. H. (1982), *A User Guide on Process Integration for the Efficient Use of Energy, Rugby, UK:* Institution of Chemical Engineers.

Manson, J. L. (1954) Heat transfer in crossflow, *Proceedings of 2nd US National Applied Mechanics Conference*, pp. 801–803.

Nusselt, W. (1911) Der Wärmeübergang im Kreuzstrom, *Zeitschrift des vereines deutscher Ingenieure*, **55**, 2021–2024.

Nusselt, W. (1930) Eine neue Formel für den Wärmedurchgang im Krewzstrom, *Technische Mechanik und Thermodynamik*, **1**, 417–422.

Sekulic, D. P. (1986) Entropy generation in a heat exchanger, *Heat Transfer Engineering*, **7**, 83–88.

Sekulic, D. P. (1990) A reconsideration of the definition of a heat exchanger, *International Journal of Heat and Mass Transfer*, **33**(12), 2748–2750.

Shah, R. K. (1981) Compact heat exchanger design procedures, In: *Heat Exchangers – Thermal-Hydraulic Fundamentals and Design*, Eds. S. Kacac, A. E. Bergles and F. Mayinger, Washington: Hemisphere, pp. 495–536.

Shah, R. K. (1982) Heat exchanger basic design methods, In: *Low Reynolds Number Flow Heat Exchangers*, Eds. S. Kacac, R. K. Shah and A. E. Bergles, Washington: Hemisphere, pp. 21–72.

Smith, D. M. (1934) Mean temperature difference in crossflow, *Engineering*, **138**, 479–481, 606–607.

Smith, E. M. (1994) Direct thermal sizing of plate-fin heat exchangers, *The Industrial Sessions Papers, 10th International Heat Transfer Conference, Brighton, UK, 14–18 August 1994*, Rugby, UK: Institution of Chemical Engineers.

Spang, B. and Roetzel, W. (1995) Neue Näherungsgleichung zur einheitlichen Berechhung von Wärmeübertragung (new approximate equation for uniform heat exchange design) *Heat and Mass Transfer (Wärme und Stoffübertragung)*, **30**, 417–422.

Stevens, R. A., Fernandez, J. and Woolf, J. R. (1957) Mean temperature difference in one, two and three-pass cross-flow heat exchangers, *Transactions ASME*, **79**, 287–297.

Townsend, D. W. and Linnhoff, B. (1982) Designing total energy systems by systematic methods, *Chemical Engineering*, No. 378, March, 91–97.

Underwood, A. J. V. (1934) The calculation of the mean temperature difference in industrial heat exchanger configuration *Journal of the Institute of Petroleum Technology*, **20**, 145–158.

Wylie, C. R. (1953) *Calculus*, New York: McGraw-Hill, pp. 283–289.

4

DIRECT SIZING OF PLATE-FIN EXCHANGERS

4.1 EXCHANGER LAYUP

Figures 4.1 to 4.4 show the layup of simple two-stream plate-fin heat exchangers made up of alternate layers of extended surface separated by flat plates. A complete block is usually of brazed construction, although diffusion bonding can be used with suitable materials. Careful design places half-height surfaces at the top and bottom of a well-insulated block. The exchanger may then be considered as slices, made up of a flat plate with a half-height extended surface attached to each side.

For the Cross-flow design, thermal sizing is eased by first determining the heat exchange duty of one slice (Fig. 4.2); the complete exchanger is then made up of an appropriate number of slices.

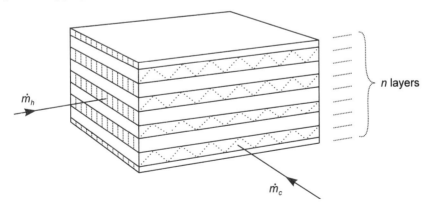

Fig. 4.1 Cross-flow plate-fin heat exchanger

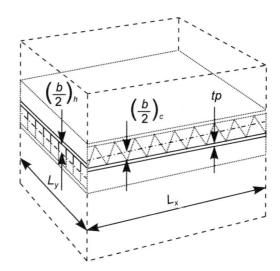

Fig. 4.2 Design geometry for cross-flow plate-fin exchanger

Flow friction and heat transfer correlations are based on the cell geometries of full-height surfaces. For correct pressure loss the task is to calculate the loss for an individual flow channel. For correct heat flow the task is to associate heat transfer coefficients with half-height fin surfaces and to relate them to the single flat-plate surface, edge length × flow length ($E_1 L_1 = E_2 L_2$).

For the contraflow design, thermal sizing is eased by mentally reconfiguring slices of the original exchanger as an equivalent single-plate heat exchanger (Fig. 4.4) with total plate surface ($S = EL$). Only the E length is shown in Fig. 4.4.

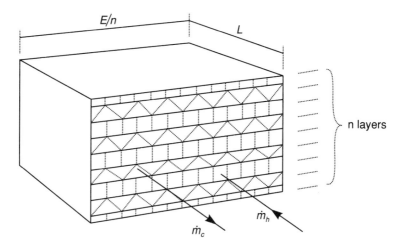

Fig. 4.3 Contraflow plate-fin heat exchanger

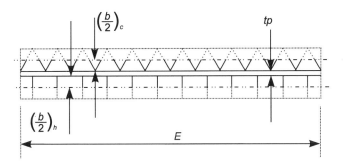

Fig. 4.4 Design geometry for contraflow exchanger

Flow friction and heat transfer correlations are based on the cell geometries of full-height surfaces. For correct pressure loss the task is to calculate the loss for an individual flow channel. For correct heat flow the task is to associate heat transfer coefficients with half-height fin surfaces and to relate them to the total plate surface $(S = EL)$.

In *two-stream* designs, direct sizing using slices of plate-fin heat exchangers implies that transverse temperature distributions in all channels are everywhere symmetric, with zero temperature gradient at the centre of each channel. This should always be the case, unless flow maldistribution exists due to other causes, e.g. variation in channel geometry through poor manufacturing techniques.

In *multi-stream contraflow* designs, transverse temperature symmetry is equally desirable, but this symmetry can be lost through other causes, even with perfect channel geometry, unless care is taken at the plant configuration stage to match longitudinal temperature profiles throughout the multi-stream exchanger.

It is not always possible to secure absolute matching, particularly when C for one or more fluids may vary along a multi-stream exchanger. When channel transverse temperature symmetry has been lost, recourse to special 'rating' procedures due to Haseler (1983) and Prasad (1993) becomes necessary in order to accommodate the 'cross-conduction' effect. However, in general, the multi-stream exchanger must first be sized before rating methods can be applied.

Several major reviews of heat exchanger design methodology have been published, e.g. London (1982), Shah (1982, 1988), Manglic and Bergles (1990), Baclic and Heggs (1985). This chapter will outline a new direct-sizing approach.

4.2 FLOW FRICTION AND HEAT TRANSFER CORRELATIONS

Plate-fin heat exchangers are conventional, largely due to the extensive research of Kays and London (1964), London and Shah (1968) and many

4 DIRECT SIZING OF PLATE-FIN EXCHANGERS

other investigators who have presented flow friction and heat transfer correlations for plate-fin surfaces. A recent development is the vortex generator surface of Brockmeier *et al.* (1993)

In direct sizing, each extended surface is referred to its base plate, but Kays and London correlations remain unchanged. For rectangular offset strip-fin surfaces, generalised explicit f- and j-correlations have been obtained by Manglic and Bergles (1990) and they permit full optimisation of heat exchanger cores. This is of particular value because the rectangular offset strip-fin surface is one of the best-performing surfaces, largely due to continuous recreation of the boundary layer on many small flat-plate fins.

Full optimisation of an exchanger core requires continuous adjustment of basic cell geometrical relationships, and the method of doing this for single- and double-cell rectagular offset strip-fin surfaces is explained below.

For every geometry it is desirable to use tabulated Re-f and Re-j data given by Kays and London (1964) and by London and Shah (1968), and to employ an interplating splinefit which allows a small weighting error for each point. For accurate design it is desirable that individual correlations should be fitted in this way, and graphical output is essential to check that results are as expected.

To confirm that the generalised Manglic and Bergles f- and j-correlations for rectangular offset strip-fin surfaces provide a good representation of original data, six London and Shah single-cell and six Kays and London double-cell surfaces were reassessed for fit, and the linear (log-log) fits presented in Figs. 4.5 and 4.6 are very close to those originally given by Manglic and Bergles. Surfaces used are set out in Table 4.4 where (α, δ, γ) are geometrical factors used by Manglic and Bergles.

In the notation of this text

$$\text{Manglic and Bergles } \alpha = \frac{\text{cell pitch}}{\text{plate spacing}} = \left(\frac{c}{b}\right)$$

$$\text{Manglic and Bergles } \delta = \frac{\text{fin thickness}}{\text{strip length}} = \left(\frac{tf}{x}\right)$$

$$\text{Manglic and Bergles } \gamma = \frac{\text{fin thickness}}{\text{cell pitch}} = \left(\frac{tf}{c}\right)$$

Correlations

$$\text{flow friction } f = 9.6243 \, (\text{Re})^{-0.7422} (\alpha)^{-0.1856} (\delta)^{0.3053} (\gamma)^{0.2659}$$
$$\times [1 + 7.669 \times 10^{-8} \, (\text{Re})^{4.429} (\alpha)^{0.920} (\delta)^{3.767} (\gamma)^{0.236}]^{0.1}$$

$$\text{heat transfer } j = 0.6522 \, (\text{Re})^{-0.5403} (\alpha)^{-0.1541} (\delta)^{0.1499} (\gamma)^{0.0678}$$
$$\times [1 + 5.269 \times 10^{-5} \, (\text{Re})^{1.340} (\alpha)^{0.504} (\delta)^{0.456} (\gamma)^{-1.055}]^{0.1}$$

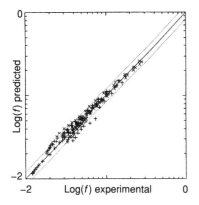

Fig. 4.5 Manglic and Bergles flow friction correlation for rectangular offset strip-fins

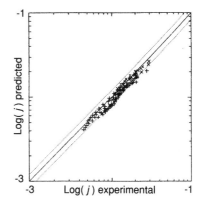

Fig. 4.6 Manglic and Bergles heat transfer correlation for rectangular offset strip-fins

4.3 PLATE-FIN SURFACE GEOMETRIES

In the Kays and London design approach, the heat transfer surface is referred to the total surface on one side of the exchanger; this is largely a function of the single-blow experimental test procedure adopted while establishing the correlations.

For design, it is desirable to refer heat transfer to the (complete) base-plate surface, and this is readily accomplished for the actual surfaces described by Kays and London and by London and Shah.

Given Kays and London data (both sides)

plate spacing	b
cell pitch	c
plate thickness	tp
fin thickness	tf
splitter thickness	ts
cell hydraulic diameter	D
(Stotal/Vtotal)	β
(Sfins/Stotal)	γ

Derived data (side 1)

alpha1 = (b1 * beta1)/(b1 + 2 * tp + b2)	(Stotal1/Vexchr)
kappa1 = (b1 * beta1)/2	(Stotal1/Splate)
lambda1 = kappa1 * gamma1	(Sfins1/Splate)
sigma1 = beta1 * D1/4	(Aflow1/Afront1)

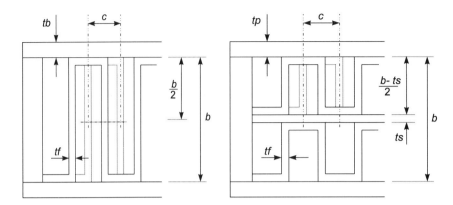

Fig. 4.7 Rectangular offset strip-fins: single-cell geometry, strip length x

Fig. 4.8 Rectangular offset strip-fins: double-cell geometry, strip length x

The procedure adopted differs from that of Soland *et al.* (1978) in which the Kays and London correlations are first regenerated so as to refer to the plate surface. There is no need to redevelop the original Kays and London correlations or the Manglic and Bergles correlations when the above parameters are used, noting that 'sigma' is defined differently to the definition used by Kays and London.

For fictitious rectangular offset strip-fin surfaces made possible by the Manglic and Bergles correlations, an approach from basic geometry becomes necessary. Two configurations are often used: no splitter (Fig. 4.7) and one central splitter (Fig. 4.8).

Providing at least one geometry is of the double-cell type, it is possible at layup time to fit a half-height layer at the block ends to provide the same pressure drop and a reasonable approximation to the correct heat transfer with half the mass flow rate.

The references scanned appear to give no information on the methods used to determine the geometrical parameters of surfaces listed by Kays and London and by London and Shah. For the double-cell geometry a problem exists as to whether the central surface areas should be considered as fin area contributing to fluid heat transfer. For symmetric transverse temperature fields, the splitter (with attached fin surface) is isothermal, and its contribution to heat transfer might be neglected. However, Kays and London include this area, which would be proper for the case of non-symmetric transverse temperature fields (Haseler 1983). But the splitter and attached surfaces do contribute to longitudinal conduction in all cases.

Table 4.4 lists relevant geometrical parameters, and Table 4.5 confirms that a good match exists with Kays and London data for the single-cell geometries, but values for double-cell geometries show somewhat less correspondence.

4.4 DIRECT SIZING OF AN UNMIXED CROSS-FLOW EXCHANGER

The best way of showing that direct-sizing works is to use the data from a rating example provided in Appendix B – Example 2 of the text by Kays and London (1964) and to design an exchanger core. It was found necessary to rework the given rating example on a computer because some round-off errors exist in the text example provided. Data used here is from the reworked case.

Values of mean bulk temperature required for evaluation of physical properties may be estimated at the mean of inlet temperatures, $(T_{h1} + T_{c1})/2$, and the accuracy improved after a first design pass. Here the same values are used as in the published example. Data are first converted to SI units. Software checks that numerical values for $Pr = C\eta/\lambda$ and $p = \rho R T_m$ remain consistent.

Input data

Performance
exchanger duty, KW $\qquad Q = 4854.82$

flow stream parameter	hot LP gas	cold HP air
mass flow, kg/s	$\dot{m}_h = 24.683$	$\dot{m}_c = 24.318$
inlet pressure, bar	$p_h = 1.02735$	$p_c = 9.1014$
allowable (core) pressure loss, N/m²	$\Delta p_h = 2859.63$	$\Delta p_c = 3562.93$
inlet temperature, K	$T_{h1} = 702.59$	$T_{c1} = 448.15$
Prandtl no.	$Pr_h = 0.670$	$Pr_c = 0.670$
sp.heat at const. press., J/(kg K)	$C_h = 1084.8$	$C_c = 1051.90$
absolute viscosity, kg/(m s)	$\eta_h = 0.000\,030\,145$	$\eta_c = 0.000\,028\,50$
thermal conductivity, J/(m s K)	$\lambda_h = 0.048\,817$	$\lambda_c = 0.044\,744$
gas constant, J/(kg K)	$R_h = 287.07$	$R_c = 287.07$
mean bulk absolute temperature, K	$T_{hm} = 611.94$	$T_{cm} = 543.05$
density, kg/m³	$\rho_h = 0.576\,79$	$\rho_c = 5.8268$

Plate material
plate thickness, mm $\qquad tp = 0.3048$
plate thermal conductivity, J/(m s K) $\quad \lambda_w = 20.77$
density, kg/m³ $\qquad \rho_w = 7030.0$

Surface geometry
warm gas $\qquad$ K-L plain 1/4-11.10 single-cell
cool air $\qquad$ K-L louver 3/8-06.06 single-cell

The cross-flow exchanger has a unique design which satisfies all constraints simultaneously. This is because the pressure loss on one side determines the

98 4 DIRECT SIZING OF PLATE-FIN EXCHANGERS

channel length and thus constrains the edge length for the other side. Only core pressure loss is used in the direct-sizing treatment. It is about 98% of the total loss in the example presented. Once the exchanger core is sized, then allowance for entrance effects, flow acceleration and exit effects can be made. Computed block dimensions obtained in direct sizing are compared in Table 4.1 with input data used in the rating method.

Table 4.1 Comparison of rating and sizing methods

Design approach	Kays and London (original sizes)		Direct sizing (m)
Units	(ft)	(m)	(m)
gas-flow path length	3.0	0.9144	0.9104
air-flow path length	6.0	1.8288	1.8199
block height	7.511*	2.2893	2.2959

*Original width adjusted to fit specified surfaces.

A slightly lower value of effectiveness was found (0.743, cf. 0.744), and the final block size was 0.63% smaller. A specific rating of 1.268 MW/m^3 existed for the original cross-flow exchanger.

4.5 DIRECT SIZING OF A CONTRAFLOW EXCHANGER

An explicit analytical solution will be discussed then an actual exchanger will be sized numerically.

Explicit analytical solution

This allows the design concept to be seen clearly among the constraints which appear in the numerical method that follows. Here we use the subscripts 1, 2 to designate sides of the exchanger.

The concept is based on treating the whole exchanger as a single plate, with half-height surfaces on each side (Fig. 4.4). Once the plate surface ($S = EL$) has been determined from flow length (L) and edge length (E), edge length E may be divided into n equal strips to create a suitable aspect ratio for the finished block exchanger.

A triangular-fin cell geometry will be employed for the simplicity of its heat transfer and flow friction correlations.

The following data are assumed known:

- from analysis of thermal performance, Q, $\Delta\theta_m$
- mass flow rate for each stream, $\dot{m}_1$, $\dot{m}_2$

4.5 DIRECT SIZING OF A CONTRAFLOW EXCHANGER

- maximum allowable core pressure losses, Δp_1, Δp_2
- individual cell flow areas, a_1, a_2
- physical properties of fluids at mean bulk temperature conditions
- physical properties of materials of construction

Explicit heat transfer and flow friction correlations for equilateral triangle cells (special case of isosceles cells) over a small Reynolds number range are given by Kays and London (1964) as

$$\text{St}(\text{Pr})^{1/3} = 2.7/\text{Re} \tag{4.1}$$

$$f = 14/\text{Re} \tag{4.2}$$

Edge length E and flow path length L (exchanger block length) are the unknowns to be determined.

Number of cells multiplied by cell pitch equals edge length on both sides

$$E = z_1 c_1 = z_2 c_2 \tag{4.3}$$

Heat transfer

Mass velocity, kg/(m² s) $\qquad G = \dfrac{\dot{m}}{az}$

Reynolds number $\qquad \text{Re} = \dfrac{DG}{\eta}$

Heat transfer coeff. from (4.1) $\qquad \alpha = \dfrac{2.7 \eta C}{(\text{Pr})^{2/3} D}$

Fin performance $\qquad \hat{m} = \sqrt{\dfrac{2\alpha}{\lambda_{fin} tf}} \qquad \phi = \dfrac{\tanh(\hat{m}Y)}{\hat{m}Y}$

The above correlations may be evaluated for stream 1 and stream 2, allowing for differing cell geometries. Referring fin surface (S_f) and exposed base surface (S_{xb}) to plate surface, the heat transfer coefficients become

$$\bar{\alpha}_1 = \alpha_1 \left[1 + \phi_1 \left(\dfrac{S_f}{S_{xb}}\right)_1\right]\left(\dfrac{base_1}{c_1}\right) \quad \alpha_w = \dfrac{\lambda_w}{tp} \quad \bar{\alpha}_2 = \alpha_2 \left[1 + \phi_2 \left(\dfrac{S_f}{S_{xb}}\right)_2\right]\left(\dfrac{base_2}{c_2}\right)$$

giving

$$\dfrac{1}{U} = \dfrac{1}{\bar{\alpha}_1} + \dfrac{1}{\alpha_w} + \dfrac{1}{\bar{\alpha}_2}$$

$$EL = \dfrac{Q}{U \Delta \theta_m} = \text{const.} \tag{4.4}$$

Design is valid *above* this rectangular hyperbola.

100   4 DIRECT SIZING OF PLATE-FIN EXCHANGERS

Pressure loss

Core loss greatly exceeds other losses, and is the only loss considered in direct sizing. Using equation (4.3) in the expression for friction factor

$$f = \frac{14\eta}{DG} = \frac{14\eta az}{D\dot{m}}$$

and inserting f in the pressure loss equation

$$\Delta p = \frac{4fG^2}{2\rho} \left(\frac{L}{D}\right)$$

then

$$L = \text{const.} \times E, \text{ say} \qquad (4.5)$$

and design is valid *to the right* of this straight line.

For streams 1 and 2

$$L_1 = Konst_1 \times E \qquad (4.6)$$

$$L_2 = Konst_2 \times E \qquad (4.7)$$

Curves (4.4), (4.6) and (4.7) may be plotted together (Fig. 4.9). The greater of the two values $E = (const./Konst_1)^{1/2}$ and $E = (const./Konst_2)^{1/2}$ is taken as the solution from which L may be determined. This identifies the controlling pressure loss. The non-controlling pressure loss for the other side has to be evaluated anew, as this is no longer the maximum permitted value.

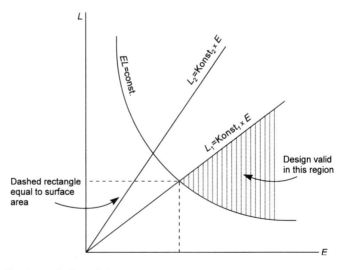

Fig. 4.9 Design solution plot

4.5 DIRECT SIZING OF A CONTRAFLOW EXCHANGER

Numerical solution

Individually splinefitted f- and j-correlations generated from data given by Kays and London and by London and Shah are preferred for accurate design. However, opportunity is now taken to demonstrate the use of the generalised Manglic and Bergles f- and j-correlations, in order to explore the effects of varying cell geometry on design.

The explicit approach above does not require construction of the solution plot, but a plot is helpful in the numerical solution. The design procedure is similar to that given for the explicit analytical solution, but care is necessary to observe the ranges of validity of the correlations. Double-cell rectangular offset strip-fins are used throughout, and the following approach proved effective for a basic two-stream design.

1. Specify exchanger duty.
2. Specify permissible controlling pressure losses. Usually it is the LP side which is controlling, and the HP pressure loss may be adjusted as necessary to maintain this condition.
3. Select surface geometry for both sides.
4. For the first fluid in steps of 50 over the range $500 < \text{Re} < 500$, determine the 'forced' values of Re for the other fluid, and determine which values fall into the validity range for each fin correlation. Then calculate the heat transfer for each Re.
5. Calculate pressure losses for each fin-cell geometry for each Re corresponding to the above restricted validity range.
6. Construct the three E-L curve one for heat transfer and two for pressure losses (typically). Fig. 4.9 is an example of a single-cell design.
7. Determine the (E, L) co-ordinates of the two intersection points if they exist, otherwise adjust the parameters appropriately and repeat from step 2.
8. Select the right-hand intersection as the controlling design point.
9. Evaluate the non-controlling pressure loss, usually the HP side.

Direct sizing is demonstrated using data provided by Campbell and Rohsenow (1992) for a proposed contraflow exchanger for a gas turbine. On the gas side of such a recuperator, it is perhaps not good practice to use rectangular offset strip-fins instead of plain fins, which can be cleaned, but we shall proceed nevertheless. We satisfy their constraint $\Sigma(\Delta P/P) = 0.08$ by introducing a one-line algorithm to calculate pressure loss on the HP side given the pressure loss for the LP side, and iterate this last value until the constraint is satisfied. Input data for HP-side pressure loss is set at a very large value so that it does not interfere with the new value calculated by the algorithm.

Herbein and Rohsenow (1988) provide details of the derivation of the expression for loss in turbine work, but their numerical results use a different gas turbine cycle.

4 DIRECT SIZING OF PLATE-FIN EXCHANGERS

$$\frac{\Delta W}{\dot{m} C \eta \gamma T_5'} = \left(\frac{P_c}{P_h}\right) \gamma \left(\frac{\Delta P_c}{P_c} + \frac{\Delta P_h}{P_h}\right) \quad (4.8)$$

Input data

Performance
exchanger duty, kW $\qquad Q = 10\,949.0$

flow stream parameter	hot LP gas	cold HP air
mass flow, kg/s	$\dot{m}_h = 49.0$	$\dot{m}_c = 49.0$
inlet absolute pressure, bar	$p_h = 1.10$	$p_c = 18.48$
allowable (core) pressure loss, N/m²	$\Delta p_h = 8800.0$	$\Delta p_c =$ floating
inlet temperature, K	$T_{h1} = 797.0$	$T_{c2} = 563.00$
outlet temperature, K	$T_{h2} = 586.0$	$T_{c1} = 774.0$
Prandtl number	$\mathrm{Pr}_h = 0.683$	$\mathrm{Pr}_c = 0.683$
sp.heat at const. press., J/(kg K)	$C_h = 1059.0$	$C_c = 1059.0$
absolute viscosity, kg/(m s)	$C_h = 0.0000\,50\,89$	$\eta_c = 0.0000\,50\,89$
thermal conductivity J/(m s K)	$\lambda_h = 0.0789$	$\lambda_c = 0.0789$
mean absolute pressure, bar	$p_{hm} = 1.056$	$p_c = 18.47$
gas constant, J/(kg K)	$R_h = 287.07$	$R_c = 287.07$
mean absolute temperature, K	$T_{hm} = 691.5$	$T_{cm} = 668.5$
density, kg/m³	$\rho_h = 0.532$	$\rho_c = 9.626$

Plate material
plate thickness, mm $\qquad tp = 0.200$
plate thermal conductivity, J/(m s K) $\qquad \lambda_w = 90.0$
density, kg/m³ $\qquad \rho_w = 8906.0$

Surface geometry
hot LP gas $\qquad$ K-L rect strip-fin 1/10-19.35 single-cell
$\qquad\qquad (b = 1.9050, c = 1.3127, x = 2,5400, tf = 0.1016\text{mm})$
cold HP air $\qquad$ K-L rect. strip-fin 1/9-24.12 single-cell
$\qquad\qquad (b = 1.9050, c = 1.0531, x = 2.8222, tf = 0.1016\text{mm})$

In direct sizing of a contraflow exchanger, a first design pass is made to determine the approximate size of the exchanger, and in order to obtain the necessary wall axial cross-section for conduction. Kroeger's method is then used to compute the LMTD reduction factor, and a second design pass made with the reduced value of LMTD.

4.5 DIRECT SIZING OF A CONTRAFLOW EXCHANGER

Running program EDGEFIN produced a first estimate of the design and showed that the exchanger was very short. Using the cross-sectional area for conduction in program LOGMEAN, the reduction factor to the applied to the LMTD to allow for longitudinal conduction was found to be 0.658. A second run of EDGEFIN produced a design allowing for longitudinal conduction with the following overall dimensions.

Core block parameters
face area, m^2 $\qquad A = 12.461$
length, m $\qquad L = 0.2797$
volume, m^3 $\qquad V = 3.4854$
Reynolds number, hot-side $\qquad Re_h = 274.49$
Reynolds number, cold-side $\qquad Re_c = 241.46$

overall Ntu, hot-side $\qquad N_h = 13.942$
overall Ntu, cold-side $\qquad N_c = 13.942$
overall heat transfer coefficient, J/(m^2 s K) $\qquad U = 436.93$

pressure loss, hot-side N/m^2 $\qquad \Delta p_h = 8620.1$
pressure loss, cold-side, N/m^2 $\qquad \Delta p_c = 601.84$

The face area is very large, the Reynolds numbers are low and the longitudinal conduction effect is unacceptable. It is obvious that the surfaces geometries must be changed to allow greater flow area for the LP warm gas (Fig. 4.10).

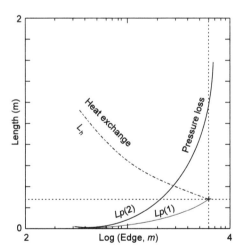

Fig. 4.10 Design solution plot for simple two-stream contraflow exchanger: hot LP gas 1.10 bar, surface K&L rect. strip 1/10-19.35(S); cold HP air 18.48 bar, surface K&L rect. strip 1/9-24.12(S)

104 4 DIRECT SIZING OF PLATE-FIN EXCHANGERS

4.6 FINE TUNING OF RECTANGULAR OFFSET-STRIP-FINS

In the design solution plot of Fig. 4.10 the two pressure loss curves were quite far apart. Anything that can be done to bring them together will improve the design. Simply increasing the plate spacing b on the LP warm side will do this, and it produces a considerable improvement in the design performance. But first a more consistent approach to optimisation will be taken, using another exchanger as the basis for experimentation.

The generalised Manglic and Bergles correlations for heat transfer and flow friction permit exploration of the effect of varying flow geometry on final core size. For the same thermodynamic performance, the optimum surface geometry can be sought for several parameters, including

- minimum block volume (overall dimensions)
- minimum block mass (excluding fluids)
- minimum frontal area

In a search, pressure loss on one side is kept constant while the other side is allowed to float, and the search arrangement is applied to both sides of the exchanger. For this comparative exercise, LMTD reduction for longitudinal conduction is not applied because it is sufficient to look merely at trends. A cryogenic N_2/N_2 block heat exchanger with single-cell surfaces on both sides was used as the model.

Heat exchanger specification

Performance
exchanger duty, kW $Q = 20.0$

flow stream parameter (nitrogen)	hot HP gas	cold LP gas
mass flow rate, kg/s	$\dot{m}_h = 0.2936$	$\dot{m}_c = 0.3425$
allowable (core) pressure loss, N/m²	$\Delta p_h = 100.0$, say*	$\Delta p_c = 600.0$, say*
inlet temperature, K	$T_{h1} = 210.00$	$T_{c2} = 141.9$
outlet temperature, K	$T_{h2} = 147.00$	$T_{c1} = 197.7$
Prandtl number	$Pr_h = 0.7698$	$Pr_c = 0.7493$
specific heat, J/(kg K)	$C_h = 1081.22$	$C_c = 1046.54$
absolute viscosity, kg/(m s)	$\eta_h = 0.000\ 011\ 66$	$\eta_c = 0.00\ 001\ 17$
thermal conductivity, J/(m s K)	$\lambda_h = 0.01637$	$\lambda_c = 0.015602$
absolute pressure, bar	$p_h = 6.0$	$p_c = 1.0$
gas constant, J/(kg K)	$R_h = 296.6$	$R_c = 296.6$
mean absolute temperature, K	$T_{hm} = 178.5$	$T_{cm} = 169.8$
density, kg/m³	$\rho h = 11.3329$	$\rho_c = 1.9856$

*Fixed pressure losses are set at approximately the mid-point of the heat transfer curve. Other values may well be more practical.

4.6 FINE TUNING OF RECTANGULAR OFFSET-STRIP-FINS

Plate material
plate thickness, mm $\quad tp = 2.00$
fin thickness, mm $\quad tf = 0.15$
splitter thickness, mm $\quad ts = 0.15$
thermal conductivity, J/(m s K) $\quad \lambda_w = 20.77$

Surface geometry
HP-side, single-cell offset rect strip-fin $\quad b, c, x =$ variable geometry
LP-side, single-cell offset rect strip-fin $\quad b, c, x =$ variable geometry

The effects of changing the fin thickness might be explored, but it was thought that the credibility of the Manglic and Bergles correlations might be pushed too far. Keeping cell width flow area constant, it was found that varying the HP fin thickness had virtually no effect on the surface area. A small LP fin thickness helped to minimise the surface area. The result is inconclusive because the works of Kelkar and Patankar (1989) and of Hesselgreaves (1993) need further study, but do note that thin fins also cause less longitudinal conduction.

Surface geometry was varied according to the scheme outlined below and in Table 4.2.

nominal sizes for both sides
$(b = 5.00$ mm, $\quad c = 2.0$ mm, $\quad x = 6.00$ mm$)$

variation about nominal (one dimension at a time)

Table 4.2 Range of geometrical parameters

Plate spacing, b (mm)	Cell pitch, c (mm)	Strip length, x (mm)
2.0	1.0	2.0
3.0	1.50	3.0
4.0	2.00	4.0
5.0	2.50	5.0
6.0	3.00	6.0
7.0	3.50	7.0
8.0	4.00	8.0

Results

Primary design parameters of interest are block volume, block mass, block length and block frontal area. Four possible graphs exist for each parameter. Reproduced in Appendix C, they show the effect of vaying dimensions b, c, x for the following four cases:

106   4 DIRECT SIZING OF PLATE-FIN EXCHANGERS

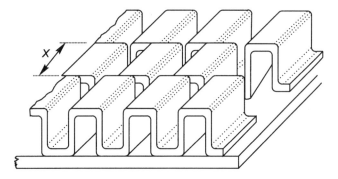

Fig. 4.11   Rectangular offset strip-fins

HIGH PRESSURE SIDE ⇔ LP-side fixed pressure loss
LOW PRESSURE SIDE ⇔ LP-side fixed pressure loss

HIGH PRESSURE SIDE ⇔ HP-side fixed pressure loss
LOW PRESSURE SIDE ⇔ HP-side fixed pressure loss

Certain figures show that adjustment in the strip length (x) on the 'controlling' side has a significant effect on the block length, whereas it hardly affects the block volume or the block mass. This makes 'tuning' of the strip length (x) on the controlling side a possible means of adjusting the block length without changing the thermal performance.

The strip length (x) depends on the distance between cuts in the original plate before it is formed. It therefore seems possible to have machinery produce customised surfaces of the rectangular offset strip-fin type which meet a critical requirement without too much difficulty (Fig. 4.11).

Observations concerning all dimensional parameters stem from validity of the Manglic and Bergles correlations, and the scatter of data should be noted in Figs. 4.5 and 4.6. And consider the approach of varying one parameter at a time then selecting a combination to optimise against a particular requirement. Although it may find the general area of best performance, it may miss the true optimum configuration. Variants of the simplex hill-climbing optimisation technique may do better, but would meet the problem of changing limits on Reynolds number validity at each iteration.

4.7 OPTIMISATION OF A CONTRAFLOW EXCHANGER

Returning to optimisation of the Campbell and Rohsenow design, but using LMTD reduction to allow for longitudinal conduction during optimisation, it is now possible to decide in which directions the surface geometry should be changed.

4.7 OPTIMISATION OF A CONTRAFLOW EXCHANGER

For the 'test' heat exchanger discussed in Section 4.6, different effects emerge for LP and HP sides of the exchanger because the controlling LP pressure loss is fixed and the HP pressure loss is allowed to float. Two plots from the sequence in Appendix C are presented in Figs. 4.12 and 4.13.

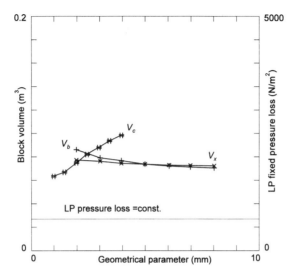

Fig. 4.12 LP-side: block volume (V) and pressure loss against geometrical parameters: $(+)b_2$, $(\#)c_2$, $(\times)x_2$

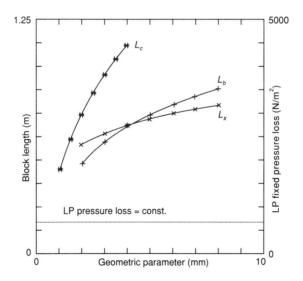

Fig. 4.13 LP-side: block length (L) and pressure loss against geometrical parameters: $(+)b_2$, $(\#)c_2$, $(\times)x_2$

It is particularly useful that the strip length (x) of the LP-side fin could be varied to change the block length without significant change in the block's volume or mass (but not its frontal area). This makes the LP-strip length dimension a useful means for final adjustment of the exchanger block length.

Returning to the Campbell and Rohsenow example of Section 4.5, a modest core block volume was achieved, but this was at the price of having a very large face area. Examination of Figs. 4.12 and 4.13 suggests that the volume can be further reduced and the flow length increased by increasing the plate spacing (b) on the low pressure side of the exchanger. Figure 4.14 shows the variation of exchanger volume, block length and LMTD reduction factor against the plate spacing (b).

Only one dimension will be changed in this exercise, leaving a more complete optimisation for those who wish to try it themselves.

It was not planned that the LMTD reduction factor should tend to unity as minimum volume was approached – it is good practice to seek only one optimum at a time. A little thought suggests that these two parameters should behave as found, but this coincidence should not in general be anticipated (see Section 3.10).

A design with a smaller volume and a smaller face area is obtained with $b = 5.00$ mm and $\Delta p_c = p_c \times 10^5 [0.08 - \Delta p_h/(p_h \times 10^5)]$ by adjusting Δp_h until the pressure loss curves coincide while conforming to $\Sigma(\Delta p/p) = 0.08$. The final configuration with LMTD reduction factor of 0.917 has the following dimensions:

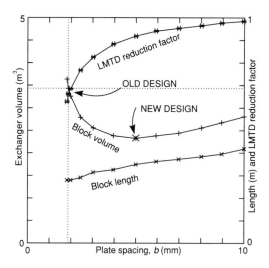

Fig. 4.14 Effect of varying plate spacing (b) *on hot LP* side of Campbell and Rohsenow design

4.7 OPTIMISATION OF A CONTRAFLOW EXCHANGER

Core block dimensions

face area, m²	$A = 7.131$
length, m	$L = 0.334$
volume,	$V = 2.384$
specific rating, MW/m³	$Q/V = 4.593$
Reynolds number, hot-side	$Re_h = 409.0$
Reynolds number, cold-side	$Re_c = 732.1$
velocity (hot LP gas), m/s	$u_h = 20.88$
velocity (cold HP air), m/s	$u_c = 3.20$
overall Ntu, hot-side	$N_h = 10.00$
overall Ntu, cold-side	$N_c = 10.00$
overall heat transfer coefficient, J/(m²sK)	$U = 795.39$
pressure loss, hot-side N/m²	$\Delta p_h = 8620.1$
pressure loss, cold-side, N/m²	$\Delta p_c = 3010.4$

hot LP gas modified K-L rect, strip-fin, 1/10-19.35 simple cell ($b = 5.000$, $c = 1.3127$, $x = 2.5400$, $tf = 0.1016$ mm)

cold HP air original K-L rect, strip-fin, 1/9-24.12 simple cell ($b = 1.9050$, $c = 1.0531$, $x = 2.8222$, $tf = 0.1016$ mm)

These results contradict the conclusions of Campbell and Rohsenow (1992) regarding plate spacing, and also show that Ntu by itself is not a reliable concept for assessing performance of compact plate-fin heat exchangers.

Only plain fins should be used on the gas side of the exchanger; this is because plain fins can be effectively cleaned whereas rectangular offset strip-fins cannot (Webb 1994). It is desirable to produce the same universal correlations for plain fins as did Manglic and Bergles for rectangular offset strip-fins. The asymptotic correlation method developed by Churchill and Usagi (1972), and used by Manglic and Bergles, is described in great clarity in the text by Webb, so it will not be repeated here.

In passing, it may be remarked that a similar technique was used by Clarke in 1966 to correlate creep curves for a nimonic nickel-based alloy, and asymptotic correlation is seen to be of wider interest.

It is worth taking a look at the final solution plot (Fig. 4.15), which indeed shows that the two pressure loss curves are now coincident. Adjusting the surface dimensions until these curves coincide is a quick way of finding an approximate optimum.

Returning to equations (4.4), (4.6) and (4.7) it can be seen that there are three simultaneous equations with only two unknowns. For a solution which is not overdetermined it is desirable that equations (4.6) and (4.7) provide the same solution at the design point. This implies that ($Konst_1 - Konst_2$) and it quickly follows that

110 4 DIRECT SIZING OF PLATE-FIN EXCHANGERS

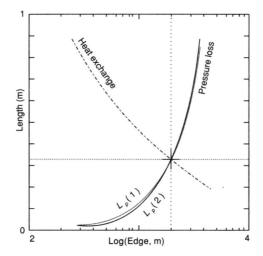

Fig. 4.15 Solution plot for Campbell and Rohsenow design with increased value of plate spacing ($b = 5.0$ mm on LP side, showing coincident pressure loss curves

$$\frac{\Delta p_1}{\Delta p_2} = \left[\frac{(a_2/c_2)D_2^2}{(a_1/c_1)D_1^2}\right] \times const \qquad (4.9)$$

Thus the ratio of useable pressure losses is fully determined for any given pair of compact surfaces. When this heat exchanger constraint can be combined with a system pressure loss contraint such as equation (4.8), viz. $\sum(\Delta p/p) = const$, then pressure losses need not be specified as input data. The design message is:

- make the pressure loss curves coincide on the $(E-L)$ plot.

4.8 DISTRIBUTION HEADERS

It is not practicable to design a contraflow plate-fin heat exchanger which does not require distribution headers. The problems introduced by this extra surface include:

- allowing for additional pressure loss
- allowing for additional heat transfer.

Haseler and Fox (1995) provide information on methods of evaluating pressure losses for seven basic types of distributor. With careful design proper allowance can be made for additional pressure losses.

In certain two-strem contraflow heat exchangers, it is possible that distribution headers for one stream may lie completely in contact within the other flow stream. Additional heat transfer surface then exits in the distribution headers, which will affect assumed entry and exit conditions for the true contraflow 'core'. Solutions to this problem seem not yet to have reached the open literature.

4.9 MULTI-STREAM DESIGN (CRYOGENICS)

It is possible to extend the contraflow design method to sizing of simple multi-stream exchangers. The case of three streams is straightforward; it is only necessary to ensure that the same pressure loss exists in both parts of the stream that is split, and that separate sections have the same length. When using rectangular offset strip-fins length adjustment is achieved by varying strip length (x).

Correction for longitudinal conduction is incorporated by adjustment of LMTD in the way described, but the problem of transverse conduction to non-adjacent streams will arise unless stream temperature profiles have already been matched in the earlier design process. This implies careful layout of cryogenic plant at the system design stage, as problems can be reduced through proper attention to matching terminal temperatures and choice of streams. However, when stream temperature profiles do not match along the length of the exchanger, recourse must be made to rating design approaches like those of Haseler (1983), Prasad and Gurukul (1992) and Prasad (1993) (see Chapter 11).

4.10 BUFFER ZONE OR LEAKAGE PLATE 'SANDWICH'

Many aspects of hardware design have not been addressed in this volume. Taylor (1987) has edited a guide to plate-fin heat exchangers which discusses mechanical construction, including headering and pressure limitations. Shah (1990) has discussed brazing methods. Haseler and Fox (1995) have considered distributor models. Imperfections in construction lead to maldistribution and loss of performance; they have been assessed by Weimer and Hartzog (1972).

One mechanical feature not previously discussed, and which directly affects thermal performance, is the leakage plate 'sandwich' used to prevent cross-contamination of two fluid streams. At lay up each separating plate is replaced by two separating plates between which is placed a shallow plain surface. No end bars are fitted to the sandwich, so that any leakage may be to the external environment or to a leak detection system (MacDonald 1995). Both the plate spacing (b) and pitch (c) are small, whereas the 'fin' thickness (t) of the shallow plain surface is as large as practicable.

In thermal design it is a simple matter to treat the leakage plate sandwich as a single plate, and proceed with direct sizing as indicated earlier. The problem is to determine an equivalent thermal conductivity for the new barrier to heat flow. The following simple treatment provides an approach which may prove useful when more accurate data is not available.

Assume that the geometry of the shallow plain surface of thickness t is in the form of a sinusoid of pitch c and amplitude b. The staggered brazing better guarantees that no cross-leakage can occur. The surface may be represented by

$$y = \left(\frac{b-t}{2}\right)\sin\left(\frac{2\pi x}{c}\right) \tag{4.10}$$

By taking the derivative at $x = 0$, $y = 0$, horizontal distance across the shallow plain surface can be found. In any single pitch c there are two such horizontal distances. Mentally removing the metal surface, the air-gaps may be slid together horizontally to give an equivalent air-gap length which is easier to handle. The vertical heat flow length is $l_2 = b$ and the air-gap width is area per unit length of exchanger A_2, given by

$$A_2 = c - 2t\sqrt{1 + \frac{c^2}{\pi^2(b-t)^2}} \tag{4.11}$$

For the metal surface the heat flow path is not at right angles to the separating plates. There are two heat flow paths of width t in any cell pitch c, hence the angled heat flow width per unit length of the exchanger is $A_1 = 2t$. Estimate conduction length using the gradient of the sinusoid at $x = 0$, $y = 0$ to obtain ℓ_1.

$$\ell_1 = b\sqrt{1 + \frac{c^2}{\pi^2(b-t)^2}} \tag{4.12}$$

To simplify notation, replace the square-root expression by the single symbol χ in equations (4.11) and (4.12), and represent each heat flow path by a lumped form of Fourier's law $Q = \lambda A(\Delta\theta/\ell)$, then

$$Q = \lambda_1 A_1(\Delta\theta/\ell_1) + \lambda_2 A_2(\Delta\theta/\ell_2) = \left[\lambda_1\left(\frac{2t}{\chi c}\right) + \lambda_2\left(1 - \frac{2t\chi}{c}\right)\right]\frac{\Delta\theta}{b}$$

In practical case $\chi \cong 1$ hence the equivalent conduction of the *gap* between the two 'leakage plates' becomes

$$\hat{\lambda} = \lambda_1\left(\frac{2t}{c}\right) + \lambda_2\left(1 - \frac{2t}{c}\right) \tag{4.13}$$

This is intuitively acceptable and is simple to incorporate in computer calculations. It follows that large values of t and small values of b and c are desirable, which is a manufacturing constraint.

Greater longitudinal conduction has now been built into the exchanger. For contraflow, the direction of the sinusoids should be arranged at right angles to the fluid flow directions. The cross-section for conduction in the single plate design of Fig. 4.4 is then for two separating plates and one narrow plate, namely $A = E(2tp + t)$.

For cross-flow there will be increased conduction in one direction due to the increased sinusoid length, which may be calculated as shown in Section

3.8. Another possibility is to arrange for the waviness of the shallow plain surface to be at 45° to both flow directions and perpendicular to the spine in Fig. 3.15, then simply to average the lengths of the sinusoidal edge and the straight edge. This may also help spread the temperature-difference spine sideways by conduction, improving performance by a second-order effect. As the heat transfer coefficient across the leakage plate sandwich will usually be higher than the two fluid coefficients referred to the plate surface, it may not usually be worthwhile to seek a more accurate solution for a 45° arrangement.

4.11 CONSISTENCY IN DESIGN METHODS

It is essential to have confidence in the design method selected. To this end the author wrote four direct-sizing programs and checked them against each other and against two rating programs.

Two direct-sizing programs were written for cross-flow. The first employed tabulated data for *individual* correlations given by Kays and London (1964) and by London and Shah (1968) to create individual interpolating splinefits, and used geometrical parameters for surfaces given by these authors. The second employed the *generalised* correlations for rectangular offset strip-fins (ROSF) given by Manglic and Bergles (1990), further discussed by Webb (1994), and used basic dimensions of the geometries to generate the geometrical parameters. Two further direct-sizing programs were written for contraflow in a similar manner.

In support of these four direct-sizing programs, two rating programs were also written. The first of these reworked the example given in Kays and London for a cross-flow exchanger, except that it was done on the computer to remove any round-off errors produced in the original hand calculation. The second was a contraflow rating program written in a similar manner. The definitions in Table 4.3 were used to generate the geometrical parameters.

Table 4.3 Geometrical parameters for side 1 of the exchanger

Description	Parameter	Kays and London	This book
Stotal1/Vexchr	alpha1	$b_1.beta1/(b1+2.tp+b2)$	$b_1.beta1/(b1+2.tp+b2)$
Stotal1/Splate	beta1	GIVEN	use Table 4.4
Sfins1/Stotal1	gamma1	GIVEN	use Table 4.4
Stotal1/Splate	kappa1	$b1.beta1/2$	use Table 4.4
Sfins 11/Splate	lambda1	$kappa1.gamma1$	use Table 4.4
Aflow1/Afront1	sigma1*	$beta1.D1/4$	$beta1.D1/4$
Splate/Vexchr	omega1	$alpha1/kappa1$	$alpha1/kappa1$

*Note that the definition of parameter sigma differs from that given by Kays and London. The present form was found to be more convenient in programming. The value of parameter *omega* should be the same for both sides of an exchanger, and its evaluation provides a useful check on the software coding. Detailed geometrical specifications are given in Appendix B.4.

When direct-sizing programs were run with the same input data, it was found that the predicted size of the exchanger might differ by about 1% between programs. Differences were finally traced to slight discrepancies in the dimensions used for local surface geometry. One source of the problem was found to be the two values of hydraulic diameter quoted in both feet and in inches in Table 9.3 of Kays and London (1964), and due to round-off these values do not quite correspond.

Another source is the different definitions used for hydraulic diameter in generating the heat transfer and flow friction correlations. Several different definitions used by different authors are to be found in the paper by Manglic and Bergles (1990), and the correctness of the obtained heat transfer coefficient and friction factor depends on using the same definition as the original author(s). This of course is messy.

It is possible to do something about this if sufficient original information is available. While there is no explicit definition of hydraulic diameter in Kays and London, the value used by these workers has been given in the paper by Manglic and Bergles, who modified the original Kays and London and London and Shah data to suit an improved definition before computing their generalised correlations.

Thus we may have some faith in the correctness of the generalised correlations except that, as might be anticipated with generalised correlations, they exhibit scatter. Nevertheless, the care taken to adjust the original data must have produced improvements in the consistency of individual correlations, which is desirable.

Careful manufacturers will find it necessary to study the twin problems of

- accurate dimensional data for surface geometries
- correct geometrical evaluation of hydraulic diameter, and other parameters

It is outside the scope of this book to undertake revision of existing correlations on a substantial scale, but some pointers towards the usefulness of individual surfaces may help to establish an order of priority in such work.

It is not to be expected that exactly the same results will be obtained when comparing designs using individually splinefitted correlations against the universal Manglic and Bergles correlations, which have to accommodate scatter.

The best-performing surface for clean conditions is probably the rectangular offset strip-fin. This is because the small strips continuously recreate the boundary layer and provide high heat transfer coefficients, and because the discontinuous surface helps to reduce effects of longitudinal conduction. The wavy fin may show slightly better heat transfer and flow friction performance, but it lacks the ability to reduce the effects of longitudinal conduction.

For fouling conditions, air-lancing will clean plain surfaces but is ineffective with rectangular offset strip-fins (Webb 1994). Thus plain fins are also prime candidates for consideration.

4.12 CONCLUSIONS

1. For each exchanger type, appropriate maximum values of Ntu are

   parallel flow   Ntu = 2.0
   one-pass unmixed cross-flow   Ntu = 4.0
   two-pass unmixed cross-flow   Ntu = 7.0
   contraflow   Ntu > 10.0

2. Geometrical relationships for single-cell and double-cell rectangular strip-fin surfaces have been presented, permitting exploration of surfaces for optimum performance.
3. A method of adjusting LMTD values to allow for longitudinal conduction in the design of a class of contraflow exchangers has been demonstrated.
4. Demonstration of direct sizing of a cross-flow exchanger confirms the precision of the method.
5. Direct sizing of a contraflow exchanger has been achieved with a better design than was achieved in the example give in the literature, while conforming to $\Sigma(\Delta p/p) = 0.08$.
6. Brief discussion of the extension of direct sizing to multi-stream exchangers has been presented.
7. Low values of Reynolds number do not imply low values of flow velocity in compact heat exchangers.

(continued on p 117)

Table 4.4 Surfaces used to check Manglic and Bergles f-and j-correlations*

Geom no.	b (mm)	c (mm)	x (mm)	tf (mm)	ts (mm)	β (1/mm)	γ	rh (mm)	Surface designation
01	6.350	1.627	3.170	0.102	–	1.549	0.809	0.596	1/8–15.61 (S)
02	7.645	1.120	2.820	0.102	–	2.067	0.885	0.434	1/9–22.68 (S)
03	1.905	1.053	2.822	0.102	–	2.831	0.665†	0.302	1/9–24.12 (S)
04	5.080	1.016	2.819	0.102	–	2.359	0.850	0.373	1/9–25.01 (S)
05	1.905	1.313	2.540	0.102	–	2.490	0.611†	0.351	1/10–19.35 (S)
06	6.350	0.940	2.540	0.102	–	2.467	0.887	0.356	1/10–27.03 (S)
07	6.020	2.127	12.70	0.152	0.152	1.512	0.796	0.567	1/2–11.94 (D)
08	8.966	2.085	4.521	0.102	0.152	1.386	0.847	0.659	1/6–12.18 (D)
09	7.722	1.613	3.629	0.102	0.152	1.726	0.895	0.517	1/7–15.75 (D)
10	6.477	1.588	3.175	0.152	0.152	1.803	0.843	0.466	1/8–16.00 (D)
11	5.207	1.282	3.175	0.102	0.152	2.231	0.841	0.385	1/8–19.82 (D)
12	5.015	1.266	3.175	0.102	0.152	2.290	0.845	0.373	1/8–20.06 (D)

*Data from Kays and London, London and Shah.
†Values quoted in Kays and London (1984) are incorrect, and the above values are taken from London and Shah (1968).

4 DIRECT SIZING OF PLATE-FIN EXCHANGERS

Table 4.5 Geometries for rectangular offset strip-fin cells*

Parameter	Single-cell	Double-cell	Notes
Sbase/x	$2(c-tf)$	$2(c-tf)$	
Splate/x	$2c$	$2c$	
Vtotal/x	bc	bc	
Sfins/x	$2(b-tf)$ $+2(b-2tf)tf/x$ $+2(c/2)tf/x$	$4[(b-ts)/2 - tf]$ $+4[(b-ts)/2 - 2tf]tf/x$ $+4(c/2)tf/x$ $+2(c-tf)$	fin sides fin ends base ends splitter
Stotal/x	$2(b-tf)$ $+2(b-2tf)tf/x$ $+2(c/2)tf/x$ $+2(c-tf)$	$4[(b-ts)/2 - 2tf]$ $+4[(b-ts)/2 - 2tf]tf/x$ $+4(c/2)tf/x$ $+2(c-tf)$ $+2(c-tf)$	fin sides base ends exposed plate splitter
Y	$(b-tf)/2$ $\{b/2\}$	$(b-ts)/2 - tf$ $\{b/2\}$	fin-height {or approx}
Per (one cell)	$2(b-tf)$ $+2(c-tf)$ $+2(b-2tf)tf/x$ $+2(c/2)tf/x$	$2[(b-ts)]/2 - tf]$ $+2(c-tf)$ $+2[(b-ts)/2 - 2tf]tf/x$ $+2(c/2)tf/x$	cell sides cell ends fin ends base ends
Aflow (one cell)	$(c-tf)(b-tf)$	$(c-tf)[(b-ts)/2 - tf]$	cell flow area
Afront	bc (one cell)	bc (two cells)	cell frontal area

Caution: Manglic and Bergles (1990) give a value for hydraulic diameter (D) of rectangular offset strip-fins as

$$D = \frac{4shl}{2(sl + hl + th) + ts}$$

where $s = c - tf$, $h = b - tf$, $l = x$, $lt = tf$ in the notation of Table 4.4. When the hydraulic diameter is evaluated from expressions in Table 4.4, a very slightly different value for D is obtained. In Manglic and Bergles notation this would be

$$D = \frac{4shl}{2(sl + hl + th) + ts - t^2}$$

Since the Manglic and Bergles generalised heat transfer and flow friction correlations are corrected to their value of hydraulic diameter, it is necessary to use their definition of D to recover the heat transfer coefficients and flow friction factors.

Table 4.6 Comparison of 'given' and calculated parameters for single-cell (S) and double-cell (D) surfaces

Geom no	β (1/mm) given	β (1/mm) calc.	γ given	γ calc.	rh (mm) given	rh (mm) calc.	Surface designation	L&S paper
01	1.549	1.546	0.809	0.810	0.596	0.597	1/8–15.61 (S)	104(S)
02	2.067	2.067	0.885	0.885	0.434	0.434	1/9–22.68 (S)	103(S)
03	2.830	2.827	0.665*	0.664	0.302	0.303	1/9–24.12 (S)	105(S)
04	2.359	2.359	0.850	0.850	0.373	0.374	1/9–25.01 (S)	101(S)
05	2.490	2.486	0.611*	0.610	0.351	0.351	1/10–19.35 (S)	106(S)
06	2.467	2.464	0.887	0.886	0.356	0.356	1/10–27.03 (S)	102(S)
07	1.512	1.500	0.796	0.794	0.567	0.572	1/2–11.94 (D)	–
08	1.386	1.371	0.847	0.845	0.659	0.667	1/6–12.18 (D)	–
09	1.726	1.708	0.859	0.858	0.517	0.523	1/7–15.75 (D)	–
10	1.803	1.797	0.843	0.845	0.466	0.468	1/8–16.00 (D)	–
11	2.231	2.218	0.841	0.841	0.385	0.387	1/8–19.82 (D)	–
12	2.290	2.248	0.845	0.840	0.373	0.381	1/8–20.06 (D)	–

*values quoted in Kays and London (1984) are incorrect, and the above values are taken from the London and Shah paper (1968).

8. Heat exchanger duty densities up to 5 MW/m^3 appear possible when surface geometries can be tuned.
9. Emphasis was placed on the desirability of careful checking of published geometrical parameters of surfaces. Any discrepancies found may influence perception of the accuracy of associated heat transfer and flow friction correlations.

REFERENCES

Baclic B. S. and Heggs, P. J. (1985), On the search for new solutions of the single-pass crossflow heat exchanger problem, *International Journal of Heat and Mass Transfer*, **28**(10), 1965–1976.

Barron, R. F. (1985), *Cryogenic Systems*, Oxford: OUP, p. 121.

Brockmeier, U., Guentermann, Th. and Fiebig, M. (1993) Performance evaluation of a vortex generator heat transfer surface and comparison with different high performance surfaces, *International Journal of Heat and Mass Transfer*, **36**(10), 2575–2587.

Campbell, J. F. and Rohsenow, W. M. (1992) Gas turbine regenerators: a method for selecting the optimum plate-finned surface pair for minimum core volume, *International Journal of Heat and Mass Transfer*, **35**, (12), 3441–3450.

Chapman, A. J. (1974) *Heat Transfer*, London: Macmillan, p. 521.

Chiou, J. P. (1978). The effect of longitudinal heat conduction on crossflow heat exchanger, *ASME Journal of Heat Transfer*, **100**, 346–351.

Chiou, J. P. (1980) The advancement of compact heat exchanger theory considering the effects of longitudinal heat conduction and flow nonuniformity, In: *Compact Heat Exchangers – History, Technological Advancement and Mechanical Design Problems*, Eds. R. K. Shah, C. F. MacDonald, C. P. Howard, New York: ASME, HTD-10, pp. 101–121.

Churchill, S. W. and Usagi, R. (1972) A general expression for the correlation of rates of transfer and other phenomena, *American Institute of Chemical Engineers Journal*, **8**(6), 1121–1128.

Clarke, J. M. (1966) A convenient representation of creep strain data for problems involving time-varying stresses and temperatures, *National Gas Turbine Establishment, Pyestock, Hants, NGTE Report No. R284*, September 1966.

Grassman, P. and Kopp, J. (1957) Zur gunstigen Wahl der Temperaturdifferenz und der Wärmeübergagszahl in Wärmeaustauschern, *Kaltetechnik*, **9**(10), 306–308.

Haseler, L. E. (1983) Performance calculation methods for multi-stream plate-fin heat exchangers, *Heat Exchangers: Theory and Practice*, Eds. J. Taborek, G. F. Hewitt and N. Afgan, New York: Hemisphere/McGraw-Hill, pp. 495–506.

Haseler, L. E. and Fox, T. (1995) Distributor models for plate-fin heat exchangers, *4th UK National Heat Transfer Conference, Manchester, 26–27 September 1995*, pp. 449–456.

Herbein, D. S. and Rohsenow, W. M. (1988) Comparison of entropy generation and conventional method of optimising a gas turbine regenerator, *International Journal of Heat and Mass Transfer*, **31**(2), 241–244.

Hesselgreaves, J. E. (1993) Optimising size and weight of plate-fin heat exchangers, *Proceedings of First International Conference on Heat Exchanger Technology, Palo Alta, CA, 15–17 February 1993*, Eds. R. K. Shah and A. Hashemi, New York: Elsevier, pp. 391–399.

Incropera, F. P. and DeWitt, D. P. (1981) *Fundamentals of Heat Transfer*, New York: Wiley, p. 521.

Kays, W. and London A. L. (1964) *Compact Heat Exchangers*, 2nd Edn, New York: McGraw-Hill, (3rd edn. 1984)

Kelkar, K. M. and Patankar, S. V. (1989) Numerical prediction of heat transfer and fluid flow in rectangular offset-fin arrays, *International Journal of Computer Methodology, Part A Applied: Numerical Heat Transfer* **15**(2), 149–164.

Kreith, F. (1965) *Principles of Heat Transfer*, 2nd Edn, Saranton, PA: International Textbook Co., p. 497.

Kroeger, P. G. (1966) Performance deterioration in high effectiveness heat exchangers due to axial conduction effects, *Proceedings of the 1966 Cryogenic Engineering Conference, Advances in Cryogenic Engineering*, **12**, Paper E-5, 363–372.

London, A. L. (1982) Compact heat exchangers – design methodology, In: *Low Reynolds Number Flow Heat Exchangers*, Eds. S. Kakac, R. K. Shah and A. E. Bergles, Washington: Hemisphere, pp. 815–844.

London, A. L. and Shah, R. K. (1968) Offset rectangular plate-fin surfaces – heat transfer and flow-friction characteristics, *ASME Journal of Engineering for Power*, **90**, 218–228.

MacDonald, C. F. (1995) Compact buffer zone plate-fin IHX—the key component for high temperature nuclear process heat realisation with advanced MHR, *Applied Thermal Engineering*, **16**(1), 3–32.

Manglic, R. M. and Bergles, A. E. (1990) The thermal hydraulic design of the rectangular offset strip-fin compact heat exchanger, In: *Compact Heat Exchangers – A*

Festschrift for A. L. London, Eds. R. K. Shah, A. D. Kraus and D. Metzger, Washington: Hemisphere, pp. 123–149.

Patankar, S. V. and Prakash, C. (1981) An analysis of the effect of plate thickness on laminar flow and heat transfer in interrupted plate passages, *International Journal of Heat and Mass Transfer*, **24**(11), 1801–1810.

Prasad, B. S. V. (1993) The performance prediction of multistream plate-fin heat exchangers based on stacking pattern, *Heat Transfer Engineering*, **12**(4), 58–70.

Prasad, B. S. V. and Gurukul, S. M. K. A. (1992) Differential methods for the performance prediction of multistream plate-fin heat exchangers, *ASME Journal of Heat Transfer*, **114**, 41–49.

Shah, R. K. (1982) Compact heat exchanger surface selection, optimisation and computer-aided thermal design, In: *Low Reynolds Number Flow Heat Exchangers*, Eds. S. Kakac, R. K. Shah and A. E. Bergles, Washington: Hemisphere, pp. 845–876.

Shah, R. K. (1988) Plate-fin and tube-fin heat exchanger design procedures, In: *Heat Transfer Equipment Design*, Eds. R. K. Shah, E. C. Subbarao and R. A. Mashelkar, Washington: Hemisphere, pp. 256–266.

Shah, R. K. (1990) Brazing of compact heat exchangers, In: *Compact Heat Exchangers – A Festschrift for A. L. London*, Eds. R. K. Shah, A. D. Kraus and D. Metzger, Washington: Hemisphere, pp. 491–529.

Smith, E. M. (1994) Direct thermal sizing of plate-fin heat exchangers, *The Industrial Sessions Papers, 10th International Heat Transfer Conference, Brighton, UK, 14–18 August 1994*, Rugby, UK: Institution of Chemical Engineers.

Soland, J. G., Mack, W. M. and Rohsenow, W. M. (1978) Performance ranking of plate-fin heat exchanger surfaces, *ASME Journal of Heat Transfer*, **100**, 514–519.

Taylor, M. A. (ed) (1987) *Plate-Fin Heat Exchangers – Guide to Their Specification and Use*, Harwell, UK: HTFS. (Amended 1990.)

Webb, R. L. (1994) *Principles of Enhanced Heat Transfer*, Chichester, UK: John Wiley & Son.

Weimer, R. F. and Hartzog, D. G. (1972) Effects of maldistribution on the performance of multistream multipassage heat exchangers, *Proceedings of the 1972 Cryogenic Engineering Conference, Advances in Cryogenic Engineering*, **17**, Paper B-2, 52–64.

5

DIRECT SIZING OF HELICAL-TUBE EXCHANGERS

5.1 DESIGN FRAMEWORK

Theoretical expressions are developed for the geometrical arrangement of the tube-bundle in a simple helical-tube multi-start coil heat exchanger, and in exchangers with central ducts. Consistent geometry provides uniform helix angles, uniform transverse and longitudinal tube pitches, and identical tube lengths throughout the bundle.

Sizing of a contraflow exchanger begins after determination of the mean temperature difference $\Delta\theta_m$ and the product US of the overall heat transfer coefficient and the surface area. Given the tube geometry and the tube-side and shell-side pressure losses, a method is presented for arriving at an optimal tube-bundle configuration for the heat exchanger with single-phase fluids.

In developing the direct-sizing method, simplified tube-side flow friction and heat transfer correlations for straight tubes are employed to permit a clean solution. This starts from knowledge of local tube and pitching geometry, and when the 'design window' is open (see Fig. 5.10) we arrive at an optimum tube-bundle configuration satisfying specified shell-side and tube-side heat transfer and pressure loss constraints.

However, tube-curvature has an effect on heat transfer and pressure loss. For design-critical conditions, once the exchanger has been sized, it is practicable to fine-tune the design by tube coil length adjustment so that the same pressure loss occurs everywhere across the shell-side and also across the tube-side.

In this chapter a fully explicit design approach can be demonstrated because all correlations for heat transfer and pressure loss are available as algebraic expressions. A numerical design approach is also possible, and perhaps

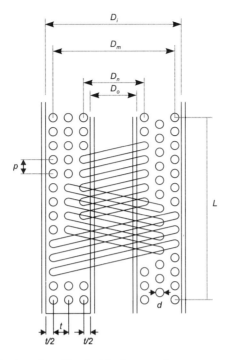

Fig. 5.1 Helical tube-bundle with start factor $r = 1$

preferable for practical design purposes. Setting up the numerical solution is left as an exercise.

The helical-tube multi-start coil heat exchanger (Fig. 5.1) has no internal baffle leakage problems. It permits uninterrupted cross-flow through the tube bank for high local heat transfer coefficients, and provides advantageous counterflow terminal temperature distribution in the overall exchanger.

Some modification to LMTD is required when the number of tube-turns is less than about 10, and this analysis has been presented by Hausen in his German text (1950) and his English texts (1983).

The design is largely restricted to non-fouling fluids, and is particularly useful when exchange is required between high pressure, low volume flow and low pressure, high volume flow, often encountered in cryogenics. But flow areas on both sides may be usefully varied. Thermal expansion can be accommodated by deflection of the ends of the coiled tube-bundle.

This type of exchanger was patented by Hampson (1895) and subsequently repatented by L'Air Liquide (1934). However, formal geometry of the helical-tube multi-start coil heat exchanger does not seem to have been given before 1960, when it was presented in an industrial report (Smith 1960). A very brief note outlining the principal results was published by Smith (1964).

Since then programmes of experimental work on heat transfer in helical-coil tube-bundles have been published: Gilli (1965), Smith and Coombs (1972),

Abadzie (1974), Smith and King (1978), Gill *et al.* (1983). Further geometrical results have been derived, and a direct method of arriving at the design of the tube-bundle has been obtained, both of which are included in this chapter.

A substantial amount of international work has been done on the helical-coil design. It has been applied in gas-cooled nuclear reactor plant, in marine and land-based PWRs and in cryogenic applications, including LNG plant. Weimer and Hartzog (1972) have preferred the helical-coil heat exchangers for LNG service, as the design is less sensitive to flow maldistribution.

In heat exchanger sizing, both LMTD-Ntu and ε-Ntu methods deliver the product of the overall heat transfer coefficient, and the related surface area (US), leaving the design configuration to be determined by other methods. This chapter describes an approach to direct sizing that starts from the product US and the LMTD.

The method applies to tube arangements in which the local geometry of the bundle is independent of the number of tubes in the exchanger, i.e. the shell-side area for flow along a single tube can be determined, and true counterflow is achieved without the use of redirecting baffles. A minimum value of $y = 10$, the number of times that shell-side fluid crosses a tube-turn, is desirable; see also Section 9.5 on the number of tube rows required. Before proceeding to thermal design, certain geometrical expressions for the helical-coil geometry have to be developed.

5.2 BASIC GEOMETRY

Start factor (r)

As a simplification, the effect of tube curvature on heat transfer and pressure loss through the tube is neglected. For the shell-side fluid, each parallel flow path should therefore have the same axial configuration. For the tube-side fluid, each tube should have the same length.

The simplest method of satisfying the above conditions is to give every tube the same helix angle, and to adopt an annular arrangement where the central coil has one tube, the second coil has two tubes, the third coil three tubes and so on. The mean coil diameters are selected so that the shell-side fluid everywhere passes over exactly the same number of tube-turns in traversing the bundle. This layout will be especially satisfactory when a small area for flow in the tube-bundle is required compared with the shell-side flow area.

It is possible to generalise the above case by multiplying the number of tubes in all coils by a constant factor r which is an integer and which may take the values 1, 2, 3, etc. This increases the number of tubes in the exchanger and the area for flow on the tube-side r times. For the same heat transfer surface, it reduces the required length of individual tubes and increases the helix angle of the tube coils.

In the expressions given below, the outermost coil is denoted as the *m*-th coil and contains *rm* tubes, whereas an intermediate coil is denoted as the *z*-th coil and contains *rz* tubes. For complete generality, a central axial cylinder is introduced (Fig. 5.1); this creates an innermost coil denoted as the *n*-th coil and containing *rn* tubes.

Mean diameter of the z-th coil (D_z)

This parameter is required for finding shell-side flow area. Noting that $p > d/\cos \phi$ always, and $t > d$ always, then for every tube in the exchanger (Fig. 5.2)

$$\tan \phi = (prz)/(\pi D_z)$$

$$D_z - D_{z-1} = (2t) = (pr)/(\pi \tan \phi)$$

$$D_z = 2zt \qquad (5.1)$$

Helix angle of coil (ϕ)

$$\tan \phi = (rp)/(2\pi t) \qquad (5.2)$$

Length of the tube bundle (L)

For every tube in the exchanger (Fig. 5.2)

$$\sin \phi = L/\ell \qquad (5.3)$$

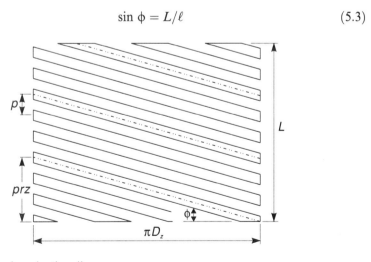

Fig. 5.2 Developed *z*-th coil

5 DIRECT SIZING OF HELICAL-TUBE EXCHANGERS

then, using relationships (5.2) and (5.3)

$$\left(\frac{\ell}{L}\right)^2 = 1 + \left(\frac{2\pi t}{rp}\right)^2 \qquad (5.4)$$

from which L may be obtained.

Number of tubes in exchanger (N)

The z-th coil contains rz tubes, so that

$$N = r[n + (n+1) + (n+2) + \cdots + (m-1) + m] \quad \text{hence}$$

$$N = r(m+n)(m-n+1)/2 \qquad (5.5)$$

Times that shell-side fluid crosses a tube turn (y)

$$Y = L/p \qquad (5.6)$$

Length of tubing in one longitudinal tube pitch (ℓ_c)

Knowledge of the dimension ℓ_c is required in heat transfer design for condensation.

$$y\ell_c = \text{total length of tubing} = N\ell$$

thus using (5.6)

$$\ell_c = Np(\ell/L) \qquad (5.7)$$

Tubing in a projected transverse cross-section (ℓ_p)

This parameter is required in evaluation of shell-side minimum area for flow (Fig. 5.3). Clearly

$$\ell_p = \pi \sum_m^n D_z$$

hence from equation (5.1)

$$\ell_p = 2\pi t[n + (n+1) + \cdots + (m-1) + m] = \pi t(m+n)(m-n+1)$$

5.2 BASIC GEOMETRY

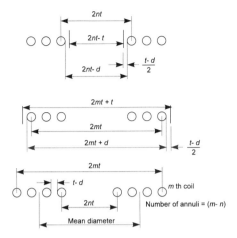

Fig. 5.3 Shell-side area for flow: area = $\Sigma[\pi \times$ (mean dia) $\times$ (no. annuli) $\times$ (width annulus)]

$$\ell_p = (2\pi Nt)/r \quad (5.8)$$

Shell-side minimum area for axial flow (A_m)

This is required for axial cross-flow through the tube-bundle. From equation (5.1) and Fig. 5.1, the outside diameter of the central axial cylinder (core mandrel) is given by

$$D_o = D_n - t = (2n-1)t \quad (5.9)$$

Similarly the inside diameter of the exchanger shell (or bundle wrapper) is

$$D_i = D_m + t = (2m+1)t \quad (5.10)$$

Considering smooth tubes only, the shell-side projected face area for flow is

$$\ell_p(t-d) = 2\pi Nt(t-d)/r$$

hence the face area for axial flow, shell-side is

$$A_s = 2\pi Nt(t-d)/r$$

Using equations (5.5), (5.9) and (5.10), or proceeding directly from Fig. 5.3

$$A_s = \pi(m+n)(m-n+1)t(t-d) = \pi(D_i^2 - D_o^2)(1-d/t)/4$$

5 DIRECT SIZING OF HELICAL-TUBE EXCHANGERS

Denoting annular area between the central axial cylinder and the exchanger shell as

$$A_a = \pi(D_i^2 - D_o^2)/4 \qquad (5.11)$$

it follows that the correction for face area is

$$A_s/A_a = 1 - d/t \qquad (5.12)$$

For flow friction and heat transfer correlations, the fluid velocities in staggered and in-line tube-bundle arrangements are generally taken at the point of minimum gap between adjacent tubes. The use of alternate right-and left-hand coils in a multi-start helical-tube heat exchanger ensures a homogeneous mixture of all cross-flow geometries between the two extremes, in-line and staggered, in the tube-bundle; this is independent of any axial displacement of individual coils (Fig. 5.1).

The effective minimum shell-side flow area (A_m) will be greater than the the minimum 'line-of-sight' flow area (A_s). The value of A_m for a multistart coil helical-tube heat exchanger is found by considering Fig. 5.4. Figure 5.4a gives a three-dimensional view of a portion of the tube-bundle, which is developed to give straight tubes. AB represents the distance between the centre-lines of adjacent rows of tubes when the tube-bundle may be considered as in-line and FG represents the same distance when the tube-bundle is staggered. This geometry is reproduced in two dimensions in Fig. 5.4b.

Because of symmetry it is sufficient to calculate the effective minimum area for flow which lies between AB and FG (A_m), and to compare this with the corresponding face area for flow (A_s). At any distance from AB, the vertical distance between K and L is pbr. Thus

$$KL = \left(\frac{pr}{\pi t}\right)\sqrt{b^2 + \{\pi t^2/(pr)\}^2}$$

and assuming that the tube cross-section can be taken as circular, the gap between the tubes is

$$\left(\frac{pr}{\pi t}\right)\sqrt{b^2 + \{\pi t^2/(pr)\}^2} - d$$

There is a slight error in the above expression due to the assumption that the tube cross-sections are circular instead of elliptical.

The minimum area for flow between AB and FG (a_m) is found by integrating the above expression between the limits $b = 0$ and $b = \pi t/2r$. The area is

$$a_m = \int_0^{\pi t/2r} \left[\left(\frac{pr}{\pi t}\right)\sqrt{b^2 + \{\pi t^2/(pr)\}^2} - d\right] db$$

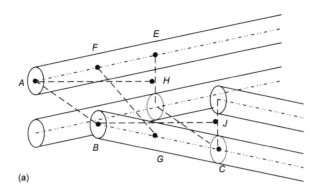

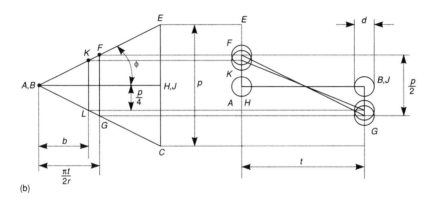

Fig. 5.4 Minimum area for flow: (a) three-dimensional view and (b) two-dimensional view

This expression is of the form

$$\int \left(\sqrt{x^2 + A^2} - B\right) dx$$

and has the solution

$$\frac{1}{2}\left[x\sqrt{x^2 + A^2} + A^2 \ln\left(x + \sqrt{x^2 + A^2}\right)\right] - Bx$$

hence

$$a_m = \left(\frac{\pi t^2}{4r}\right) ROOT + \left(\frac{\pi t^3}{2pr}\right) \ln\left[\left(\frac{p}{2t}\right) + ROOT\right] - \frac{d\pi t}{2r}$$

where $\quad ROOT = \sqrt{\left(\dfrac{p}{2t}\right)^2 + 1}$

The corresponding face area between AB and FG is $a_s = (\pi t)(t-d)/2r$, thus the correction factor for helix angle in the bundle is

$$A_m/A_s = a_m/a_s = \left(\frac{1}{t-d}\right)\left[\left(\frac{t}{2}\right)ROOT + \left(\frac{t^2}{p}\right)\ln\left(\frac{p}{2t} + ROOT\right) - d\right] \tag{5.13}$$

The shell-side area for flow is determined from equations (5.11), (5.12) and (5.13) as

$$A_m = A_a \left(\frac{A_s}{A_a}\right)\left(\frac{A_m}{A_s}\right) \tag{5.14}$$

$$A_m = \left(\frac{2\pi N t^2}{r}\right)\left(\frac{A_s}{A_a}\right)\left(\frac{A_m}{A_s}\right) \tag{5.15}$$

This expression will not apply when there is no spacing, or very little spacing between the tubes, for then an entirely new shell-side geometry is created, and correlations may be better expressed in terms of tube-bundle porosity (see below).

Tube-side area for flow (A_t)

$$A_t = N(\pi/4)d^2 \tag{5.16}$$

Shell-side to tube-side flow area ratio (a_r)

This parameter is independent of the number of tubes, which permits direct sizing of helical-tube multi-start coil heat exchangers.

$$a_r = \left(\frac{A_m}{A_t}\right) = \left(\frac{8}{r}\right)\left(\frac{t}{d_i}\right)^2 \left(\frac{A_s}{A_a}\right)\left(\frac{A_m}{A_s}\right) \tag{5.17}$$

independent of the number of tubes (N).

Shell-side porosity (P_y)

This parameter is used in correlating friction factor data for shell-side flow in the helical-coil tube-bundle (Smith and Coombs 1972, Smith and King 1978).
The volume taken up by the tubes is

$$V_t = N(\pi/4)d^2\ell$$

The annular volume between the exchanger shell and the internal duct is

$$V_a = (\pi/4)(D_i^2 - D_o^2)L = \pi(m+n)(m-n+1)t^2L = \left(\frac{2\pi}{r}\right)Nt^2L$$

Thus the shell-side porosity is $P_y = 1 - $(tube volume)/(annulus volume)

$$P_y = 1 - \left(\frac{r}{8}\right)\left(\frac{d}{t}\right)^2\left(\frac{\ell}{L}\right) \qquad (5.18)$$

Exchanger with central duct

A central bypass duct with a flow-control valve may be an advantage when close control of fluid temperature is necessary. It is convenient first to establish the leading dimensions of a simple exchanger to obtain the length of the bypass duct required. The inside diameter of the central duct is then selected on the basis of pressure loss in the duct, and the outside diameter will then determine the value of n for the innermost coil using equation (5.9).

5.3 SIMPLIFIED GEOMETRY

When a wire space of diameter $(t-d)$ is used to space tubes in both transverse and longitudinal directions, then several of the geometrical relationships are simplified as follows:

$$\cos \phi = t/p \qquad (5.19a)$$

$$\sin \phi = r/2\pi \qquad (5.19b)$$

Using equation (5.13)

$$\frac{L}{t} = \frac{r}{2\pi} \qquad (5.20)$$

and in equation (5.18)

$$P_y = 1 - \left(\frac{\pi}{4}\right)\left(\frac{d}{t}\right)^2 \qquad (5.21)$$

Substituting (5.20) in (5.4)

$$\left(\frac{t}{p}\right)^2 = 1 - \left(\frac{r}{2\pi}\right)^2 \qquad (5.22)$$

We note that r may take only integer values (1 to 6) for real values of t/p; see Table 5.1.

For axial flow in the bundle it is best if the angle of inclination does not exceed 20° if heat transfer correlations are to remain valid, thus $r = 1$ or 2.

For radial flow exchangers it may be preferable to have the tubes of succesive coils set at right angles to each other, thus $r = 4$ or 5

Table 5.1 Simplified geometry for tube-bundle

r	1	2	3	4	5	6
p/t	1.013	1.055	1.138	1.297	1.651	3.369
ℓ/L	6.283	3.142	2.094	1.571	1.257	1.047
ϕ	9.158	18.56	28.52	39.54	52.73	72.74

In obtaining expression (5.15) for shell-side minimum area for axial flow, certain assumptions were made concerning the location of minimum flow area.

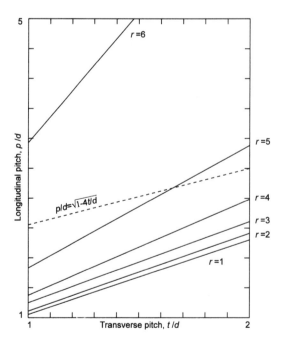

Fig. 5.5 Location of shell-side minimum area for flow

It is necessary to show that at the locally staggered section of the bundle, the minimum flow area will always lie in the diagonal direction.

With the simplified geometry, when the intermediate coil of the other coiling hand is in the 'middle' position

transverse area for flow $= (2t - d)$

diagonal area for flow $= 2[t\sqrt{(1 + 4t/d)} - d\,]$

if $p/d \geqslant \sqrt{(1 + 4t/d)}$ then minimum flow area is in transverse direction

if $p/d \leqslant \sqrt{(1 + 4t/d)}$ then minimum flow area is in diagonal direction

For values of r up to 4, Fig. 5.5 shows that the assumption made earlier is certainly valid for the simplified geometry. The assumption should probably be checked for values of r greater than 2 for more general tube-pitching arrangements. Again, when either the transverse pitch ratio (t/d) or the longitudinal pitch ratio (p/d) is unity, the shell-side flow area will need to be reassessd. When both ratios are equal to unity, the relevant parameter will be the porosity of the tube bundle.

5.4 THERMAL DESIGN

Input data

To illustrate the design method, data for one of the OECD Dragon helium/steam heat exchangers was modified to provide a single-phase problem. Constant (mean) fluid properties are employed, but the technique can be extended to piecewise calculation of exchangers in which change in fluid properties is significant. Terminal temperatures, log mean temperature difference $\Delta\theta_{lmtd}$ and the exchanger duty Q are known data, which will provide the product US.

Exchanger performance
exchanger duty $\qquad\qquad\qquad\qquad\qquad\qquad Q = 1500$
log mean temp difference, K $\qquad\qquad\qquad \Delta\theta_{lmtd} = 39.152$

Tube-side fluid (steam)
mass flow rate, kg/s $\qquad\qquad\qquad\qquad\quad \dot{m}_t = 1.750$
specific heat, J/(kg K) $\qquad\qquad\qquad\qquad\quad C_t = 6405.0$
density, kg/m³ $\qquad\qquad\qquad\qquad\qquad\quad \rho_t = 88.00$
thermal conductivity, J/(m s K) $\qquad\qquad\quad \lambda_t = 0.1040$
absolute viscosity, kg/(m s) $\qquad\qquad\qquad \eta_t = 0.000\ 029\ 78$
Prandtl number $\qquad\qquad\qquad\qquad\qquad\; Pr_t = 1.484$

Shell-side fluid (helium)
mass flow rate, kg/s $\qquad\qquad\qquad\qquad\quad \dot{m}_s = 1.500$

specific heat, J/(kg K) $C_s = 5120.0$
density, kg/m³ $\rho_s = 1.200$
thermal conductivity, J/(m s K) $\lambda_s = 0.256$
absolute viscosity, kg/(m s) $\eta_s = 0.000\,038.\,50$
Prandtl number $\text{Pr}_s = 0.770$

Local geometry
tube external diameter, m $d_o = 0.022$
tube internal diameter, m $d_i = 0.018$
optimised tube spacing, m $(t - d) = 0.00761^*$
tube minimum coiling diameter, m $D_m = 0.200$
tube thermal conductivity, J/(m s K) $\lambda_w = 190.0$
coiling start factor $r = 1$

Correlations and constraints

Tube-side correlations

$$\text{heat transfer} \quad \text{Nu} = 0.023(\text{Re})^{0.8}(\text{Pr})^{0.4} \tag{5.23}$$

$$\text{flow friction} \quad f = 0.046(\text{Re})^{-0.2} \tag{5.24}$$

Shell-side correlation

$$\text{heat transfer} \quad \text{Nu} = 0.0559(\text{Re})^{0.794} \tag{5.25}$$

$$\text{friction factor} \quad f = P_y \times 0.26(\text{Re})^{0.117} \tag{5.26}$$

Equation (5.23) is the standard result for turbulent flow in a straight tube with the viscosity term omitted for simplicity. Equation (5.24) follows from (5.23) using the Reynolds analogy. The Dean number correlation for flow in curved tubes is omitted as it would introduce complications in the first optimisation. The design may be varied later using correlations selected from papers by Ito (1959), Mori and Nakayama (1965, 1967a,b), Ozisik and Topakoglu (1968), Jensen and Bergles (1981), Yao (1984) and Gnielinski (1986). These papers contain references to further publications.

Equations (5.25) and (5.26) were obtained during an experimental programme of work on investigation of shell-side flow in helical-coil heat exchangers (Smith and Coombs 1972, Smith and King 1978) and the constants

---

*The optimised tube spacing corresponds to $t/d = 1.346$. This can only be obtained after the computational runs required to construct Figs. 5.8 and 5.9, and it corresponds to maximum utilisation of available pressure losses. Its use at this point avoids extensive listing of data which does not correspond to the design point. The t/d ratio is also a constraint in that it must lie within the range of values used in the test programme that established the shell-side correlations $(1.125) \leqslant t/d \leqslant 1.500$).

can be modified to include the effects of Prandtl number, which Abadzic (1974) found to be $(Pr)^{0.36}$ with $Pr = 0.71$. More recent work is reported by Kanevets and Politykina (1989).

Each correlation (5.23) to (5.26) has a range of validity of Reynolds number. In addition, limiting velocities may exist for erosion (upper bound) and for fouling (lower bound). The problem of flow-induced tube vibration may also have to be considered (Chen 1978). Maximum desired pressure losses are specified for both tube-side and shell-side flow.

The experimental Reynolds number ranges for equations (5.25) and (5.26) are not great. Abadzic (1974) examined heat transfer data from several sources and recommended three generalised equations for an extended range of Reynolds number:

$$Nu = 0.332(Re)^{0.6} (Pr)^{0.36} \text{ for } (1 \times 10^3 < Re < 2 \times 10^4)$$
$$Nu = 0.123(Re)^{0.7} (Pr)^{0.36} \text{ for } (2 \times 10^4 < Re < 2 \times 10^5)$$
$$Nu = 0.036(Re)^{0.8} (Pr)^{0.36} \text{ for } (2 \times 10^5 < Re < 9 \times 10^5)$$

These are straight-line (log-log) segments, and such multiple correlations can now be replaced by interpolating splinefits of data with weighted errors.

Most of the data correlated by Abadzic correponded to helix angles of around 9°, but the data are not for uniform helical-coil tube-bundles. A comparison of the correlations presented by Smith and Coombs (1972) and

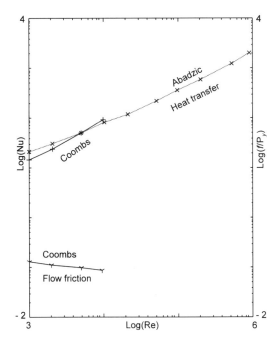

Fig. 5.6 Shell-side correlations

5 DIRECT SIZING OF HELICAL-TUBE EXCHANGERS

Abadzic (1974) is presented in Fig. 5.6. Unfortunately, Abadzic does not report friction factor correlations. Shell-side correlations used in this paper are those which were developed specifically for helical-coil tube-bundles and of which the author has personal knowledge.

A comprehensive review of available heat transfer and flow friction correlations for helical-tube multi-start coil heat exchanger bundles was given by Le Feuvre (1986), but not all of the winding geometries reported are consistent. When data outwith this extended range is required the reader might turn to the extensive work on normal crossflow for in-line and staggered geometries by Zukauskas (1987), Zukauskas and Ulinskas (1988), or to Bejan's summary of these results (1993).

Tube-side constraints
Re for heat transfer and pressure loss 10^4 to 10^6
velocity for fouling/erosion, m/s $1.5 < u_{fe} < 6.0$
maximum design pressure loss, N/m² 2000.0

Shell-side constraints
Re for heat transfer and pressure loss 10^3 to 10^4
velocity for tube vibration Chen (1978)
maximum design pressure loss, N/m² 5000.0

The design approach has to work within the limits of the above envelope.

Local geometry
From relations (5.19) to (5.22) p, ℓ/L, P_y and t/p are evaluated numerically. Then number of tubes in exchanger, equation (5.5)

$$N = (r/2)(m+n)(m-n+1) = \text{unknown} \tag{5.27}$$

Tube-side area for flow, equation (5.16)

$$A_t = N(\pi/4)d_i^2 = 0.000\,254\,47\ m^2 \tag{5.28}$$

Correction for face area, equation (5.12)

$$A_s/A_a = 0.257\,01 \tag{5.29}$$

Correction for helix angle, equation (5.13)

$$A_M/A_s = 1.160\,46 \tag{5.30}$$

Flow area ratio (shell-side/tube-side), equation (5.17)

$$a_r = \left(\frac{8}{r}\right)\left(\frac{t}{d}\right)^2 \left(\frac{A_s}{A_a}\right)\left(\frac{A_m}{A_s}\right) = 6.456\,51 \tag{5.31}$$

which is independent of N, i.e. the shell-side area for flow corresponding to a single tube is determined. Hence the shell-side area for flow is

$$A_s = a_r A_t = 0.00164299 N \quad m^2 \tag{5.32}$$

Velocity constraints
Because tube size is specified, the tube-side velocity bounds on Reynolds number and on fouling/erosion may be used to determine the number of tubes required. However, it is important to recognise that they are not the only considerations to be taken into account.

$$A_t = N(\pi/4)d_i^2 \quad \dot{m}_t = \rho_t A_t u_t \quad Re = \rho_t u_t d_n/\eta_t$$

Re upper bound
$V_{max} = 18.80$ m/s
$A_{min} = 0.0010578$ m^2
$N_{min} = 5$ tubes

Re lower bound
$V_{min} = 0.188$ m/s
$A_{max} = 0.10578$ m^2
$N_{max} = 415$ tubes

Erosion upper bound
$V_{max} = 6.0$ m/s
$A_{min} = 0.003314$ m^2
$N_{min} = 14$ tubes

Fouling lower bound
$V_{min} = 0.1$ m/s
$A_{max} = 0.1989$ m^2
$N_{max} = 781$ tubes

Only restrictions on Reynolds number are known for shell-side flow, and they are handled in a slightly different manner.

$$A_s = a_r A_t \quad G_s = \dot{m}_s/A_s \quad Re_s = dG_s/\eta_s$$

Re upper bound
$N_{min} = 53$ tubes

Re lower bound
$N_{max} = 521$ tubes

Overall heat transfer is now considered, carying forward the unknown number of tubes (N) and the unknown tube length (l), and referring all heat transfer coefficients to the outside of the tube.

Heat transfer constraints
The tube-side heat transfer coefficient is determined from correlation (5.23) as follows:

$$G_t = \dot{m}_t/A_t = 6877.0/N$$
$$Re_t = d_i G_t/\eta_t = 4\,156\,705.6/N$$
$$Nu_t = 0.023(Re_t)^{0.8}(Pr_t)^{0.4} = 5312.52/(N)^{0.8}$$
$$\alpha_t = (\lambda_t/d_i)Nu_t = 30\,694.6/(N)^{0.8}$$

5 DIRECT SIZING OF HELICAL-TUBE EXCHANGERS

and referring to tube outside diameter, multiplier (d_i/d)

$$\alpha_t = 25\,113.7/(N)^{0.8} \quad \text{J}/(\text{m}^2\text{sK}) \tag{5.33}$$

The shell-side heat transfer coefficient is determined from correlation (5.25) as follows,

$$G_s = \dot{m}_s/A_s = \dot{m}_s/(a_r A_t) = 912.971/N$$

$$\text{Re}_s = dG_s/\eta_s = 521\,697.5/N$$

$$\text{Nu}_s = 0.0559(\text{Re}_s)^{0.794} = 0.063\,235(\text{Re}_s)^{0.794}(\text{Pr}_s)^{0.36} = 1936.611/(N)^{0.794}$$

$$\alpha_s = (\lambda_s/d)\text{Nu}_s = 22\,535.1/(N)^{0.794} \quad \text{J}/(\text{m}^2\text{sK}) \tag{5.34}$$

The tube-wall heat transfer coefficient is determined as

$$\alpha_w = \frac{2\lambda_w}{d\ln(d/d_i)} = 86\,075.0 \ \text{J}/(\text{m}^2\text{sK}) \tag{5.35}$$

The expression for overall heat transfer coefficient becomes

$$U = 22\,535.1/[0.8973(N)^{0.8} + 0.261\,808 + (N)^{0.794}] \ \text{J}/(\text{m}^2\text{sK}) \tag{5.36}$$

The given product $(US) = (Q/\Delta\theta_{lmtd}) = 38\,311.93$ may now be employed

$$38\,311.9 = [U \text{ from eqn}(5.36)]N\pi d \tag{5.37}$$

Thus

$$24.598 = \frac{N\ell}{0.897\,32(N)^{0.8}} + 0.261\,81 + (N)^{0.794} \tag{5.38}$$

Equation (5.38) has to be satisfied to ensure correct heat transfer. Values of N already obtained under 'velocity constraints' could now be substituted to obtain several values for the tube length (ℓ), but this would not necessarily guarantee performance because pressure losses have not yet been considered. All bounding values for N must be found before taking the design decision.

Pressure loss constraints

Two further equations exist involving N and ℓ. These are related to tube-side and shell-side pressure losses, respectively.

For the tube-side

$$G_t = \dot{m}_t/A_t = 6877.065/N$$
$$\text{Re}_t = d_i G_t/\eta_t = 4\,156\,721.9/N$$
$$f_t = 0.046(\text{Re})^{-0.2} = 0.002\,1828(N)^{0.2} \text{ from equation (5.24)}$$

$$\Delta p_t = \left(\frac{4f_t G_t^2}{2\rho_t}\right)\left(\frac{\ell}{d_i}\right) \tag{5.39}$$

and with the desired pressure drop of 2000.0 N/m²

$$\ell = (N)^{1.8}/65.171\,64 \tag{5.40}$$

For the shell side

$$G_s = \dot{m}_s/A_s = 912.971/N$$
$$\text{Re}_s = dG_s/\eta_s = 521\,697.5/N$$
$$P_y = 1 - \left(\frac{\pi}{4}\right)\left(\frac{d}{t}\right)^2 = 0.5676$$

$$f_s = P_y \times 0.26(\text{Re})^{-0.117} = 0.031\,628(N)^{0.117} \text{ from equation (5.26)}$$

$$\Delta p_s = \left(\frac{fG^2}{2\rho}\right)\left(\frac{S}{A_s}\right) = \left(\frac{4fG^2}{2\rho}\right)\left(\frac{\ell}{d_i}\right)\left(\frac{d}{a_r d_i}\right) \tag{5.41}$$

and with the desired pressure loss of 5000.0 N/m²

$$\ell = (N)^{1.883}/92.416\,56 \tag{5.42}$$

Equations (5.41) and (5.42) each deliver the minimum number of tubes to satisfy respective pressure drops. Each is solved in turn with the heat transfer equation (5.39) to obtain numerical values for N.

Tube-side

Eqns (5.38) and (5.40)
N_{min} 55 tubes

Shell-side

Eqns (5.38) and (5.42)
N_{min} 55 tubes

These values are identical because the design has already been optimised to satisfy tube-side and shell-side pressure losses simultaneously. If any other constraint had entered into the design consideration, these two figures would be different.

138 5 DIRECT SIZING OF HELICAL-TUBE EXCHANGERS

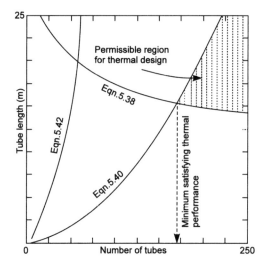

Fig. 5.7 Direct sizing of helical-coil exchanger

Following solution of these equations, the design region satisfying thermal design may be identified as in Fig. 5.7, in which the tube-side curve is shifted to the right for the purpose of illustration. In the present design, the shell-side and tube-side curves are almost coincidental, and both cut the heat transfer curve at the same point because the optimised tube spacing (t–d) has been used.

5.5 COMPLETION OF THE DESIGN

Values of N obtained in earlier sections are summarised in Table 5.2.

Table 5.2 Design 'window' based on tubes

Tube-side		
Reynolds no.	5 (high Re)	415 (low Re)
Fouling/erosion	14 (erosion)	731 (fouling)
Heat transfer	55 (max Δp)	–
Shell-side		
Reynolds no.	53 (high Re)	521 (low Re)
Vortex shedding	no data	no data
Heat transfer	55 (max Δp)	–

The minimum surface area for which the exchanger design is viable would occur with the largest value of N_{min}, namely N=55 tubes. This will not necessarily prove to be the number of tubes in a helical-coil exchanger as further geometrical constraints have to be satisfied to ensure that every coiling station in the tube-bundle is filled.

The size of the central duct is known from exchanger bypass requirements, which determines the minimum coiling diameter. The smallest possible integer value of n is first obtained from the minimum coiling diameter of 0.200 m, by rounding up the value obtained from

$$0.200 = 2nt \tag{5.43}$$

and a first estimate of the integer m is obtained by using

$$N = r(m+n)(m-n+1)/2 \tag{5.44}$$

Equation (5.44) is employed again in reverse using these values of m and n, to obtain a first 'helical-coil' value for N.

If the new value of N is greater than 55, the value of n is increased by 1 progressively until the closest match above 55 is obtained. In unusual circumstances it may be appropriate to decrease the value of m by 1 and adjust n to get the best match.

If the new value of N is below 55 then m is increased by 1, and n is again increased progressively by 1 until the closest match above 55 is obtained. The nearest approach to the desired value of 55 is subsequently taken. Optimising the bundle to the smallest number of tubes greater than 55 gives the following configuration:

$N = 56$	$(t/d) = 1.3459$
$m = 11$	$D_i = (2m+1)t = 0.6808$ m
$m = 5$	$D_o(2n-1)t = 0.2664$ m

Using (5.27), individual tube length $\ell = 20.400$ m

Using (5.6), length of coiled bundle $L = 3.248$ m

The chosen value of $t/d = 1.3459$ makes almost maximum use of both allowable pressure losses and provides a close approach to the smallest practicable tube-bundle volume. The design method might be extended by adding an iterative loop to vary the tube spacing within the range of validity of the shell-side correlations. However, it was thought advisable to proceed by separate calculation for each value of $(t-d)$ because the several design constraints may invoke alternative limiting conditions, conditions which the designer should be aware of.

140 5 DIRECT SIZING OF HELICAL-TUBE EXCHANGERS

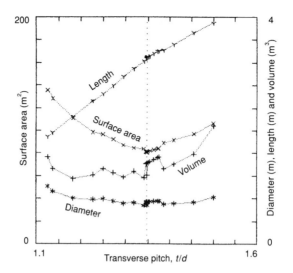

Fig. 5.8 Optimised design of helical-coil exchanger

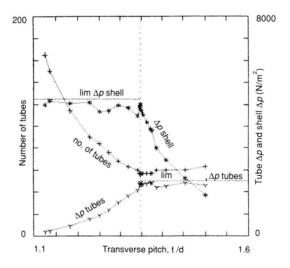

Fig. 5.9 Optimised design of helical-coil exchanger

Figure 5.8 illustrates the effect of varying transverse tube pitching. The jagged curves are a consequence of adjusting tube numbers to satisfy helical-coil tube geometry requirements. Limiting conditions for shell-side and tube-side pressure drop are shown in Fig. 5.9, but one or other, or both, could easily be replaced by maximum Re or erosion/fouling velocity limitations for different input data.

5.6 THERMAL DESIGN FOR $(t/d) = 1.346$

The results, in this section were obtained from a program written in Pascal.

surface area for heat transfer, m²	$S = 78.988$
tube length, m	$\ell = 20.408$
bundle length, m	$L = 3.248$
core mandrel outside diameter, m	$D_o = 0.266$
bundle wrapper inside diameter, m	$D_i = 0.681$
face area of bundle, m²	$A_a = 0.3643$
volume of bundle, m³	$V_a = 1.1832$
specific performance, MW/m³	$Q/V_a = 1.268$
tube heat transfer coefficient, J/(m² s K)	$\alpha_{\_t} = 1003.14$
shell heat transfer coefficient, J/(m² s K)	$\alpha_s = 949.47$
wall heat transfer coefficient, J/(m² s K)	$\alpha_w = 86\,075.0$
overall heat transfer coefficient, J/(m² s K)	$U = 485.04$
tube-side Reynolds number	$Re_t = 74\,227.2$
shell-side Reynolds number	$Re_s = 9316.1$
tube-side velocity, m/s	$u_{\_t} = 1.396$
shell-side velocity, m/s	$u_s = 13.586$
tube-side friction factor	$f_t = 0.004\,883$
shell-side friction factor	$f_s = 0.050\,55$
tube-side pressure loss, N/m²	$\Delta p_t = 1897.4$
shell-side pressure loss, N/m²	$\Delta p_s = 4806.1$
tube-side area for flow, m²	$A_t = 0.01434$
shell-side area for flow, m²	$A_s = 0.920$
shell/tube area ratio	$a_r = 6.456$
exchanger duty $(US\Delta\theta_{lmtd})$, kW	$Q = 1500.0$

The optimised helical-tube design provides a useful standard against which competing designs may be assessed (Fig. 5.10). This multi-start helical-tube design is a preferred configuration when one fluid is evaporating, because the shell-side is fully interconnected for that fluid.

5.7 FINE TUNING

Fine tuning of the design will now be developed to avoid thermodynamic mixing losses which degrade exchanger performance. Gnielinski's (1986) review of tube-side correlations for curved tubes omitted the significant earlier work of Ito and of Mori and Nakayama. The Gnielinksi correlations are

142 5 DIRECT SIZING OF HELICAL-TUBE EXCHANGERS

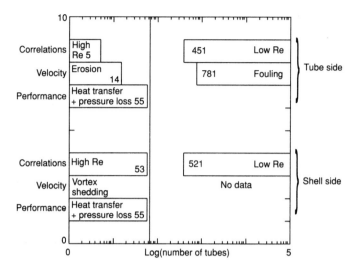

Fig. 5.10 Design window for Dragon-type heat exchanger

compared with the earlier work of Ito (1959) and of Mori and Nakayama (1965, 1967a, b), and a selection is made.

Resume of correlations

All the correlations, in this section are consistent with the definition of friction factor for laminar flow in straight tubes leading to $f = 16/\text{Re}$.

Transition Reynolds number
The following expressions were examined by Gnielinski (1986):

$$\text{Ito}: \quad \text{Re} = 2000(d/D)^{0.32} \tag{5.45}$$

$$\boxed{\text{Schmidt}: \text{Re} = 2300[1 + 8.6(d/D)^{0.45}]} \tag{5.46}$$

When plotted in the ranges $1000 < \text{Re} < 100\,000$ and $10 < D/d < 10\,000$, the Ito correlation is a straight line but the Schmidt correlation is curved. There is, however, little to choose between them up to a value of $D/d = 100$, which is already a fairly large diameter heat exchanger. The Schmidt correlation is preferred as it correctly tends towards the straight tube transition at Re = 2300 for large values of Dean number De = D/d (Fig. 5.11a, b).

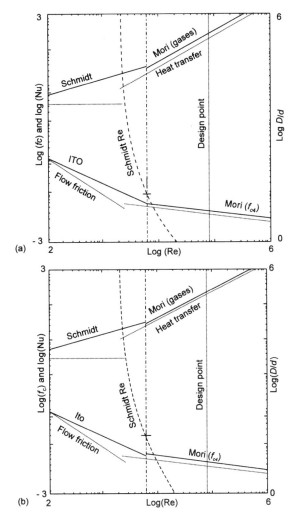

Fig. 5.11 Correlations for flow in curved tubes with (a) D/d = 16.5 and (b) D/d = 36.2

Laminar flow friction factor

Ito

$$\frac{f_c}{f_s} = \frac{21.5\text{De}}{[1.56 + \log(\text{De})]^{5.73}} \qquad (5.47)$$

valid for $(13.5) \leqslant \text{De } 2000$

Gnielinski

$$\frac{f_c}{f_s} = [1 + 0.33(\text{De})^4]\left(\frac{\eta_{wall}}{\eta}\right)^{0.27} \qquad (5.48)$$

5 DIRECT SIZING OF HELICAL-TUBE EXCHANGERS

Mori

$$\frac{f_{c2}}{f_s} = \frac{0.1080\sqrt{De}}{(1 - 3.253/\sqrt{De})} \tag{5.49}$$

valid for $(1 \leqslant Pr \leqslant \infty)$

When plotted for values of $D/d = 5, 10, 20, 50, 100, 200$, the Gnielinski and and Ito correlations were practically identical within the ranges $100 < Re < 100\,000$ and $0.001 < f_c < 1$. The Mori correlation is purely theoretical and was not compared. The Ito correlation is preferred in this analysis because it does not require knowledge of bulk-to-wall properties.

Laminar flow heat transfer
Of the flow regimes, laminar heat transfer is least well understood and further experimental investigation is required.

Schmidt

$$m = 0.5 + 0.2903(d/D)^{0.194}$$

$$Nu = \left\{ 3.65 + 0.08 \left[1 + 0.8 \left(\frac{d}{D}\right)^{0.9} \right] \right\} (Re)^m (Pr)^{0.33} \left(\frac{Pr}{Pr_{wall}} \right)^{0.14} \tag{5.50}$$

Mori and Nakayama

$$\text{for } Pr \leqslant 1.0 \text{ (gases) } \xi = \frac{1}{5}\left[2 + \sqrt{10/(Pr)^2 - 1} \right]$$

$$\text{for } Pr \geqslant 1.0 \text{ (liquids) } \xi = \frac{2}{11}\left[1 + \sqrt{77/(4Pr^2)} \right]$$

$$Nu = 0.864 \frac{\sqrt{De}}{\xi}\left[1 + \frac{2.35}{\sqrt{De}} \right]$$

The last correlation was recommended in the third paper by Mori and Nakayama as being for practical use. Two other theoretical correlations developed in their earlier papers involved a great number of terms, and did not find favour.

When plotted for values of $D/d = 5, 10, 20, 50, 100, 200$, large differences were found in Nu values within the ranges $100 < Re < 100\,000$ and $1 < Nu < 100$ as predicted by Schmidt and by Mori and Nakayama. When very large values of D/d were tried, in an effort to recover the straight-tube correlation at low Reynolds numbers, the Schmidt correlation exhibited convergence towards the theoretical value, $48/11 = 4.36$, for uniform heat flux.

5.7 FINE TUNING

The Schmidt correlation is based on experimental results and seems the better of the two, but requires knowledge of bulk-to-wall properties.

Turbulent flow friction factor

Ito

$$f_c = \frac{\sqrt{d/D}}{4}\left\{0.029 + 0.304\left[\text{Re}(d/D)^2\right]^{-0.25}\right\}$$

valid for $0.034 < \left[\text{Re}(d/D)^2\right] < 300$, i.e. moderate Re

Gnielinski

$$f_c = \frac{0.3164/(\text{Re})^{0.25} + 0.03\sqrt{(d/D)}[\eta_{wall}/\eta]^{0.27}}{4}$$

Mori and Nakayama

$$f_{c4} = \frac{0.305\left\{1 + \frac{0.112}{[\text{Re}(d/D)^2]}\right\}\sqrt{(d/D)}}{4\left[\text{Re}(d/D)^2\right]^{0.2}} \qquad (5.51)$$

moderate Re

$$f_{c5} = \frac{0.192\left\{1 + \frac{0.068}{[\text{Re}(d/D)^{2.5}]^{0.167}}\right\}\sqrt{(d/D)}}{4\left[\text{Re}(d/D)^2\right]^{0.167}}$$

large Re

The Ito and Gnielinski correlations are both experimental and give identical results with $\eta_{wall}/\eta = 1$. The Mori correlations are theoretical; the Mori f_{c4} correlation corresponds closely with Ito over the ranges $1000 < \text{Re} < 100\,000$ and $0.001 < f_c < 0.1$. The integer 4 which appears in each of the above correlations adjusts the friction factor to the definition for straight tubes given earlier. The Mori f_{c4} correlation is employed.

Turbulent flow heat transfer

Gnielinski The correlation makes use of the Gnielinski turbulent flow friction factor (f_c) from above.

$$\text{Nu} = \frac{(f_c/2)\text{Re}(\text{Pr})^{0.4}(\text{Pr}/\text{Pr}_{wall})^{0.14}}{1 + 12.7[\text{Pr}^{0.666} - 1]\sqrt{(f_c/2)}}$$

Mori and Nakayama

$$\text{Nu} = \frac{\Pr(\text{Re})^{0.8}(d/D)^{0.1}\left[1 + \frac{0.098}{\text{Re}(d/D)^2}\right]^{0.2}}{26.2(\Pr)^{0.666} - 0.074} \quad (5.52)$$

for Pr < 1.0 (gases)

$$\text{Nu} = \frac{\Pr^{0.4}(\text{Re})^{5/6}(d/D)^{1/12}\left[1 + \frac{0.061}{\{\text{Re}(d/D)^{2.5}\}^{0.167}}\right]}{41.0}$$

for Pr ⩾ 1.0 (liquids)

The Mori and Nakayama gas correlation gives virtually identical results to the Gnielinski correlation in the ranges $1000 < \text{Re} < 100\,000$ and $1 < \text{Nu} < 1000$; it is preferred because it does not require knowledge of the bulk-to-wall properties.

Preferred correlations for curved tubes are shown as solid lines in Fig. 5.11a, b for $D/d = 16.5$ and $D/d = 36.2$ with $\Pr = 1.484$; matching of the transition between laminar and turbulent regions is remarkably good. The dashed line is the correlation for transition Reynolds number for flow in curved tubes. Flow friction and heat transfer correlations for straight tubes are also shown. Transition between laminar and turbulent flow for the curved tube is marked by a vertical chain-dotted line, and the exchanger tube-side design condition is marked by a vertical dotted line in the turbulent region.

5.8 DESIGN FOR CURVED TUBES

Straight-tube correlations

The reference design employed straight-tube turbulent heat transfer and flow friction correlations for tube-side flow

$$\text{Nu} = 0.023(\text{Re})^{0.8}(\Pr)^{0.4} \quad (5.53a)$$

$$f = 0.046/(\text{Re})^{0.2} \quad (5.53b)$$

The resulting tube-bundle configuration satisfying all constraints has

inner coil tube count	$n = 5$
outer coil tube count	$m = 11$
total number of tubes	$N = 56$

inner coil Dean number (D/d)	$De_n = 16.45$
outer coil Dean number (D/d)	$De_m = 36.19$

Severest conditions occur at the innermost coiling diameter. Equations (5.53a) and (5.53b) may therefore be replaced by the straight-tube equivalents of equations (5.51) and (5.52) using $t/d = 1.346$ found in the earlier optimisation. For the innermost coil, $n = 5$ gives $D/d = 13.46$, which provides

$$\mathrm{Nu} = 0.0326(\mathrm{Re})^{0.7922}(\mathrm{Pr})^{0.4} \qquad (5.54a)$$

$$f = 0.0661/(\mathrm{Re})^{0.2089} \qquad (5.54b)$$

This ensures that all constraints are properly taken into consideration. The design program is run again with the following results:

inner coil tube count	$n = 6$
outer coil tube count	$m = 12$
total number of tubes	$N = 63$
inner coil Dean number (D/d)	$De_n = 19.74$
outer coil Dean number (D/d)	$De_m = 39.48$

Values of tube-side and shell-side pressure drops obtained are within the required limits

tube-side pressure drop, N/m²	$\Delta p_t = 1797.1$
shell-side pressure drop, N/m²	$\Delta p_s = 3455.8$

Fine tuning with curved-tube correlations

The approach to fine tuning is determined by examination of the equations governing thermal performance for a single tube

$$Q = US\Delta\theta_{lmtd} = \dot{m}_t C_t (T_{t1} - T_{t2}) = \dot{m}_s C_s (T_{s1} - T_{s2}) \qquad (5.55)$$

If a cryogenic exchanger is involved, thermodynamic mixing losses are to be avoided as temperature differences of a few kelvins become important. Applying the same philosophy of having constant exit temperatures everywhere in this high temperature exchanger, design LMTD and design terminal temperatures have to be maintained at each coiling diameter of the helical-coil design, thus

$$R_m = (\dot{m}_t/\dot{m}_s) = \mathrm{const.} \qquad (5.56)$$

5 DIRECT SIZING OF HELICAL-TUBE EXCHANGERS

Tube-side pressure loss is reduced with increased coiling diameter. Adjustment of tube-side flow by orificing becomes necessary in order to maintain constant terminal temperatures. Since the ratio $R_m = \dot{m}_t/\dot{m}_s$ must also remain constant; the corollary is an increase in the length of the shell-side flow path with increasing coil diameter.

Individual coil design

The number of coils is known, and thermal performance of a single tube may be determined at each coiling diameter.

Solution involves simultaneous equations with two unknowns: tube length (ℓ) and shell-side mass flow rate ($\dot{m}_s$). Track is kept of these unknowns in the equations which follow by enclosing them in angle brackets $<\#>$ and by assigning the parameter symbol K to that part which can be evaluated numerically.

Shell-side pressure loss

This is constant at each coiling diameter.

$$G_s = \left(\frac{\dot{m}_s}{A_s}\right) = \left[\frac{4}{\pi a_r d_i^2}\right]\langle \dot{m}_s \rangle$$

where a_r is dependent only on local geometry.

$$\mathrm{Re}_s = \left[\frac{dG_s}{\eta_s}\right]\langle \dot{m}_s \rangle$$

$$f_s = P_y \times 0.26(\mathrm{Re}_s)^{-0.117}\langle \dot{m}_s^{-0.117}\rangle$$

$$\Delta p_s = \frac{f_s G_s^2}{2\rho_s}\left(\frac{L}{r_h}\right) \quad \text{with} \quad r_h = \frac{AL}{S}$$

where $s = \pi d \ell$ and $A = a_r \pi d_i^2/4$

Then

$$\langle \ell \dot{m}_s^{(2-0.117)}\rangle = \frac{2\rho_s a_r \Delta p_s}{4 f_s d}\left(\frac{d_i}{G_s}\right)^2 \tag{5.57}$$
$$= K1, \text{ say}$$

Shell-side heat transfer

$$\mathrm{Nu}_s = 0.063\,235(\mathrm{Re}_s)^{0.794}(\mathrm{Pr}_s)^{0.36}$$

$$\alpha_s = (\lambda_s/d) \times 0.063\,235(\mathrm{Re})_s^{0.794}(\mathrm{Pr}_s)^{0.4}\langle \dot{m}_s^{0.794}\rangle \tag{5.58}$$

$$= K2 \times \langle \dot{m}_s^{0.794}\rangle, \text{ say}$$

Tube-side heat transfer coefficient

$$\mathrm{Nu}_t = \frac{\mathrm{Pr}_t(\mathrm{Re}_t)^{0.8}(d_i/D)^{0.1}\left\{1+\dfrac{0.098}{[\mathrm{Re}(d_i/D)^2]^{0.2}}\right\}}{\left[26.2(\mathrm{Pr}_t)^{2/3}-0.074\right]} \quad (5.59)$$

where Pr_t and d_i/D are specified.

$$G_t = \frac{\dot{m}_t}{A_t} = \frac{4R_m}{\pi d_i^2}\langle \dot{m}_s\rangle$$

$$\mathrm{Re} = \frac{d_i G_t}{\eta_t}\langle \dot{m}_s\rangle$$

$$\alpha_t = K3 \times (\mathrm{Re}_t)^{0.8}\langle \dot{m}_s^{0.8}\rangle\left[1+\frac{K4}{(\mathrm{Re}_t)^{0.2}\langle \dot{m}_s^{0.2}\rangle}\right], \quad \text{say} \quad (5.60)$$

$$\text{where} \quad K3 = \left(\frac{\lambda_t}{d_i}\right)\left[\frac{\mathrm{Pr}_t}{26.2(\mathrm{Pr}_t)^{2/3}-0.074}\right]\left(\frac{d_i}{D}\right)^{0.1}$$

$$\text{and} \quad K4 = \frac{0.098}{(d_i/D)^{0.4}}$$

Referring α_t to the outside diameter

$$\alpha_t = K5 \times \langle \dot{m}_s^{0.8}\rangle + K6 \times \langle \dot{m}_s^{0.6}\rangle \quad (5.61)$$

$$\text{where} \quad K5 = K3 \times (\mathrm{Re}_t)^{0.8}\left(\frac{d_i}{d}\right)$$

$$\text{and} \quad K6 = K3 + K4 \times (\mathrm{Re}_t)^{0.6}$$

Overall heat transfer coefficient

$$\frac{1}{U} = \left[\frac{1}{\alpha_t}+\frac{1}{\alpha_w}+\frac{1}{\alpha_s}\right] = \frac{1}{K5\langle \dot{m}_s^{0.8}\rangle + k6\langle \dot{m}_s^{0.6}\rangle} + \frac{1}{\alpha_w} + \frac{1}{K2\langle \dot{m}_s^{0.794}\rangle} \quad (5.62)$$

Heat transfer (referred to outside surface)

$$U(\pi d\ell)\Delta\theta_{lmtd} = \dot{m}_s C_s(T_{s1}-T_{s2})$$
$$\langle \ell/\dot{m}_s\rangle = K7/U, \quad \text{say} \quad (5.63)$$

where $K7 = \dfrac{C_s(T_{s1} - T_{s2})}{\pi d \Delta\theta_{lmtd}}$

The tube length (ℓ) is eliminated from simultaneous equations (5.57) and (5.63), and $\dot{m}_s$ is found by binary search. A good first estimate for $\dot{m}_s$ is available from the reference design. Once $\dot{m}_s$ is determined, ℓ and $\dot{m}_s$ can be found and the tube-side pressure loss may be evaluated.

Tube-side pressure loss (coiled)

Tube performance in each coiling diameter may now be now evaluated

$$G_t = \dfrac{4\dot{m}_t}{\pi d_t^2} \quad \mathrm{Re}_t = \dfrac{d_i G_t}{\eta_t} \quad f_{c4} = \text{equation (5.51)}$$

$$\Delta P_{coil} = \dfrac{4 f_{c4} G_t^2}{2\rho}\left(\dfrac{\ell}{d_i}\right)$$

Required tube-orifice pressure drops are obtained as $\Delta P_t - \Delta P_{coil}$.

Variations in mass flow rates

The results of fine tuning are given in Table 5.3. The variation in mass flow rates across the bundle can be seen. When the program was first run, all constraints were satisfied except for the cumulative mass flow rates, which were 104.3% of the desired values. This is a consequence of using curved-tube heat transfer and pressure drop correlations for the innermost coil, to ensure that all flow constraints were observed when calculating the reference configuration. It is also because the number of tubes proved to be greater.

Table 5.3 Flow rates and tube-side pressure loss across the tube-bundle

Coil no.	$\dot{m}_s$ (kg/m²)	$\dot{m}_s$ (kg/m²)	ℓ(m)	L (m)	Δp_{coil} (N/m²)	$\sum S$ (m²)
6	0.023 995	0.027 994	18.052	2.873	1761.5	7.486
7	0.023 930	0.027 918	18.146	2.888	1739.3	16.265
8	0.023 873	0.027 852	18.226	2.901	1720.4	26.343
9	0.023 824	0.027 794	18.298	2.912	1704.1	37.725
10	0.023 779	0.027 743	18.362	2.922	1689.8	50.416
11	0.023 740	0.027 696	18.420	2.932	1677.1	64.419
12	0.023 704	0.027 654	18.472	2.940	1665.7	79.740

The constraints to be satisfied are as follows:
cumulative tube-side flow, kg/s   cum $\dot{m}_t = 1.75$
cumulative shell-side flow, kg/s   cum $\dot{m}_s = 1.50$

tube-side pressure loss, N/m² $\Delta p_t = 1797$
shell-side pressure loss, N/m² $\Delta p_s = 3456$

To accommodate the variation, a small reduction is applied to the Nusselt number in the reference design calculation only; this is achieved using a multiplier. The multiplier is adjusted until correct values of cumulative mass flow appears in the fine-tuning calculation, which takes account of the changing thermal performance across the bundle. As the two programs are run in sequence, this takes only a few moments.

5.9 DISCUSSION

Actual exchangers for the Dragon reactor have $n=6$ and $m=12$ with a start factor of $r=1$; all tubes have a right-handed helix angle of 16° (Gilli 1965). This would lead to a ratio of $p/t=1.800$, which is much greater than in the design presented here.

For the present design, Fig 5.12 shows the variation in tube-side flow rates across the bundle. Note that the right-hand scale has a suppressed zero. Shell-side mass flow rates follow exactly the same trend as tube-side mass flow rates. It indicates the extent of tube-side orificing required. Figure 5.13 presents the outline of the fine-tuned bundle, illustrating the required variation in tube length against coiling diameter.

inner coil tube count	$n = 6$
outer coil tube count	$m = 12$
total number of tubes	$N = 63$
helix angle of tubes, degrees	$\phi = 9.158$
inner mean coiling diameter $(2nt)$, m	$D_n = 0.355$
outer mean coiling diameter $(2mt)$, m	$D_m = 0.711$
inner bundle length, m	$L_n = 2.874$
outer bundle length, m	$L_m = 2.940$
core outer diameter $(2n-1)t$, m	$D_o = 0.326$
shell inner diameter $(2m+1)t$, m	$D_i = 0.740$

The terminal temperatures used were as follows:

shell-side inlet temperature (helium),°C	$T_{s1} = 600.0$
shell-side outlet temperature (helium),°C	$T_{s2} = 404.7$
tube-side outlet temperature (steam),°C	$T_{t1} = 522.4$
tube-side inlet temperature (steam),°C	$T_{t2} = 388.5$
thermal effectiveness	$(T_{s1} - T_{s2})/T_{span} = 0.923$

152 5 DIRECT SIZING OF HELICAL-TUBE EXCHANGERS

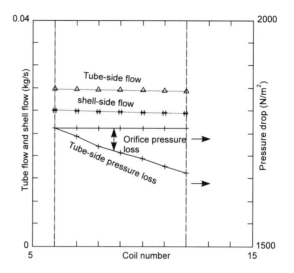

Fig. 5.12 Fine tuning of Dragon-type exchanger

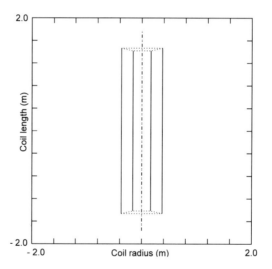

Fig. 5.13 Aspect of fine-tuned head exchanger tube bundle

The real Dragon primary heat exchangers were designed for boiling on the steam side (tube side), consequently the LMTD was also considerably different (ENEA 1960) and the present results cannot be directly compared, although the number of tubes in the present exchanger is exactly the same as for the Dragon exchangers.

5.10 PART-LOAD OPERATION WITH BYPASS CONTROL

Each Dragon heat exchanger was provided with a central bypass duct to control exit gas temperature on the shell side of the exchanger during part-load operation. Under these conditions, pressure loss in the central duct+control valve is equal to the pressure loss in the tube-bundle. The two pressure loss equations can be used, together with the mixing equation at exit, to solve for the mass flow rates and the exit temperature. Heat transfer and flow friction correlations for straight tubes are adequate for the purpose, as the control valve makes any necessary adjustment.

5.11 CONCLUSIONS

1. Geometry relevant to the design of helical-coil exchangers has been presented.
2. Because the flow area ratio (shell-side/tube-side) is independent of the number of tubes in the exchanger, direct sizing of the tube-bundle becomes possible.
3. A simple example illustrating the method of thermal desing has been presented. This highlights the constraining factor which may then be scrutinised.
4. Design optimisation is possible by varying tube spacing $(t - d)$. Full optimisation to minimise any selected parameter (e.g. bundle volume, face area, total tube length) may be carried out by repeating the process for each commercially available tube size.
5. Correlations published by different authors for flow friction factor and heat transfer in curved tubes show consistency of prediction, except for heat transfer in laminar flow.
6. Flow friction and heat transfer correlations for flow in curved tubes match well at the transition between laminar and turbulent regions, compared with those for straight tubes (Fig. 5.11).
7. Curved-tube correlations for tube-side flow should be used for fine tuning of the design when thermodynamic mixing losses are to be avoided. For exacting applications, adjustment of tube length may be required across the tube-bundle. Orificing pressure loss may be allowed for in extended tube 'tails'.
8. The number of tubes in the Dragon primary heat exchangers is confirmed, even though coiling directions and helix angles are different, and steam-side heat transfer and LMTD are different.
9. The final configuration is represented in the 'design window' of Fig. 5.10 as a chain-dotted line.

REFERENCES

Abadzic, E. E. (1974), Heat transfer on coiled tubular matrix, *ASME Winter Annual Meeting, New York, 1974, ASME Paper 74-WA/HT-64*.

Bejan, A. (1993) *Heat Transfer*, New York: Wiley.

Chen, Y. N. (1978), General behaviour of flow induced vibrations in helical tube bundle heat exchangers, *Sulzer Technical Review, Special Number 'NUCLEX 78'*, 59–68.

ENEA (1960) *OECD High Temperature Reactor Project (Dragon)*, Annual Reports, 1960–64.

Gill, G. M., Harrison, G. S. and Walker, M. A. (1983), Full scale modelling of a helical boiler tube, *International Conference on Physical Modelling of Multi-Phase Flow, BHRA Fluid Engineering Conference, April, 1983, Paper K4*, pp. 481–500.

Gilli, P. V. (1965), Heat transfer and pressure drop for crossflow through banks of multistart helical tubes with uniform inclinations and uniform longitudinal pitches, *Nuclear Science and Engineering*, **22**, 298–314.

Gnielinski, V. (1986), Heat transfer and pressure drop in helically coiled tubes, *8th International Heat Transfer Conference, San Francisco*, Vol. 6, 1986, pp. 2847–2854.

Hampson, N. (1895) Improvements relating to the progressive refrigeration of gases, *British patent 10165*.

Hausen, H. (1950), *Warmeubertragung im Gegenstrom, Gleichstrom und Kreuzstrom*, Berlin: Springer Verlag, pp. 213–228.

Hausen, H. (1983), *Heat Transfer in Counterflow, Parallel Flow and Cross Flow*, New York: McGraw-Hill, pp. 232–248.

Ito, H. (1959) Friction factors for turbulent flow in curved pipes, *ASME Journal of Basic Engineering*, **81**, 123–129.

Jensen, M. K. and Bergles, A. E. (1981), Critical heat flux in helically coiled tubes, *ASME Journal of Heat Transfer*, **103**, 660–666.

Kanevets, G. Ye. and Politykina, A. A. (1989), Heat transfer in crossflow over bundles of coiled heat exchanger tubes, *Applied Thermal Sciences*, **2**(1), 38–41.

L'Air Liquide (1934), Improvements relating to the progressive refrigeration of gases, *British patent 416. 096*.

Le Feuvre, R. F. (1986) A method of modelling the heat transfer and flow resistance characteristics of multi-start helically-coiled tube heat exchangers, *8th International Heat Transfer Conference, San Francisco*, Vol. 6, pp. 2799–2804.

Mori, Y. and Nakayama, W. (1965) Study on forced convective heat transfer in curved pipes (first report, laminar region), *International Journal of Heat and Mass Transfer*, **8**, 67–82.

Mori, Y. and Nakayama, W. (1967a), Study on forced convective heat transfer in curved pipes (second report, turbulent region), *International Journal of Heat and Mass Transfer*, **10**, 37–59.

Mori, Y. and Nakayama, W. (1967b), Study on forced convective heat transfer in curved pipes (third report, theoretical analysis under the condition of uniform wall temperature and practical formulae), *International Journal of Heat and Mass Transfer*, **10**, 681–695.

Ozisik, M. N. and Topakoglu, H. (1968), Heat transfer for laminar flow in a curved pipe, *Heat Transfer*, August, 313–318.

Smith, E. M. (1960), The geometry of multi-start helical coil heat exchangers, *Internal Report NRC 60–121*, Nuclear Research Centre, C. A. Parsons & Co. Ltd., December, pp. 1–17.

Smith, E. M. (1964), Helical-tube heat exchangers, *Engineering*, 7 February, 232.

Smith, E. M. (1986) Design of helical-tube multi-start coil heat exchanger, *ASME Winter Annual Meeting, Anaheim, CA, 7–12 December 1986, ASME Publication HTD* **66**, 95–104.

Smith, E. M. and Coombs, B. P. (1972), Thermal performance of cross-inclined tube bundles measured by a transient technique, *Journal of Mechanical Engineering Science*, **14**, 205–220.

Smith, E. M. and King, J. L. (1978), Thermal performance of further cross-inclined in-line and staggered tube banks, *6th International Heat Transfer Conference, Toronto, Paper HX-14*, pp. 267–272.

Weimer, R. F. and Hartzog, D. G. (1972), Effects of maldistribution on the performance of multistream heat exchangers, *Proceedings of de 1972 Cryogencies Engineering Conference, Advances in Cryogenic Engineering*, **18**, Paper B–2, 52–64.

Yao, L. S. (1984), Heat convection in a horizontal curved pipe, *ASME Journal of Heat Transfer*, **106**, 71–77.

Zukauskas, A. A. (1987) Convective heat transfer in cross flow, *Handbook of Single-Phase Convective Heat Transfer*, Eds. S. Kabac, R. K. Shah and W. Aung, New York: Wiley, Chap. 6.

Zukauskas, A. A. and Ulinskas, R. (1988) *Heat Transfer in Tube Banks in Crossflow*, Washington/Berlin: Hemisphere/Springer-Verlag.

6

DIRECT SIZING OF BAYONET-TUBE EXCHANGERS

6.1 ISOTHERMAL SHELL-SIDE CONDITIONS

Explicit design of the bayonet-tube heat exchanger is practicable when the shell-side fluid is essentially isothermal, i.e. for some condensing and evaporating conditions, and for isothermal cross-flow. Analytical expressions and dimensionless plots are presented for the four possible configurations, giving full temperature profiles, exchanger effectiveness, position of closest shell to tube-side temperature approach, and direct determination of exchanger length.

In designing bayonet-tube heat exchangers for the case when the shell-side fluid is essentially isothermal (e.g. condensing, evaporating, or isothermal cross-flow), a modified version of Hurd's (1946) theoretical approach appears to be necessary; there exist explicit design conditions. Four configurations, A, B, C, D in Fig. 6.1, will be examined in turn, two for evaporation and two for the condensing condition.

Notation is awkward for the bayonet-tube exchanger because fluid in the bayonet tube enters and exists from the same end, and because each pass of that fluid requires separate identification. As only overall heat transfer coefficients will be involved in the analysis which follows, the symbols (α, β) can be used for parameters in the solution. The concepts of LMTD and mean TD are not useful.

It was found convenient to introduce the concept of 'perimeter transfer units' (P, $\hat{P}$) equivalent to 'Ntu per unit length' of the exchanger surface. These parameters arise quite naturally in the differential equations, and conversion to Ntu values (N, $\hat{N}$) is straightforward once the solutions have been obtained.

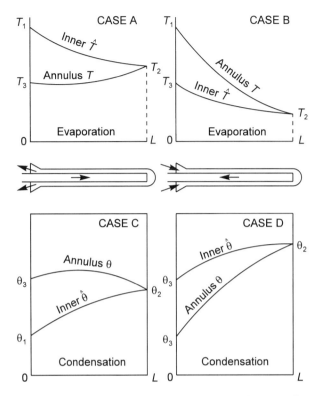

Fig. 6.1 Alternative exchanger configurations: condensation is reflected evaporation

$$P = \frac{N}{L} = \frac{UZ}{\dot{m}C} \text{ where } Z \text{ is perimeter of outer tube}$$

$$\hat{P} = \frac{\hat{N}}{L} = \frac{\hat{U}\hat{Z}}{\dot{m}C} \text{ where } \hat{Z} \text{ is perimeter of inner tube}$$

In the solutions which follow, all physical parameters remain constant.

6.2 EVAPORATION

Case A

An energy balance can be written for a differential length dx of the tube (Fig. 6.2).

Inner tube

$$\left\{\begin{matrix}\text{energy entering} \\ \text{with fluid}\end{matrix}\right\} - \left\{\begin{matrix}\text{energy leaving} \\ \text{with fluid}\end{matrix}\right\} - \left\{\begin{matrix}\text{heat transferred} \\ \text{to annulus}\end{matrix}\right\} = \left\{\begin{matrix}\text{energy stored} \\ \text{in fluid}\end{matrix}\right\}$$

158 6 DIRECT SIZING OF BAYONET-TUBE EXCHANGERS

$$\dot{m}C\hat{T} - \dot{m}C\left(\hat{T} + \frac{d\hat{T}}{dx}\delta x\right) - \hat{U}(\hat{Z}\delta x)(\hat{T} - T) = 0 \qquad (6.1)$$

Annulus

$$\left[\dot{m}C\left(T + \frac{dT}{dx}\delta x\right) + U(Z\delta x)(\hat{T} - T)\right] - [\dot{m}CT + U(Z\delta x)(T - 0)] = 0 \quad (6.2)$$

giving respectively

$$-\frac{d\hat{T}}{dx} = +\hat{P}(\hat{T} - T) \qquad (6.3)$$

$$+\frac{dT}{dx} = -\hat{P}(\hat{T} - T) + PT \qquad (6.4)$$

Eliminating T from equations (6.3) and (6.4) produces

$$\frac{d^2\hat{T}}{dx^2} - P\frac{d\hat{T}}{dx} - P\hat{P}\hat{T} = 0 \qquad (6.5)$$

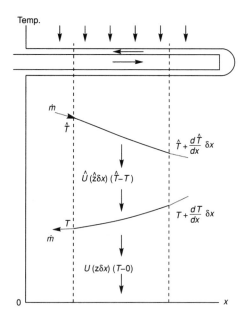

Fig. 6.2 Differential energy balance for case A: origin at flow entry and exit

which has the solution

$$\hat{T} = A_i\, e^{\alpha x} + B_i\, e^{\beta x} \tag{6.6}$$

with

$$\alpha = \frac{P}{2}\left[1 + \sqrt{(1 + 4\hat{P}/P)}\right]$$
$$\beta = \frac{P}{2}\left[1 - \sqrt{(1 + 4\hat{P}/P)}\right] \tag{6.7}$$

$$P = (\alpha + \beta) \quad \text{and} \quad \frac{\hat{P}}{P} = \frac{-\alpha\beta}{(\alpha+\beta)^2} \tag{6.8}$$

An identical result exists for the other unknown temperature profile

$$T = A_o\, e^{\alpha x} + B_o\, e^{\beta x} \tag{6.9}$$

Annulus temperature profile (T)
Two boundary conditions are required, but only $T = T_3$ at $x = 0$ is immediately available. A second condition is obtained by noting that the overall energy balance must be satisfied.

$$Q = \dot{m}C(T_1 - T_3) = \int_{x=0}^{L} [U(Z\delta x)(T - 0)]$$

$$(T_1 - T_3) = \left(\frac{UZ}{\dot{m}C}\right)\int_{x=0}^{L} T\,\delta x = P\int_{x=0}^{L} T\,\delta x = (\alpha + \beta)\int_{x=0}^{L} T\,\delta x \tag{6.10}$$

Inserting boundary conditions $T = T_3$ at $x = 0$ in equation (6.9)

$$T_3 = A_0 + B_0$$

Substituting in (6.9)

$$T + (T_3 - B_0)e^{\alpha x} + B_0\, e^{\beta x} \tag{6.11}$$

and then in (6.10)

$$(T_1 - T_3) = (\alpha + \beta)\left[(T_3 - B_o)\left(\frac{e^{\alpha L} - 1}{\alpha}\right) + B_o\left(\frac{e^{\beta L}}{\beta}\right)\right]$$

160 6 DIRECT SIZING OF BAYONET-TUBE EXCHANGERS

from which B_o may be found for reintroduction in (6.11). Equation (6.11) is then solved for T_2 at $x = L$ and following some algebra too extensive to reproduce

$$T_2/T_3 = (termX + termY + termZ)/denom$$

where

$$termX = +\alpha^2 e^{\alpha L}(e^{\beta L} - 1)$$
$$termY = +\alpha\beta(e^{\beta L} - e^{\alpha L})(T_1 - T_3)$$
$$termZ = -\beta^2 e^{\beta L}(e^{\alpha L} - 1)$$
$$denom = (\alpha + \beta)\left[\alpha(e^{\beta L} - 1) - \beta(e^{\alpha L} - 1)\right]$$

giving

$$T_2/T_3 = \phi[\alpha, \beta, L, (T_1/T_3)] \tag{6.12}$$

Inner temperature profile ($\hat{T}$)

Again two boundary conditions are required, but only $\hat{T} = T_1$ at $x = 0$ is immediately available. From equation (6.3) at $x = L$, $\hat{T} = T$; thus $d\hat{T}/dx = 0$. Three results from equation (6.6) are then obtained.

$$\begin{aligned} T_1 &= A_i + B_i \\ T_2 &= A_i e^{\alpha L} + B_i e^{\beta L} \\ 0 &= \alpha A_i e^{\alpha L} + \beta B_i e^{\beta L} \end{aligned} \tag{6.13}$$

Solving the first two for A_i and B_i, and inserting in the third condition

$$\boxed{\frac{T_2}{T_1} = \frac{(\beta - \alpha)e^{(\alpha+\beta)L}}{\beta e^{\beta L} - \alpha e^{\alpha L}}} \tag{6.14}$$

Combining this result with equation (6.12)

$$\frac{T_1}{T_3} = \frac{T_1}{T_2} \cdot \frac{T_2}{T_3} = \frac{(\beta e^{\beta L} - \alpha e^{\alpha L})}{(\beta - \alpha)e^{(\alpha+\beta)L}} \phi[\alpha, \beta, L, (T_1/T_3)]$$

from which, after substantial algebraic reduction, there emerges

$$\boxed{\frac{T_1}{T_3} = \frac{\beta e^{\beta L} - \alpha e^{\alpha L}}{\beta e^{\alpha L} - \alpha e^{\beta L}}} \tag{6.15}$$

providing the explicit result

$$L = \frac{1}{(\alpha + \beta)} \ln \left[\frac{\beta + \alpha(T_1/T_3)}{\alpha + \beta(T_1/T_3)}\right] \quad (6.16)$$

Equation (6.15) delivers $\lim(T_1/T_3) = (-\alpha/\beta)$ as $L \to \infty$, thus the restriction $1 < (T_1/T_3) < (-\alpha/\beta)$ applies to (6.16).

From Fig. 6.1 it is evident that a minimum value may exist in the profile for T. This is readily found from equation (6.9) by solving

$$T_3 = A_o + B_o$$
$$T_2 = A_o e^{\alpha L} + B_o e^{\beta L} \quad (6.17)$$
$$0 = \alpha A_0 e^{\alpha X_{min}} + \beta B_o e^{\beta X_{min}}$$

giving the position of the minimum as

$$X_{min} = \frac{1}{(\beta - \alpha)} \ln \left[\frac{\alpha e^{\beta L} - (T_2/T_3)}{\beta e^{\alpha L} - (t_2/T_3)}\right] \quad (6.18)$$

(T_2/T_3) is found from (6.14) and (6.15) as

$$\frac{T_2}{T_3} = \frac{(\beta - \alpha)e^{(\alpha+\beta)L}}{\beta e^{\alpha L} - \alpha e^{\beta L}} \quad (6.19)$$

Inserting this result in (6.18), the locus of X_{min} is obtained as a straight line

$$X_{min} = L + \frac{1}{(\beta - \alpha)} \ln \left[(\alpha/\beta)^2\right] \quad (6.20)$$

Minimum annulus temperature T_{min} is obtained by inserting X_{min} in the expression for the annulus temperature profile (T/T_3).

$$\frac{T_{min}}{T_3} = \frac{\beta e^{\alpha L} e^{\beta X_{min}} - \alpha e^{\beta L} e^{\alpha X_{min}}}{\beta e^{\alpha L} - \alpha e^{\beta L}} \quad (6.21)$$

This completes the analysis for Case A, but simplification is possible, noting from equation (6.7) that

6 DIRECT SIZING OF BAYONET-TUBE EXCHANGERS

$$\alpha = aP \quad \text{where} \quad a = \frac{1}{2}\left[1 + \sqrt{(1 + 4\hat{P}/P)}\right]$$
$$\beta = bP \quad \text{where} \quad b = \frac{1}{2}\left[1 - \sqrt{(1 + 4\hat{P}/P)}\right] \tag{6.22}$$

Table 6.1 offers a selection of values for (a, b) covering most applications. The summary below recasts the above relationships in terms of (a, b) and $Ntu = N = PL$.

Summary of results for Case A

Exchanger length

$$N = \frac{1}{(\beta - \alpha)} \ln\left[\frac{b + a(T_3/T_1)}{a + b(T_3/T_1)}\right] \tag{6.16a}$$

Position of minimum (T_{min} if it exists)

$$N_{min} = N + \frac{1}{(b - a)} \ln\left[\left(\frac{b}{a}\right)^2\right] \tag{6.18a}$$

Open-end temperature ratio

$$\frac{T_3}{T_1} = \frac{be^{aN} - ae^{bN}}{be^{bN} + ae^{aN}} \quad \text{with} \quad \lim(T_3/T_1) = (-b/a) \text{ as } N \to \infty \tag{6.15a}$$

Bayonet-end temperature ratio

$$\frac{T_2}{T_1} = \frac{(b - a)e^{aN}e^{bN}}{be^{bN} - ae^{aN}} \tag{6.14a}$$

Minimum annulus temperature (T_{min} if it exists)

$$\frac{T_{min}}{T_1} = \frac{be^{aN}e^{bN(X_{min}/L)} - ae^{bN}e^{aN(X_{min}/L)}}{be^{bN} - ae^{aN}} \tag{6.21a}$$

6.2 EVAPORATION 163

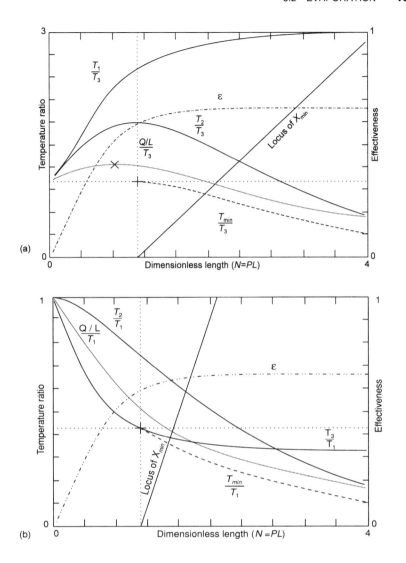

Fig. 6.3 Variation of parameters for Case A with a = +}/2 = −1/2: (a) and T_1/T_3, T_2/T_3, $(Q/L)/T_3$, X_{min} and ϵ vs. $N = PL$; (b) T_3/T_1, T_2/T_1, T_{min}/T_1, $(Q/L)/T_1$, X_{min} and ϵ vs. $N = PL$

Inner temperature profile

$$\frac{\hat{T}}{T_1} = \frac{be^{bN} e^{aN(x/L)} - ae^{aN} e^{bN(x/L)}}{be^{bN} - ae^{aN}}$$

164 6 DIRECT SIZING OF BAYONET-TUBE EXCHANGERS

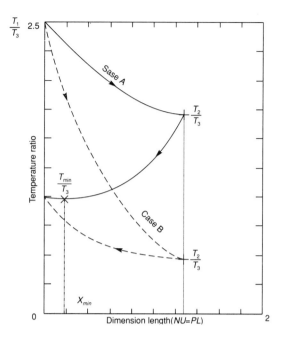

Fig. 6.4 Actual temperature profiles for Case A ($a = +3/2$, $b = -1/2$) and Case B ($a = +1/2$, $b = 3/2$); a minimum exists for Case A

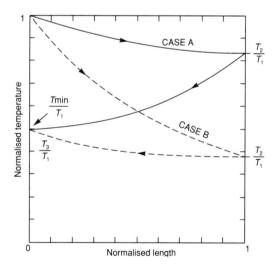

Fig. 6.5 Normalised temperature profies for Case A ($a = +3/2$, $b = -1/2$) and Case B ($a = +1/2$, $b = 3/2$); there is a minimum at the fluid exit for Case A

Annulus temperature profile

$$\frac{\hat{T}}{T_1} = \frac{be^{aN}e^{bN(x/L)} - ae^{bN}e^{aN(x/L)}}{be^{bN} - ae^{aN}}$$

Limiting effectiveness

$$\epsilon_{\lim} = 1 + \frac{b}{a}$$

Case B

Case B is simpler to analyse than Case A because the lower temperature profile does not include an intermediate minimum. Following a similar procedure, the basic differential equations are

$$\frac{d\hat{T}}{dx} = \hat{P}(\hat{T} - T) \qquad (6.23)$$

$$\frac{dT}{dx} = \hat{P}(\hat{T} - T) - PT \qquad (6.24)$$

Eliminating T as before

$$\frac{d^2\hat{T}}{dx^2} = P\frac{d\hat{T}}{dx} - P\hat{P}\hat{T} = 0 \qquad (6.25)$$

which has the solution

$$\hat{T} = A_i e^{\alpha x} + B_i e^{\beta x} \qquad (6.26)$$

with

$$\alpha = \left(\frac{P}{2}\right)\left[-1 + \sqrt{(1 + 4\hat{P}/P)}\right]$$

$$\beta = \left(\frac{P}{2}\right)\left[-1 - \sqrt{(1 + 4\hat{P}/P)}\right] \qquad (6.27)$$

and $\qquad P = -(\alpha + \beta) \qquad (6.28)$

166   6 DIRECT SIZING OF BAYONET-TUBE EXCHANGERS

An identical result exists for the other unknown temperature profile

$$T = A_o\, e^{\alpha x} + B_o\, e^{\beta x} \tag{6.29}$$

From the annulus temperature profile and overall heat balance

$$T_2/T_1 = (termX + termY + termZ)/denom$$

where

$$termX = +\alpha^2\, e^{\alpha L}(e^{\beta L} - 1)$$
$$termY = +\alpha\beta(e^{\beta L} - e^{\alpha L})(T_3/T_1)$$
$$termZ = -\beta^2\, e^{\beta L}(e^{\alpha L} - 1)$$
$$denom = (\alpha + \beta)[\alpha(e^{\beta L} - 1) - \beta(e^{\alpha L} - 1)]$$

giving

$$T_2/T_1 = \phi[\alpha,\ \beta,\ L,\ (T_3/T_1)] \tag{6.30}$$

Boundary conditions for the lower (inner) temperature profile, equation (6.24), again give $\hat{T} = T_3$ at $x = 0$ and $d\hat{T}/dx = 0$ at $x = L$.

$$T_1 = A_i + B_i$$
$$T_2 = A_i\, e^{\alpha L} + B_i\, e^{\beta L} \tag{6.31}$$
$$0 = \alpha A_i\, e^{\alpha L} + \beta B_i\, e^{\beta L}$$

which delivers

$$\boxed{\dfrac{T_2}{T_3} = \dfrac{(\beta - \alpha)e^{(\alpha+\beta)L}}{\beta e^{\beta L} - \alpha e^{\alpha L}}} \tag{6.32}$$

As before, using (6.30) and (6.32)

$$\dfrac{T_1}{T_3} = \dfrac{T_1}{T_2} \cdot \dfrac{T_2}{T_3} = \dfrac{\beta e^{\alpha L} - \alpha e^{\beta L}}{\beta e^{\beta L} - \alpha e^{\alpha e^{\alpha L}}} \tag{6.33}$$

allowing the explicit result

$$\boxed{L = \dfrac{1}{(\alpha + \beta)} \ln\left[\dfrac{\alpha + \beta(T_1/T_3)}{\beta + \alpha(T_1/T_3)}\right]} \tag{6.34}$$

Here equation (6.34) delivers $\lim(T_1/T_3) = (-\beta/\alpha)$ as $L \to \infty$, thus the restriction $1 < (T_1/T_3) < (-\beta/\alpha)$ applies to (6.34).

6.2 EVAPORATION

This completes the analysis for Case B, which is a reversal of flow direction of the tube-side fluid for Case A. Again simplification is possible, recasting the above relationships in terms of (a, b) and $Ntu = N = PL$.

Summary of results for Case B

Exchanger length

$$N = \frac{1}{(b-a)} \ln\left[\frac{a + b(T_3/T_1)}{b + a(T_3/T_1)}\right] \quad (6.34a)$$

Open end temperature ratio

$$\frac{T_3}{T_1} = \frac{be^{bN} - ae^{aN}}{be^{aN} - ae^{bN}} \quad \text{with} \quad \lim(T_3/T_1) = (-a/b) \quad \text{as} \quad N \to \infty \quad (6.33a)$$

Bayonet end temperature ratio

$$\frac{T_2}{T_1} = \frac{(b-a)e^{aN} e^{bN}}{be^{aN} - ae^{bN}} \quad (6.30a)$$

Inner temperature profile

$$\frac{\hat{T}}{T_1} = \frac{be^{bN} e^{aN(x/L)} - a e^{aN} e^{bN(x/L)}}{be^{aN} - ae^{bN}}$$

Annulus temperature profile

$$\frac{T}{T_1} = \frac{be^{aN} e^{bN(x/L)} - a e^{bN} e^{aN(x/L)}}{be^{aN} - ae^{bN}}$$

Limiting effectiveness

$$\epsilon_{\lim} = 1 + \frac{a}{b}$$

A design solution with the same $\hat{P}/P$ ratio as Case A is presented in Fig. 6.4. But note that the values for (a, b) are reversed, i.e. $a = +1/2$ and $b = -3/2$.

168 6 DIRECT SIZING OF BAYONET-TUBE EXCHANGERS

Evaluation of equations (6.33) and (6.34) produces $T_1/T_2 = 2.5$ at $x = 0$ and $N = 1.2825$, identical with Case A. Heat transfer and tube length are unaffected by direction of flow of the tube-side fluid, but temperature profiles are different.

6.3 CONDENSATION

Case C

Symbols $(\theta, \hat{\theta})$ are adopted for temperatures so as to permit transformation to $(T, \hat{T})$ later in the analysis. Proceeding as for Case A, energy balances are again written for differential lengths of tube.

Inner tube

$$\left\{ \begin{array}{c} \text{energy entering} \\ \text{with fluid} \end{array} \right\} - \left\{ \begin{array}{c} \text{energy leaving} \\ \text{with fluid} \end{array} \right\} - \left\{ \begin{array}{c} \text{heat transferred} \\ \text{from annulus} \end{array} \right\} = \left\{ \begin{array}{c} \text{energy stored} \\ \text{in fluid} \end{array} \right\}$$

$$\dot{m}C\hat{\theta} - \dot{m}C\left(\hat{\theta} + \frac{d\hat{\theta}}{dx}\delta x\right) + \hat{U}(\hat{Z}\,\delta x)(\theta - \hat{\theta}) = 0 \qquad (6.35)$$

Annulus

$$\left[\dot{m}C\left(\theta + \frac{d\theta}{dx}\delta x\right) + \hat{U}(\hat{Z}\,\delta x)(\theta_c - \theta)\right] - \left[\dot{m}C\theta + U(Z\,\delta x)(\theta - \hat{\theta})\right] = 0 \qquad (6.36)$$

giving respectively

$$+\frac{d\hat{\theta}}{dx} = +\hat{P}(\theta - \hat{\theta}) \qquad (6.37)$$

$$-\frac{d\theta}{dx} = +P(\theta_c - \theta) - \hat{P}(\theta - \hat{\theta}) \qquad (6.38)$$

Putting

$$\hat{T} = \theta_c - \hat{\theta}, \quad d\hat{T} = -d\hat{\theta}$$
$$T = \theta_c - \theta, \quad dT = -d\theta$$

equations (6.37) and (6.38) become identical with equations (6.3) and (6.4) of Case A.

Case D

In a manner similar to Case B, it may be shown that the governing equations for Case D are identical to equations (6.23) and (6.24).

6.4 DESIGN ILLUSTRATION

In the absence of pressure loss data for the bayonet end and the consequent impossibility of calculating total pressure loss in the bayonet-tube at this time, it is presently only possible to design on the basis of heat transfer alone. Figure 6.6 shows the variation of ϵ against Ntu with $(\hat{P}/P = \hat{N}/N)$ as parameter. Inspection of these curves leads to formulation of a direct method of design:

1. For a given cross-section of the exchanger calculate
   - both overall heat transfer coefficients $(U, \hat{U})$
   - both perimeter transfer units $(P, \hat{P})$
   - obtain (a, b), say $a = +3/2$, $b = -1/2$

2. Calculate the limiting value of the temperature ratio
$$\lim(T_1/T_3) = (-a/b) = 3$$

3. Calculate the limiting effectiveness
$$\epsilon_{\lim} = 1 - (T_3/T_1) = 2/3$$

4. Take a design fraction of this, say $f = 0.9$, to obtain the actual effectiveness
$$\epsilon = 0.6$$

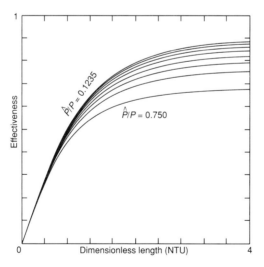

Fig. 6.6 Effectiveness and limiting effectiveness: see Table 6.1 for results 7 to 14

170 6 DIRECT SIZING OF BAYONET-TUBE EXCHANGERS

5. Solve for the actual temperature ratio (T_1/T_3), which is a measure of the heat transfer possible

$$(T_1/T_3) = 1/(1 - \epsilon) = 5/2$$

6. Solve equation (6.16) explicitly to obtain $N = 1.2825$. The resulting temperature profiles for the design selected are shown as solid lines in Fig. 6.4.

Table 6.1 Useful values of $\hat{P}/P$, a, b for Case A*

No.	$\hat{P}/P$	$\sqrt{(root)}$	a	b	1+a/b
01	16/225 = 0.0711	17/15	16/15	−1/15	15/16
02	15/196 = 0.0765	16/14	15/14	−1/14	14/15
03	14/169 = 0.0828	15/13	14/13	−1/13	13/14
04	13/144 = 0.0903	14/12	13/12	−1/12	12/13
05	12/121 = 0.0992	13/11	12/11	−1/11	11/12
06	11/100 = 0.1100	12/10	11/10	−1/10	10/11
07	10/81 = 0.1235	11/9	10/9	−1/9	9/10
08	9/64 = 0.1406	10/8	9/8	−1/8	8/9
09	8/49 = 0.1633	9/7	8/7	−1/7	7/8
10	7/36 = 0.1944	8/6	7/6	−1/6	6/7
11	6/25 = 0.2400	7/5	6/5	−1/5	5/6
12	5/16 = 0.3125	6/4	5/4	−1/4	4/5
13	4/9 = 0.4444	5/3	4/3	−1/3	3/4
14	3/4 = 0.7500	4/2	3/2	−1/2	2/3
15	10/9 = 1.1111	7/3	5/3	−2/3	3/5
16	21/16 = 1.3125	10/4	7/4	−3/4	4/7
17	50/25 = 2.0000	15/5	10/5	−5/5	5/10
18	91/36 = 2.5278	20/6	13/6	−7/6	6/13
19†	45/16 = 2.8125	7/2	9/4	−5/4	4/9
20	144/49 = 2.9388	25/7	16/7	−9/7	7/16
21†	28/9 = 3.1111	11/3	7/3	−4/3	3/7
22	240/64 = 3.7500	32/8	20/8	−12/8	8/20
23	360/81 = 4.4444	39/9	24/9	−15/9	9/24
24†	77/16 = 4.8125	9/2	11/4	−7/4	4/11
25	600/100 = 6.0000	50/10	30/10	−20/10	10/30
26	900/121 = 7.4380	61/11	36/11	−25/11	11/36
27	1260/144 = 8.7500	72/12	42/12	−30/12	12/42
28	1764/169 = 10.4379	85/13	49/13	−36/13	13/49
29	2451/196 = 12.5051	100/14	57/14	−43/14	14/57
30	3484/225 = 15.4844	119/15	67/15	−52/15	15/67

*$\sqrt{root} = \sqrt{(1 + 4\hat{P}/P)}$, $a = [1 + \sqrt{root}]/2$, $b = [1 - \sqrt{root}]/2$, $\epsilon_{\lim} = 1 + (a/b)$ For heat transfer design, possibly onely results in the range 1–14 may be of interest.
† Entries seemingly out of natural sequence.

Helical annular flow between inner and outer tubes

For effective operation it is important that the heat transfer coefficient in the annulus be significantly higher than the heat transfer coefficient in the inner tube. With single-phase flow it becomes necessary to provide helical fins between the outside diameter of the inner tube so that the annulus fluid is forced to flow in a helical path. This provides a smaller characteristic dimension and a smaller flow area than for the simple concentric annulus. This same condition applies also to the design configurations which follow. For a narrow annulus a wire-wrap might be considered.

6.5 NON-ISOTHERMAL SHELL-SIDE CONDITIONS

An explicit design solution exists for the bayonet-tube heat exchanger with non-isothermal shell-side conditions only for the special case of equal water equivalents. Four possible flow configurations exist, each having four (reflected) temperature profiles. For the non-explicit solutions, selection of an appropriate configuration before numerical evaluation is eased when expected temperature profiles can be examined.

Exchangers are classified as *contraflow* if tube-side annulus flow and shell-side flow are in the opposite sense, and *parallel flow* if in the same sense. Each class has four possible cases (A, B and their reflections) corresponding to those for isothermal shell-side conditions; they are set out in Table 6.2.

Table 6.2 Non-isothermal configurations

Exchanger class	A	B	C	D
Contraflow	type 1	type 3	type 2	type 4
Parallel flow	type 5	type 7	type 6	type 8

For non-isothermal shell-side flow conditions, the origin is placed at the bayonet end because this simplifies the analysis. In Figs. 6.9 to 6.12, presented later, dotted lines represent flow in the inner tube of the bayonet. Narrow band shading is used to denote heat transfer between shell-side fluid and annulus tube-side fluid.

Figure 6.7 illustrates the heat balances used to obtain the three coupled governing equations for Contraflow, Case A (type 1). Note that all equations are for $x = 0$ at the bayonet end; selection of the origin differs from the isothermal solution.

Temperatures $(T, \hat{T})$, mass flow rate $(\dot{m}_b)$, and specific heat (C_b) will be used for the bayonet-tube fluid.

Temperature (T_e), mass flow rate $(\dot{m}_e)$, and specific heat (C_e) will be used for the external fluid.

6 DIRECT SIZING OF BAYONET-TUBE EXCHANGERS

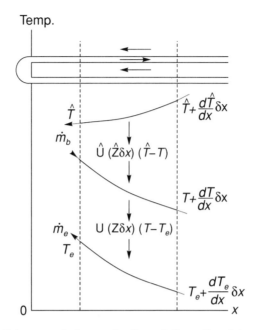

Fig. 6.7 Differential energy balances for Case A (type 1); origin at bayonet end.

Inner

$$\left\{\begin{array}{c}\text{energy entering}\\ \text{with fluid}\end{array}\right\} - \left\{\begin{array}{c}\text{energy leaving}\\ \text{with fluid}\end{array}\right\} - \left\{\begin{array}{c}\text{heat transferred}\\ \text{to annulus}\end{array}\right\} = \left\{\begin{array}{c}\text{energy stored}\\ \text{in fluid}\end{array}\right\}$$

$$\dot{m}_b\, C_b \left(\hat{T} + \frac{d\hat{T}}{dx}\delta x\right) - \dot{m}_b\, C_b\, \hat{T} - \hat{U}(\hat{Z}\,\delta x)(\hat{T} - T) = 0$$

giving

$$+\frac{d\hat{T}}{dx} = +\hat{P}(\hat{T} - T) \quad \text{where} \quad \hat{P} = (\hat{U}\hat{Z})/(\dot{m}_b\, C_b) \qquad (6.39)$$

Annulus similarly

$$-\frac{dT}{dx} = -\hat{P}(\hat{T} - T) + P(T - T_e) \quad \text{where} \quad P = (UZ)/(\dot{m}_b\, C_b) \qquad (6.40)$$

External similarly

$$-\frac{dT_e}{dx} = +P_e(T - T_e) \quad \text{where} \quad P_e = (UZ)/(\dot{m}_e C_e) \qquad (6.41)$$

The perimeter transfer units $(P, \hat{P}, P_e)$ are positive; so for all configurations, when temperature differences are kept positive, the sign of the gradient gives the slope of the temperature profile.

At the bayonet end $(\hat{T} = T)$, then

$$+\frac{d\hat{T}}{dx}\bigg|_{x=0} = 0 \qquad (6.39a)$$

$$-\frac{dT}{dx}\bigg|_{x=0} = P_e(T - T_e) \qquad (6.40a)$$

$$-\frac{dT_e}{dx}\bigg|_{x=0} = P_e(T - T_e) \qquad (6.41a)$$

When $\hat{T} = T$ and $x = 0$ are specified at the bayonet end then all gradients are known at $x = 0$, allowing generation of the temperature profile using finite differences from that end. See Section 6.8 on non-explicit solutions.

6.6 EXPLICIT SOLUTION

The set of partial differential equations (6.39) to (6.41) are similar to those which Kroeger (1966) used in exploring the problem of longitudinal conduction in contraflow heat exchangers. Kroeger found an explicit solution only for the special case of equal water equivalents, which is also the situation here.

Substitute (6.39) and (6.41) in (6.40)

$$\frac{dT}{dx} = \frac{d\hat{T}}{dx} + \left(\frac{P}{P_e}\right)\frac{dT_e}{dx}$$

$$\frac{d(\hat{T} - T)}{dx} = \left(\frac{-P}{P_e}\right)\frac{dT_e}{dx} \qquad (6.42)$$

$$\hat{T} - T = \left(\frac{-P}{P_e}\right)T_e + \text{const.}$$

Boundary conditions

At $x = 0$, $T_e = T_{e2}$ and $\hat{T} - T = 0$

$$0 = \left(\frac{-P}{P_e}\right) T_{e2} + \text{const}, \quad \text{thus} \quad \text{const} = \left(\frac{P}{P_e}\right) T_{e2}$$

At $x = L$, $T_e = T_{e1}$ and $\hat{T} - T = T_1 - T_3$

$$T_1 - T_3 = \left(\frac{P}{P_e}\right)(T_{e2} - T_{e1})$$

or

$$\dot{m}_b C_b (T_1 - T_3) = \dot{m}_e C_e (T_{e2} - T_{e2}) = Q \tag{6.43}$$

which is the correct energy balance, and we do not gain another equation.
From (6.42) and the first boundary condition

$$\hat{T} - T = \left(\frac{P}{P_e}\right)(T_{e2} - T_e) \tag{6.44}$$

From (6.44)

$$T_e = T_{e2} - \left(\frac{P_e}{P}\right)(\hat{T} - T) \tag{6.45}$$

For the inner temperature profile $(\hat{T})$, substitute (6.45) in (6.40)

$$-\frac{dT}{dx} = (P_e - \hat{P})T_e + (\hat{P} + P - P_e)T - PT_{e2} \tag{6.46}$$

From (6.39)

$$T = \hat{T} - \left(\frac{1}{\hat{P}}\right)\frac{d\hat{T}}{dx} \tag{6.47}$$

Differentiate

$$\frac{dT}{dx} = \frac{d\hat{T}}{dx} - \left(\frac{1}{\hat{P}}\right)\frac{d^2\hat{T}}{dx^2} \tag{6.48}$$

Substitute (6.47) and (6.48) in (6.46)

$$\frac{d^2\hat{T}}{dx^2} + (P - P_e)\frac{d\hat{T}}{dx} - \hat{P}P\hat{T} = -\hat{P}PT_{e2} \tag{6.49}$$

The solution of the homogeneous equation is of the form $\hat{T} = e^{dx}$, thus

$$b = \frac{-(P - P_e) \pm \sqrt{(P - P_e)^2 + 4\hat{P}P}}{2} \qquad (6.50)$$

The roots are real but one is negative. And the internal temperature profile is

$$\hat{T} = e^{b1\,x} + e^{b2\,x} + \text{const.} \qquad (6.51)$$

Boundary conditions

$$\hat{T} = e^{b1\,x} + e^{b2\,x} + \text{const.} \qquad (6.52)$$

subject to $\hat{T} = T_1$ at $x = L$

$$\frac{d\hat{T}}{dx} = b_1 e^{b1\,x} + b_2 e^{b2\,x} \qquad (6.53)$$

subject to $d\hat{T}/dx = 0$ at $x = 0$

From (6.53) $b_1 + b_2 = 0$. If the assumption $b_1 = +b$, $b_2 = -b$ is made in order to get the special-case analytical solution, then in (6.50) this implies

$$-(P - P_e)\sqrt{root} = +(P - P_e) + \sqrt{root}$$

that is $P = P_e$, so there exists the constraint $\dot{m}_b C_b = \dot{m}_e C_e$, implying

$$(T_{e2} - T_{e1}) = (T_1 - T_3)$$

Substituting in equation (6.50)

$$b = \sqrt{\hat{P}P} \qquad (6.54)$$

In equation (6.51) const. $= T_1 - \left(e^{+bL} + e^{-bL}\right)$, which provides the explicit inner temperature profile.

6.7 COMPLETE SOLUTION

Bayonet tube

Inner profile ($\hat{T}$) Case A (type 1)

$$\hat{T} = \left[e^{+bx} + e^{-bx}\right] + \left[T_1 - \left(e^{+bL} + e^{-bL}\right)\right] \qquad (6.55)$$

6 DIRECT SIZING OF BAYONET-TUBE EXCHANGERS

Bayonet end temperature (T_2) Case A (type 1)
At $x = 0$ put $\hat{T} = T = T_2$, say

$$T_2 = 2 + \left[T_1 - \left(e^{+bL} + e^{-bL}\right)\right] \qquad (6.65)$$

Inner profile gradient, Case A (type 1)
Differentiate (6.55)

$$\frac{d\hat{T}}{dx} = b\left(e^{+bx} - e^{-bx}\right) \qquad (6.57)$$

which confirms that $d\hat{T}/dx = 0$ at $x = 0$.

Annulus

Annulus profile (T) Case A (type 1)
From (6.39) and (6.57)

$$T = \hat{T} - \left(\frac{b}{\hat{P}}\right)\left(e^{+bx} - e^{-bx}\right) \qquad (6.58)$$

and substituting from (6.55)

$$T = \left[e^{+bx} + e^{-bx}\right] + \left[T_1 - \left(e^{+bL} + e^{-bL}\right)\right] - \left(\frac{b}{\hat{P}}\right)\left[e^{+bx} - e^{-bx}\right] \qquad (6.59)$$

Check on bayonet end temperature, case A (type 1)
At $x = L$ put $T = T_3$

$$T_3 = T_1 - \left(\frac{b}{\hat{P}}\right)\left(e^{+bL} - e^{-bL}\right) \qquad (6.60)$$

At $x = 0$ put $T = \hat{T} = T_2$

$$T_2 = 2 + \left[T_1 - \left(e^{+bL} + e^{-bL}\right)\right], \quad \text{which checks with equation (6.56)}$$

Exchanger duty, Case A (type 1)
From equation (6.60)

$$(T_1 - T_3) = \left(\frac{b}{\hat{P}}\right)\left(e^{+bL} - e^{-bL}\right) \qquad (6.61)$$

$$Q = \dot{m}_b C_b (T_1 - T_3) = \dot{m}_b C_b \left(\frac{b}{\hat{P}}\right)\left(e^{+bL} - e^{-bL}\right) \qquad (6.62)$$

Annulus profile gradient, Case A (type 1)

$$\frac{dT}{dx} = b\left(e^{+bx} - e^{-bx}\right) - \left(\frac{b^2}{\hat{P}}\right)\left(e^{+bx} - e^{-bx}\right) \qquad (6.63)$$

At $x = 0$

$$\frac{dT}{dx} = -2\left(\frac{b^2}{\hat{P}}\right) \qquad (6.64)$$

which is always negative.

External profile

From equation (6.45)

$$T_e = T_{e2} - \left(\frac{P_e}{P}\right)(\hat{T} - T)$$

but $P = P_e$, thus $T_e = T_{e2} - (\hat{T} - T)$, and then use (6.58).

External profile, Case A (type 1)

$$T_e = T_{e2} - \left(\frac{b}{\hat{P}}\right)\left(e^{+bx} - e^{-bx}\right) \qquad (6.65)$$

At $x = L$ put $T_e = T_{e1}$

$$(T_{e2} - T_{e1}) = \left(\frac{b}{\hat{P}}\right)\left(e^{+bL} - e^{-bL}\right) \qquad (6.66)$$

which corresponds with equation (6.61).
Obtain a check on the energy balance

$$(T_{e2} - T_{e1}) = \left(\frac{b}{\hat{P}}\right)\left(e^{+bL} - e^{-bL}\right) = (T_1 - T_2)$$

6 DIRECT SIZING OF BAYONET-TUBE EXCHANGERS

Using (6.66) in (6.65), an alternative expression for the external profile can be obtained.

Alternative external profile, Case A (type 1)

$$T_e = T_{e1} + \left(\frac{b}{\hat{P}}\right)[(e^{+bL} - e^{-bL}) - (e^{+bx} - e^{-bx})] \quad (6.65a)$$

Exchanger length

From equation (6.60),

$$e^{+bL} - e^{-bL} = \left(\frac{\hat{P}}{b}\right)(T_1 - T_3) = \left(\frac{\hat{P}}{b}\right)(T_{e2} - T_{e1}) = \frac{\hat{P}Q}{b\dot{m}_b C_b} = a, \text{ say} \quad (6.67)$$

Exchanger length, Case A (type 1)

$$L = \left(\frac{1}{b}\right)\ln\left[\frac{a + \sqrt{a^2 + 4}}{2}\right] \quad (6.68)$$

positive root only

Non-dimensional profiles

Inner profile, Case A (type 1)

$$\frac{\hat{T} - T_{e1}}{T_1 - T_{e1}} = 1 - \frac{(e^{+bL} + e^{-bL}) - (e^{+bx} + e^{-bx})}{(T_1 - T_{e1})} \quad (6.69)$$

Annulus profile, Case A (type 1)

$$\frac{T - T_{e1}}{T_1 - T_{e1}} = \left(\frac{b}{\hat{P}}\right)\frac{(e^{+bx} - e^{-bx})}{(T_1 - T_{e1})} \quad (6.70)$$

External profile, Case A (type 1)

$$\frac{T_e - T_{e1}}{T_1 - T_{e1}} = \left(\frac{b}{\hat{P}}\right)\frac{(e^{+bx} - e^{-bx})(e^{+bL} - e^{-bL})}{(T_1 - T_{e1})} \quad 6.71)$$

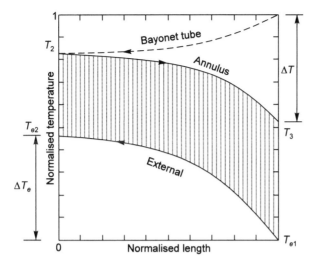

Fig. 6.8 Temperature profiles for explicit solution of the non-isothermal case with $(P=P_e)$, $(\dot{m}_b C_b = \dot{m}_e C_e)$, $(\Delta T = \Delta T_e)$

Typical temperature profiles for the only explicit solution with $(P = P_e)$, $(\dot{m} c = \dot{M} C)$ are given in Fig. 6.8.

6.8 NON-EXPLICIT SOLUTIONS

The analytical approach to solution of the same coupled PDEs has already been given by Kroeger (1966), but proceeding from equations (6.39a), (6.40a) and (6.41a) it is straightforward to set up a numerical method for generating temperature profiles. Starting from the bayonet end, it is necessary to assume a value for the bayonet-end temperature difference $(T_2 - T_{e2})$.

Table 6.3 Numerical parameters used in solution

Parameter	Internal	External
Tube mean diameter (m)	0.015	0.025
Mass flow rate (kg/s)	0.010	0.070
Specific heat (J/kg K)	4200.0	1000.0
Heat transfer coefficient (J/m² s K)	1000.0	2000.0

The resulting temperature profiles, shown in Figs. 6.9 to 6.12, were obtained by using the same arbitrary data for each configuration, given in Table 6.3. These solutions do not take into account longitudinal conduction in the tube walls. In laminar flow, a short-length bayonet-tube exchanger may not allow fully developed profiles to appear.

180 6 DIRECT SIZING OF BAYONET-TUBE EXCHANGERS

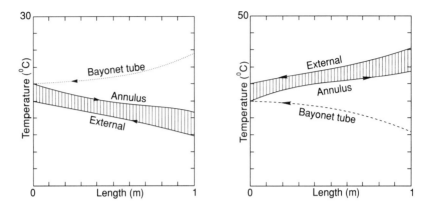

Fig. 6.9 (a) Type 1, contraflow case A and (b) Type 2, contraflow refl. A

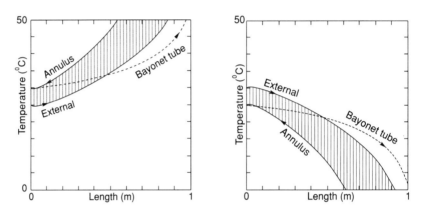

Fig. 6.10 (a) Type 3, contraflow case B and (b) Type 4, contraflow refl. B

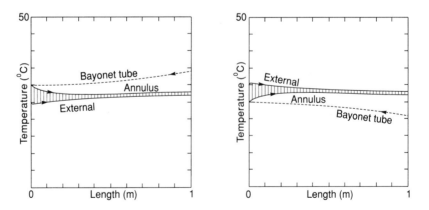

Fig. 6.11 (a) Type 5, parallel flow case C and (b) Type 6, parallel flow refl. C

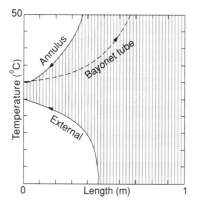 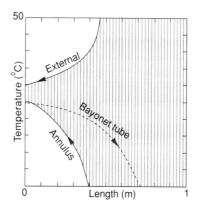

Fig. 6.12 **(a)** Type 7, parallel flow case D and **(b)** Type 8, parallel flow refl. D

6.9 PRESSURE LOSS

To obtain higher heat transfer coefficients in the annulus, it becomes essential to reduce annular flow area by forming helical channels.

It is necessary to know pressure loss in the bayonet-tube heat exchanger before direct sizing becomes possible. Although sufficient information exists to calculate pressure loss in the central tube and in the annular flow surrounding it, information about pressure loss at the bayonet end is currently sparse. Pressure loss at the bayonet end is known to depend on flow direction.

For a given outer tube diameter (D) there may be one inner tube diameter and thickness and one inner-tube/bayonet-end spacing which will provide the optimum (possibly least) pressure loss for a given reference Reynolds number. With this asumption the direct-sizing plot will be of length (L) versus external surface diameter (D). For the isothermal case, only one pressure loss curve will exist.

Some Russian data on pressure loss at the bayonet end has been published by Idelchik and Ginzburg (1968) and is reported in the textbook by Miller (1990). This information was provided by D. Graham, a senior honours student at Heriot-Watt University, Edinburgh, and further data (unpublished) for simple annular flow was obtained in an experimental programme under Dr B. Burnside.

Although adequate end-loss data may not yet be available, the question of pressure loss in the annulus can still be explored, and it seems worthwhile to obtain the correct analytical value for hydraulic diameter in the annulus under laminar flow conditions. It will then be consistent when attempting to fit heat-transfer and flow-friction data to turbulent correlations.

Simple annular flow

Steady laminar flow in a tube may be analysed using the cylindrical coordinate system to be found in most standard texts. A simple expression for flow velocity (u) along the tube is given by

$$u = \frac{-(p_1 - p_2)r}{4\eta \ell} + A \ln(r) + B$$

Inserting boundary conditions, u *finite* at $r = 0$ and $u = 0$ at $r = a$, we may obtain an expression for the velocity profile across the tube

$$u = \frac{(p_1 - p_2)}{4\eta \ell}(a^2 - r^2)$$

from which a mean velocity, $\bar{u}$ say, may be found by integration, giving for a circular tube

$$\Delta p = \frac{8\eta \bar{u} \ell}{a^2} \tag{6.72}$$

An identical analysis for pressure loss in steady laminar flow within an annulus of outer radius a and inner radius b gives

$$\Delta p = \frac{8\eta \bar{u} \ell}{(a^2 + b^2) - (a^2 - b^2)/\ln(a/b)} \tag{6.73}$$

Equations (6.72) and (6.73) may be equated to give an expression for equivalent frictional diameter of an annulus

$$d_f = \sqrt{(D^2 + d^2) - (D^2 - d^2)/\ln(D/d)} \tag{6.74}$$

As $d \rightarrow 0$ in the limit $d_f \rightarrow D$ i.e. the plain circular tube is recovered. The ratio d_f/D from equation (6.74) is plotted in Fig. 6.13 as a solid line.

When a similar analysis is made for flow in a very narrow annulus (in the limit, flow between two flat plates of spacing s), then with Cartesian coordinates the expression for pressure loss becomes

$$\Delta p = \frac{12\eta u \ell}{s^2} \tag{6.75}$$

Again by analogy with the solution for a circular tube, $12\eta \bar{u} \ell /s^2 = 8\eta \bar{u} \ell / a^2$, from which the equivalent frictional diameter for a very narrow annulus is obtained as

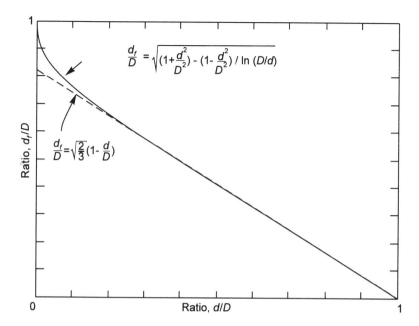

Fig. 6.13 Laminar flow friction equivalent diameter for concentric annulus

$$d_f = \sqrt{\frac{2}{3}}(D - d) \tag{6.76}$$

The ratio d_f/D from equation (6.76) is plotted in Fig. 6.13 as a dashed line, and it is remarkable how well it matches the value of d_f/D for an annulus for many of the diameter ratios. Indeed this may be seen as supporting the approximate equivalent diameter for flow friction in an annulus

$$d_f = \frac{4 \times \text{area for flow}}{\text{wetted perimeter}} = (D - d)$$

because the constant $\sqrt{2/3}$ may be assimilated in the empirical constant of a correlation.

Helical annular flow

An abstract survey covering the last 10 years suggests that published data on helical annular flow in near rectangular ducts is very sparse, and only the paper by Wang and Andrews (1995) provides the correct analysis for helical annular flow, plus references to the few papers of interest. With the additional effect of the 180° return at the bayonet-tube end, pressure loss becomes highly flow-direction dependent.

184 6 DIRECT SIZING OF BAYONET-TUBE EXCHANGERS

With the bayonet-tube fluid entering the central tube, flow at the bayonet tube end should be mainly radial and longitudinal in character. With the bayonet-tube fluid entering the helical annulus, an additional tangential component is introduced to affect flow conditions.

A programme of experimental and theoretical work seems necessary to establish exactly what is happening under these conditions, and to optimise diameter ratios and helical flow angle for the bayonet-tube exchanger.

For heat transfer, it may also be that temperature profiles derived earlier would be affected to second-order of magnitude by the slight discontinuities now introduced by helical annular flow.

6.10 CONCLUSIONS

1. The bayonet-tube exchanger transfers useful heat only from the outer tube, and the annulus should have helical channels for effective performance.
2. Pressure losses in the bayonet-tube end will be flow-direction dependent, and a research programme to determine these is needed.

Isothermal shell-side

3. Explicit temperature profiles are presented for the bayonet-tube exchanger having evaporation, condensation or isothermal cross-flow on the shell side.
4. Overall heat exchange and optimum length of exchanger are unaffected by the direction of tube-side flow.
5. Temperature profiles are significantly affected by the direction of tube-side flow, and this may be relevant in some design situations, e.g. Case A would be preferred to Case B when freezing of the tube-side fluid is to be avoided, and Case C preferred to Case D when boiling of the tube-side flow is to be avoided.
6. Possible applications include freezing of wet ground in order to stablise conditions for excavation, and ice formation around sunken objects as a means of flotation.

Non-isothermal shell-side

7. The present derivation of temperature profiles for an individual bayonet-tube exchanger assumes that a constrained external longitudinal flow exists. This design situation may arise when superheating secondary steam at the top of a PWR fuel-element channel.
8. An explicit solution has been obtained for the case of equal water equivalents ($\dot{m}c = \dot{M}C$), which provides exact exchanger temperature profiles. The explicit solution provides a check on numerical solutions.

9. For the case of unequal water equivalents, information helpful in selecting a suitable flow configuration has been provided, and sufficient information has been gathered to allow intelligent attacks on actual design problems.
10. One possible application is the use of a single vertical bayonet tube at the centre of a large cryogenic storage tank, with external natural convection. Such an exchanger provides axisymmetric cooling in the tank; it would encourage slow and controlled circulation of the contents of the tank, thus helping to inhibit 'roll-over' incidents.
11. Bayonet-tube heat exchangers are suitable for heat recovery at high temperatures where metals are not strong enough. Silicon carbide bayonet tubes are used.

REFERENCES

Guedes, R. O. C., Cotta, R. M. and Brum, N. C. L. (1991) Heat transfer in laminar flow with wall axial conduction and external convection, *Journal of Thermophysics*, **5**(2), 508–513.

Hernandez-Guerrero, A. and Macias-Machin, A. (1991) How to design bayonet heat-exchangers, *Chemical Engineering*, **79**, 122–128.

Hurd, N. L. (1946) Mean temperature difference in the field or bayonet tube, *Industrial and Engineering Chemistry*, **38** (12), 1266–1271.

Idelchik, I. E. and Ginzburg, Ya. L. (1968) The hydraulic resistance of 180° annular bends, *Thermal Engineering*, **15** (4), 109–114.

Joye, D. D. (1994) Optimum aspect ratio for heat transfer enhancement in curved rectangular channels, *Heat Transfer Engineering*, **15**(2), 32–38.

Kroeger, P. G. (1966) Performance deterioration in high effectiveness heat exchangers due to axial conduction effects, *Proceedings of the 1966 Cryogenic Engineering Conference*, Boulder, CO, 13–15 June 1966; also in *Cryogenic Engineering*, **12**, 363–372.

Miller, D. S. (1990) *Internal Flow Systems*, BHRA (Information Services), pp. 218–225.

Pagliarini, G. and Barozzi, G. S. (1991) Thermal coupling in laminar flow double-pipe heat exchangers, *Transactions ASME, Journal of Heat Transfer*, **113**, 526–534.

Smith, E. M. (1981) Optimal design of bayonet tube exchangers for isothermal shell-side conditions, *20th Joint ASME/AIChemE National Heat Transfer Conference, Milwaukee, WI, 2–5 August 1981, ASME Paper 81-HT-34*.

Todo, I. (1976) Dynamic response of bayonet-type heat exchangers, Part I – response to inlet temperature changes, *Bulletin of the Japanese Society of Mechanical Engineers*, **19**(136), 1135–1140.

Todo, I. (1978) Dynamic response of bayonet-type heat exchangers, Part II – response to flow rate changes, *Bulletin of the Japanese Society of Mechanical Engineers*, **21**(154), 644–651.

Wang, J. -W. and Andrews, J. R. G. (1995) Numerical simulation of flow in helical ducts (helical coordinate system and equations for flow in helical ducts), *AICLE Journal*, **41** (5), 1071–1080.

Ward, P. W. (1985) Ceramic tube heat recuperator – a user's experience, *Advances in Ceramics – Vol. 14, Ceramics in Heat Exchangers*, Eds. B. D. Foster and J. B. Patton, American Ceramics Society.

Zaleski, T. (1984) A general mathematical model of parallel-flow, multichannel heat exchangers and analysis of its properties, *Chemical Engineering Science*, **39** (7/8), 1251–1260. (Includes bayonet-tube exchangers.)

7

DIRECT SIZING OF RODBAFFLE EXCHANGERS

7.1 DESIGN FRAMEWORK

The direct-sizing approach in this chapter is provisional. It should be checked against the established rating method.

The RODbaffle exchanger is a better-performing shell-and-tube design than conventional tube-and-baffle designs. Design methods proposed by the originators of this exchanger type require prior knowledge of the diameter of the exchanger shell, thus they can be classed only as rating methods. Direct sizing of an exchanger becomes possible when the tube-bundle can be designed with reference to local geometry only. This chapter indicates an approach to such a method.

When the local geometry in a heat exchanger is fully representative of the whole geometry, then direct methods of thermal sizing become possible (Smith 1986, 1994). Both compact plate-fin and helical-tube heat exchangers are amenable to this approach, and this chapter proposes that the RODbaffle design may be handled in the same manner.

The paper by Gentry *et al.* (1982) presents a method for rating RODbaffle heat exchangers. This is based on test results obtained from experimental rigs on real heat exchangers. In setting out the Gentry *et al.* approach, several decisions were taken which effectively prevent their method from being used for direct sizing of RODbaffle heat exchangers:

- The exchanger inner shell surface area is incorporated in the hydraulic diameter for pressure loss on the shell side.
- Coefficients C_L and C_T in heat transfer correlations for laminar and turbulent flow include expressions for A_ℓ/A_s and L/D_{bo} each

188   7   DIRECT SIZING OF RODBAFFLE EXCHANGERS

of which requires knowledge of exchanger shell diameter (see Notation).
- Coefficients C_1 and C_2 in the pressure loss correlation for baffle-sections each require knowledge of exchanger shell diameter, (see Notation).

This chapter sets out an alternative design approach to permit direct thermal sizing of RODbaffle heat exchangers. As no experimental work has yet been carried out to confirm the approach, direct sizing should be used only for preliminary design, and the method of Gentry *et al.* should be used to complete the final design.

Minor changes to the notation used by Gentry *et al.* will be used in the interests of clarity.

7.2   CONFIGURATION OF THE RODBAFFLE EXCHANGER

The RODbaffle exchanger is essentially a shell-and-tube exchanger with conventional plate-baffles (segmental or disc-and-doughnut) replaced by grids of rods. Unlike plate-baffles, RODbaffle-sections extend over the full transverse cross-section of the exchanger.

Only square pitching of the tube-bundle is practicable with RODbaffles. To minimise blockage, one set of vertical rods in a baffle-section is placed between every second row of tubes. At the next baffle-section, the vertical rods are placed in the alternate gaps between tubes not previously filled at the first baffle-section. The next two baffle-sections have horizontal rod spacers, similarly arranged. Thus each tube in the bank receives support along its length.

7.3   APPROACH TO DIRECT SIZING

As the RODbaffle design is based on a set of four baffles, two with horizontal rods and two with vertical rods, this may not seem consistent with having constant local geometry throughout the bundle. However, fluids with no memory do not recognise when a set of four baffles begins, thus length design to at least the nearest baffle pitch becomes practicable, neglecting flow distributions between the shell nozzles and the first and last baffles. And Hesselgreaves (1988) shows that RODbaffle flow creates von Kármán vortex streets, well distributed in the shell-side fluid. Hesselgreaves took street length as the pitch between adjacent RODbaffles, but the street length may be longer.

Published correlations for heat transfer and shell-side pressure loss were assessed for direct sizing (see references) but, in the end the data presented in Figs. 6 and 8 of Gentry *et al.* (1982) for shell-side heat transfer and RODbaffle pressure loss were splinefitted to obtain data for the ARA bundle configuration.

The RODbaffle pressure loss data of Gentry *et al.* claims to take into account both loss through the plane of the baffle, and friction on the inner shell surface. This seems an awkward concept, for it implies that the baffle hydraulic diameter must change with shell diameter, which contravenes the basic concept of 'local action' in continuum mechanics.

An alternative concept of evaluating longitudinal leakage flow between the shell and the outside of the bundle might be employed. Because of the need to locate the baffle-rods, it is necessary to fit baffle-rings between the tube-bundle and the exchanger shell; this may permit leakage flow. Shell-side flow through the tube-bundle could first be evaluated using local geometry concepts, and this same pressure loss used to calculate the leakage flow between the baffles and the exchanger shell. The final outlet temperature would be the result of mixing of both streams. The necessary experimental data for presure loss due to leakage between baffle and shell is available in the papers by Bell and Bergelin (1957) and Bergelin *et al.* (1958). Dimensions for the baffle-rings are provided in the paper by Gentry (1990). The remaining problem is to convert the present trial and error procedure for bypass flow into a solution which will compute.

With the above proposal, when the shell-side flow is being heated there will be some diffusion from the shell side of the tube-bundle into the leakage stream, and an opposite effect when the shell-side fluid is being cooled. However, it is likely that the major contribution to leakage pressure loss would occur in the small clearance gaps around the baffle-rings.

The present design approach will simply assume that leakage flow losses can be included in the baffle loss coefficient (k_b). This permits the direct-sizing approach.

7.4 CHARACTERISTIC DIMENSIONS

For shell-side heat transfer in the interior of a tube-bundle, the Reynolds number can be based on local geometry only, allowing a definition of hydraulic diameter (D_s) for plain tubes

$$d_s = \frac{4 \times \text{area for flow}}{\text{wetted perimeter}} = \frac{4p^2 - \pi d^2}{\pi d} \quad (7.1)$$

For shell-side pressure loss, two characteristic dimensions are required, one for plain tubes only and one for the baffle-section. The above expression for D_s can be used for plain tubes, and an expression for the baffle-ring section may be evaluated over a tube length equal to the thickness of the baffle (i.e. $d_r = 2_r$), whence from Fig. 7.1,

$$d_b = \frac{4 \times \text{area for flow}}{\text{wetted perimeter}} = \frac{4p^2 - \pi d^2 - 4pr}{\pi(d + p/2)} \quad (7.2)$$

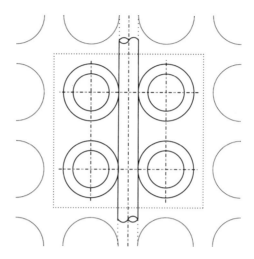

Fig. 7.1 Local geometry of tube-bundle at a RODbaffle section

7.5 FLOW AREAS

Flow areas per single tube

Tube-side

$$a_t = \frac{\pi}{4} d_i^2 \qquad (7.3)$$

Shell-side (plain tubes)

$$a_s = p^2 - \frac{\pi}{4} d^2 \qquad (7.4)$$

Shell-side (baffle-section)

$$a_b = P^2 - \frac{\pi}{4} d^2 - pr \qquad (7.5)$$

Total flow areas

Tube-side total flow area

$$A_t = Z a_t \qquad (7.6)$$

Shell-side total flow area (plain tubes)

$$A_s = Z a_s \qquad (7.7)$$

Shell-side total flow area (baffle-section)
$$A_b = Z a_b \tag{7.8}$$

7.6 DESIGN CORRELATIONS

Whenever explicit algebraic correlations for heat transfer and friction factor can be used throughout, it becomes possible to seek a direct algebraic solution for L and Z, although tracing the missing numerical values through the analysis requires some care (see Chapter 5 on helical-tube multi-start coil heat exchangers).

Here it is the intention to use the correlations provided by Gentry *et al.* in graphical form, and to spline fit the correlations for heat transfer, flow friction and baffle loss coefficient on the shell side. This obviates the need to know the exchanger shell diameter and the baffle-ring diameters before design commences. A possible case for making this simplification can be seen from the graphs of Gentry *et al.* The scatter around each correlation is within acceptable limits, suggesting that it is possible to avoid detailed building of the main correlations from subcorrelations involving shell diameters and tube-bundle length.

Different correlations would probably be necessary for different tube-bundle arrangements. This is beyond the present task, which is to establish that direct-sizing is possible, but see Appendix D.

The procedure is first to evaluate Reynolds number constraints on both shell-side and tube-side correlations. Valid Reynolds number values on the shell-side can then be scanned, and corresponding Reynolds number values on the tube side can be forced. Design within the valid envelope can then be completed.

7.7 REYNOLDS NUMBERS

Shell-side heat transfer

With an assumed value for shell-side Re_s

$$G_s = \frac{\mathrm{Re}_s \eta_s}{d_s} \qquad A_s = \frac{\dot{m}_s}{G_s} \qquad Z + \frac{A_s}{a_s}$$

and the number of tubes is determined.

Tube-side heat transfer and pressure loss

The forced tube-side Reynolds number may now be obtained

$$a_t = \frac{\pi}{4} d_i^2 \qquad A_t = Z a_t \qquad G_t = \frac{\dot{m}_t}{A_t}$$

$$\text{Re}_t = \frac{d_i G_t}{\eta_t} \tag{7.9}$$

Shell-side pressure loss

Two Reynolds numbers are involved. The plain-tube value is identical with that assumed for heat transfer. The baffle-section Reynolds number is obtained as follows.

$$G_b = \frac{\dot{m}_s}{A_b}$$
$$\text{Re}_b = \frac{D_b G_b}{\eta_s} \tag{7.10}$$

7.8 HEAT TRANSFER

Shell side

The heat transfer correlation shown in Fig. 6 of the paper by Gentry *et al.* is depicted as two straight-line segments, but in the text the curve is described as exhibiting a gradual change of slope. This feature is preserved in the splinefit of Fig. 7.2.

Nusselt numbers can be determined by assuming the viscosity ratio term is unity. The shell-side heat transfer coefficient becomes

$$\alpha_s = \left(\frac{\lambda_s}{d}\right) \text{Nu}_s \tag{7.11}$$

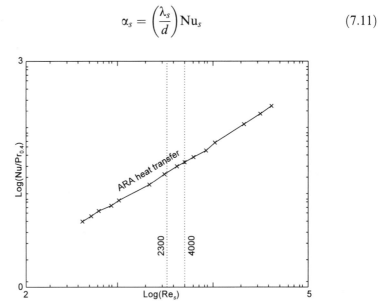

Fig. 7.2 Heat transfer correlation for configuration ARA (Adapted from Gentry *et al.* 1982)

Tube wall

The tube-wall heat transfer coefficient may be written as

$$\alpha_w = \frac{2\lambda_w}{d \ln(d/d_i)} \tag{7.12}$$

Tube side

The conventional tube-side correlation might be used

$$\text{Nu} = 0.023(\text{Re})^{0.8}(\text{Pr})^{0.33}$$

or a more comprehensive correlation due to Churchill (1977, 1988, 1992).

With the 'forced' value for Re_t, and correcting to outside diameter, we obtain

$$\alpha_t = \left(\frac{d_i}{d}\right)\left(\frac{\lambda_t}{d_i}\right)\text{Nu}_t = \left(\frac{\lambda_t}{d}\right)\text{Nu}_t \tag{7.13}$$

Overall coefficient

$$U = (1/\alpha_s + 1/\alpha_w + 1/\alpha_t)$$

Heat transfer equation

$$LZ = \frac{Q}{\pi d U \Delta\theta_{lmtd}} \tag{7.14}$$

7.9 PRESSURE LOSS TUBE-SIDE

The total pressure loss is made up of three components: one due to friction, one due to flow acceleration/deceleration and one due to entrance/exit effects. The largest is due to friction, sometimes reaching 98% of the total pressure loss. Only the frictional loss is considered in direct sizing but the other losses should be evaluated once the dimensions of the exchanger are known.

Chen (1979) provides an explicit correlation for turbulent friction factor in a pipe over the Reynolds number range $4000 < \text{Re} < 4 \times 10^8$ taking into account roughness ϵ/d in the range $5 \times 10^{-7} < \epsilon/d < 0.05$ (Fig. 7.3).

Using the standard friction factor expression

$$\frac{1}{\sqrt{f}} = -2.0 \log \left[\frac{(\epsilon/d)}{3.7065} - \frac{5.0452}{(\text{Re})} \log \left\{ \frac{(\epsilon/d)^{1.1098}}{2.8257} + \frac{5.8506}{(\text{Re})^{0.8981}} \right\} \right] \tag{7.15}$$

194 7 DIRECT SIZING OF RODBAFFLE EXCHANGERS

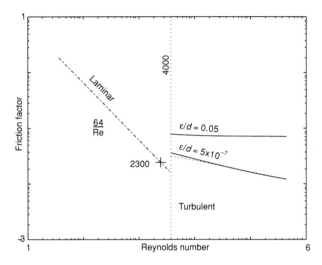

Fig. 7.3 Plain-tube flow-friction correlation: laminar (Moody), turbulent (Chen)

with the pressure loss expression

$$\Delta p_t = 4f_t \left(\frac{G_t^2}{2\rho}\right)\left(\frac{L}{d_i}\right) = \left(\frac{4f_t}{2\rho}\right)\left(\frac{\dot{m}_t^2}{d_i a_t^2}\right)\frac{L}{Z^2}$$

the pressure loss equation for the tube side may be written

$$L = Z^2 \Delta p_t \left[\left(\frac{2\rho}{4f_t}\right)\left(\frac{\dot{m}_T^2}{d_i a_t^2}\right)\right] \quad (7.16)$$

$$L = Z^2 \Delta p_t [bracket]$$

7.10 PRESSURE LOSS SHELL-SIDE

The total pressure loss is made up of four components: one due to friction on plain tubes, one due to baffle losses, one due to flow acceleration/deceleration and one due to entrance/exit effects. Only flow friction and baffle-losses are considered in direct sizing, but the other losses should be evaluated once the dimensions of the exchanger are known.

The technical literature contains papers on longitudinal flow in tube-bundles, but they are mainly concerned with testing heat transfer and pressure loss in nuclear fuel-element bundles. A nearly triangular pitch is often adopted, which makes the results of less interest for the RODbaffle exchanger. Additionally, very close spacing of the fuel rods is employed (Rehme 1992). Results for square pitching tend to be at high Reynolds numbers (Tong 1968).

In shell-side flow, pressure loss in the plain-tube section of a RODbaffle exchanger tends to be an order of magnitude less than in the rod-baffles. In the absence of exact information, plain-tube pressure loss is calculated from the correlation for flow inside smooth tubes, using an appropriate hydraulic diameter.

Plain tubes

The Chen friction factor correlation may be used again, then

$$G_s = \frac{\dot{m}_s}{A_s} = \frac{\dot{m}_s}{Za_s}$$

$$\Delta p_f = 4f_s \left(\frac{G_s^2}{2\rho}\right) \frac{L}{d_s}$$

(7.17)

Baffle-section

The baffle loss coefficient is obtained from a splinefit of k_b vs. Re_b (Fig. 7.4). then

$$G_b = \frac{\dot{m}_s}{A_b} = \frac{\dot{m}_s}{Za_b}$$

$$\Delta p_b = k_b \left(\frac{G_b^2}{2\rho}\right) \frac{L}{L_b}$$

(7.18)

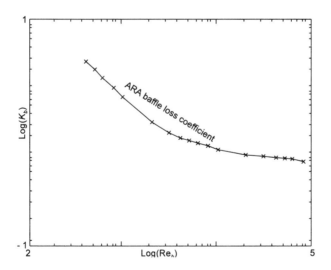

Fig. 7.4 Baffle loss coefficient (K_b) for configuration ARA (Adapted from Gentry et al. 1982)

Additive pressure loss

$$\Delta p_s = \Delta p_p + \Delta p_b = \text{(plain tube + baffle loss)}$$

$$= 4f_p \left(\frac{G_s^2}{2\rho}\right)\frac{L}{d_s} + k_b \left(\frac{G_b^2}{2\rho}\right)\frac{L}{L_b}$$

$$= \left[\frac{4f_p}{2\rho}\left(\frac{\dot{m}_s}{a_s}\right)^2 \frac{1}{d_s} + \frac{k_b}{2\rho}\left(\frac{\dot{m}_s}{a_b}\right)^2 \frac{1}{L_b}\right]\frac{L}{Z^2}$$

$$= [bracket]\frac{L}{Z^2}$$

$$L = Z^2 \frac{\Delta p_s}{[bracket]} \qquad (7.19)$$

7.11 DIRECT SIZING

To complete the design it is now appropriate to plot equations (7.14), (7.16) and (7.19), together with abscissa Z and ordinate L, producing the direct-sizing solution of Fig. 7.5. It is not known beforehand if shell-side pressure loss or tube-side pressure loss will lie to the right and be 'controlling'. Whichever curve provides the solution, tube length may be chosen such that a whole number of shell-side baffles is obtained. This constraint is perhaps not absolute because shell-side flow at entry to and exit from the bundle may be transverse to the tubes.

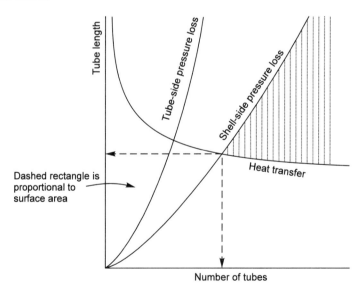

Fig. 7.5 Design solution plot

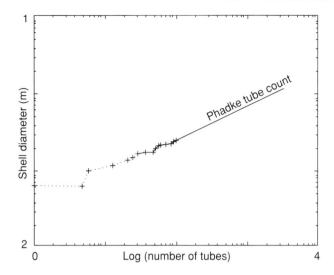

Fig. 7.6 Plot of tube count (Z) versus inside shell diameter (D_i)

7.12 TUBE-BUNDLE DIAMETER

The diameter of the tube-bundle may be determined using the results provided by Phadke (1984). Using data generated from the Phadke expressions, a log-log plot of tube count (Z) versus inside shell diameter (D_i) is very nearly a straight line for tube numbers greater than 100 (Fig. 7.6).

For accurate values it is necessary to consult the Phadke paper. In doing so, it is seen that Phadke always assumed one tube at the exact centre of the tube-plate. There exists another square-pitch tube layout in which there is no tube at the centre of the tube-plate. This may give a slightly different tube count.

It is very likely that the calculated value of Z does not provide the correct number of tubes to completely fill the tube-plate without leaving gaps. The number of tubes must now be adjusted to an appropriate value.

This may be done either by selecting appropriate values for Z and L, which lie in the shaded area of the design solution plot, and then completing a rating design, or alternatively by adjusting values of allowable pressure loss in the sizing design until suitable values of Z and L appear.

7.13 PRACTICAL DESIGN

The paper by Gentry (1990) provides information on the design of a gas-gas heat exchanger used as a feed preheater in a catalytic incinerator process for a petrochemical plant.

Although a substantial amount of information is available, there are some gaps which need filling. The physical properties of nitrogen are used because there was no information about gas composition. The specific heat of the cold shell-side fluid is found from a heat balance, allowing an estimation of specific heat for the hot gas. Terminal temperatures in the exchanger can then be found, plus Ntu values from the LMTD-Ntu relationships.

The following data was used in direct sizing of the heat exchanger surface:

Exchanger specification

Performance
exchanger duty, kW $\qquad Q = 5450.0$
log mean temperature difference, K $\qquad \Delta\theta_{lmtd} = 106.2$

Fluid properties

Shell-side fluid (gas)
bulk mean temperature, K $\qquad T_s = 467.0$
absolute viscosity, kg/(m s) $\qquad \eta_s = 0.0000245$
thermal conductivity, J/(m s K) $\qquad \lambda_s = 0.0366$
specific heat, J/(kg K) $\qquad C_s = 1027.1$
Prandtl number $\qquad \Pr_s = 0.686$

Tube-side fluid (gas)
bulk mean temperature, K $\qquad T_t = 573.0$
absolute viscosity, kg(/m s) $\qquad \eta_t = 0.0000283$
thermal conductivity, J/(m s K) $\qquad \lambda_t = 0.0426$
specific heat, J/(kg K) $\qquad C_t = 1034.8$
Prandtl number $\qquad \Pr_t = 0.685$

Local geometry

RODbaffle configuration ARA
Tube outside diameter, mm $\qquad d = 38.10$
Tube inside diameter (guesstimate), mm $\qquad d_i = 31.75$
Baffle-rod diameter, mm $\qquad d_r = 6.35$
Baffle-section spacing, mm $\qquad L_b = 150$

Pressure losses quoted by Gentry were outside the design window for direct sizing. In direct sizing, pressure loss data is for friction in the heat exchange surface only; the Gentry data are for total losses in the exchanger. On the tube side, in addition to flow friction in the tubes, there exist, tube entry and exit effects, flow acceleration and nozzle outlet losses. On the shell side, nozzle inlet and outlet losses will be present, plus flow distribution losses at the ends and flow friction in the baffled shell side.

It may seem as if a design comparison is not possible, but by using the maximum values of pressure loss, constrained by the heat transfer curve in

7.13 PRACTICAL DESIGN

direct sizing, a good approximation to the tube-bundle design described by Gentry is obtained. Since no data is used beyond the experimental limits given by Gentry, the pressure losses actually determined may be close to those of the exchanger described.

Absolute values of pressure are required to evaluate gas densities, and these are guesstimates.

shell-side pressure (guesstimate), bar	$p_s = 5.5$
tube-side pressure (guesstimate), bar	$p_t = 1.1$
shell-side pressure loss (max possible), N/m²	$\Delta p_s = 11\,305.46$
tube-side pressure loss (max possible), N/m²	$\Delta p_t = 8493.63$

Direct-sizing result

Figure 7.7 is the actual design solution plot obtained from the computer, in which the pressure loss curves are seen to be coincident.

Leading dimensions

Tube length, m	$L = 11.437$
Number of tubes	$Z = 519$
Number of RODbaffles	$B = 76$

Reynolds numbers

Shell-side, plain tubes	$\text{Re}_s = 40\,000.0$
Tube-side	$\text{Re}_t = 43\,601.9$

Heat transfer coefficients (referred to o.d.)

Shell-side heat transfer coeff. J/(m²s K)	$\alpha_s = 205.39$

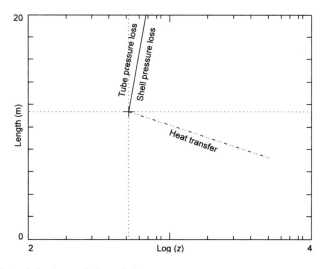

Fig. 7.7 Actual design solution plot

Tube-side heat transfer coeff. J/(m²s K) $\alpha_t = 134.98$
Tube-wall heat transfer coeff. J/(m²s K) $\alpha_w = 12\,956.3$
Overall heat transfer coeff. J/(m²s K) $U = 72.27$

Number of transfer units
Ntu shell-side $N_s = 3.34$
Ntu tube-side $N_t = 3.16$
Ntu overall $Ntu = 3.34$

Actual pressure losses
Shell-side, $\Delta p_s = 11\,305.46$
Tube-side, $\Delta p_t = 8493.63$

Phadke shell-sizing
Shell/baffle-ring clearance, m $s/b = 0.005$
Baffle-ring/tubes clearance, m $b/t = 0.005$

Nearest number of tubes to fill shell $Z = 529$
Corresponding inside shell diameter, m $D_i = 1.217$

RODbaffle performance
volume of exchanger, m³ $V = 13.304$
specific performance, MW/m³ $Q/V = 0.4097$

Table 7.1 Design comparison

Parameter	Gentry design	Direct sizing
Number of tubes	516	519
Shell diameter (m)	1.219	1.217
Number of baffles	78	76
Baffled tube length (m)	11.582	11.437
Unsupported tube length (m)	0.610	—
Total surface area (m²)	753.0	709.94[#]
Overall heat trans. coeff (J/m² s K)	61.3	72.27[#]
Shellside pressure loss (N/m²)	35700.0[*]	11305.46[†]

[#]No allowance for leakage flow
[*]Total loss
[†]Bundle only

For the results given in Table 7.1, Direct sizing assumes that RODbaffles are spaced over the complete bundle, with 76 baffles compared with 78 in the Gentry design. Some 0.610 m of tube-bundle in the Gentry design was said to be unsupported, and adding this to the length found in direct sizing gives 12.047 m, which is within 1.2% of the tube length quoted by Gentry (1990). Leakage flow past the RODbaffle rings may be the reason for lower coefficients calculated by Gentry *et al.* The shell diameter found using Phadke (1984) would accommodate 529 tubes, but some of this space is taken up by support rings and tie-rods.

Shell-side flow between the end RODbaffles and entry/exit nozzles does not necessarily relate to data for conventionally baffled shell-and-tube exchangers, and the writer could not find data for these end-effects. More information is required for shell-side distributional flow between the inlet/outlet nozzles and the end RODbaffles. This is particularly so when an annular vapour belt is provided because flow in the tube-bundle will then be radial.

Apart from the inlet/outlet discrepancy and the difference in values of the overall heat transfer coefficients, there seems to be close enough correspondence between the Gentry design and direct sizing to warrant further investigation of direct sizing.

7.14 GENERALISED CORRELATIONS

Shell-side heat transfer

The generalised correlations proposed by by Hesselgreaves (1988) interpret the results for shell-side heat transfer and pressure loss tests in terms of parameters derived from the detailed geometry of the RODbaffle exchanger. Such generalised correlations can be very useful in optimisation; see the Manglic and Bergles correlations for plate-fin exchangers. For the shell-side, Hesselgreves found that geometric valus of the ratio

$$\frac{D_{hyd}}{(L_b + L')/2} = \frac{\text{hydraulic diameter of intertube space}}{\text{mean of (baffle spacing + vortex spacing*)}}$$

were necessary parameters in his correlations. See Daugherty *et al.* (1985), Duncan *et al.* (1975) and White (1986) for discussion of the von Kármán vortex street.

Shell-side heat transfer was reasonably correlated over the whole range of Reynolds number by combining vortex street (laminar) Nusselt number (Nu_L), and turbulent Nusselt number (Nu_T), in the form

$$\boxed{\frac{Nu}{Pr^{0.4}} = \left[\left(\frac{Nu_L}{Pr^{0.4}}\right)^2 + \left(\frac{Nu_T}{Pr^{0.4}}\right)^2\right]^{1/2}}$$

where Nu_L and Nu_T are to be found from the following expressions recommended by Gnielinski (1990):

---

*Spacing on one side of the von Kármán street only.

$$\mathrm{Nu}_L = \frac{0.664}{\sqrt[6]{\mathrm{Pr}}} \sqrt{\left(\mathrm{Pe}\, \frac{d_{hyd}}{L'}\right)}$$

$$\mathrm{Nu}_T = \frac{(f/8\, \mathrm{Pr}\, (\mathrm{Re} - 100))}{1 + 12.7\sqrt{\{(f/8)(\mathrm{Pr}^{2/3} - 1)\}}} \left[1 + \left(\frac{d_{hyd}}{L'}\right)^{2/3}\right]$$

The turbulent friction factor was by Chen (1969)

$$\frac{1}{\sqrt{f}} = -2.0 \log_{10}\left(\frac{5.0452}{\mathrm{Re}} \log_{10} \frac{5.8506}{\mathrm{Re}}\right)$$

Shell-side baffle pressure loss

Hesselgreaves found that an approximate representation of the drag coefficient (C_D) could be represented by

$$C_D = 1.00 + \frac{3000}{\mathrm{Re}_b}$$

allowing the pressure loss per baffle to be written

$$\boxed{\Delta p = \frac{1}{2} \rho u_b^2\, C_D \left(\frac{A_{total}}{A_{flow}}\right)}$$

where

$$\frac{A_{total}}{A_{flow}} = \frac{4p^2}{4p^2 - \pi d^2 - 4pr}$$

may be understood by reference to Fig. 7.1, and u_b is the baffle velocity at the minimum flow area.

The attempt at a generalised correlation for baffle pressure loss was less successful; relatively high scatter in the data is evident in the figures of the paper by Hesselgreaves (1988). Further work was recommended, but there may now be sufficient information to establish the structure of an optimisation proceedure.

The generalised data fit for heat transfer appears to be as good as that of Manglic and Bergles for rectangular offset strip-fins, but for baffle pressure loss there is considerable scatter. Present designs should perhaps use experimental results for individual RODbaffle geometries which can be splinefitted.

7.15 RECOMMENDATIONS

All experimental work to date appears to involve testing of RODbaffle geometries complete with baffle-rings and RODbaffle-ring/shell-bypass leakage. If the proposed concept for dealing with separate leakage flow (Section 7.3) is to be accurately assessed, the thermal performance of RODbaffle geometry should be carried out in a square duct with no bypass leakage. This will provide the local thermal performance, without the unknown contribution due to bypass leakage.

Bell and Bergelin's (1957) method for calculating bypass leakage flow may then be applied with more accuracy, and the degradation in performance due to leakage will be clearly seen.

On the question of tube counts relating to inside shell diameter, it is relevant to enquire whether a second 'Phadke' analysis is possible, starting with four tubes placed symmetrically at the centre of the tube-plate.

From the calculated heat transfer coefficients, enhancement of tube-side heat transfer would produce significant reduction in surface area. The prospect of using twisted-tape inserts has recently been examined by Manglik and Bergles (1993) in two extensive papers, and the possibility of using internally rippled tubes seems worth considering.

7.16 OTHER SHELL-AND-TUBE DESIGNS

Direct sizing is not recommended for accurate design of conventional shell-and-tube heat exchangers because they need allowance for leakage paths in the clearance between tubes and tube-holes in the baffle-plates, and because the shell-side flow changes direction while traversing the tube-bundle.

However, direct-sizing methods could be applied, using the generalised shell-side correlation given by Kern (1950) and reported in the text by Hewitt *et al.* (1994). The Bell-Delaware method is an extension of the Kern approach; it introduces correction factors for shell-side leakage flow paths, bypassing flow and baffle shape.

The spiral shell-side flow, angle-baffled shell-and-tube heat exchanger (Næss *et al.* 1990, Austergard *et al.* 1993, 1994) is a new development which may become of interest for direct sizing once the concept is fully developed and shell-side correlations become available.

The helically-twisted flattened-tube heat exchanger discussed in Section 1.6 seems first to have been developed for aircraft applications in Russia as indicated in the texts by Danilov *et al.* (1986) and Dzyubenko *et al.* (1990). See also Ievlev *et al.* (1982). In parallel the Brown Fintube Company is developing industrial-scale shell- and tube heat exchangers (Butterworth *et al.* 1996). This geometry is equally suited to direct-sizing providing the tube pitching is given equal triangular or square pitching, and the tubes are oriented so that the tubes are aligned along their length.

A more compact design that the RODbaffle configuration is possible, but equally the same uncertainties exist regarding transverse mixed flow at the ends of the bundle near shell-side nozzles or annular inlet/outlet connections. The problem of filling a circular shell with exactly the right number of tubes is the same (Phadke 1984), and the same problem exists of having extra shell-side flow space next the shell.

7.17 CONCLUSIONS

1. A method of direct sizing of RODbaffle exchangers has been proposed and results from *one* example have been compared with the design approach recommended by Gentry *et al.*
2. It would be of considerable assistance in design to have dimensionless plots of baffle loss coefficient and heat transfer correlation for each RODbaffle configuration. Tables of the recommended smoothed data used in constructing these curves would also be useful, as in the text on compact heat exchangers by Kays and London.
3. Existing design information on shell-side flow between the entry/exit nozzles and the end-baffles needs to be assessed and incorporated into the design.
4. Shell-side leakage flow should be examined to assess the prospect of handling losses in other methods besides that recommended by Gentry *et al.*
5. In its present form, direct sizing seems to provide a reasonable method for arriving at the shell diameter, so it is possible to proceed with the original method of design recommended by Gentry *et al.*

REFERENCES

Austergard, A., Næss, E. and Sonju, O. K. (1993) Three dimensional flow modelling of shell-side flow in a novel helical flow shell-and-tube heat exchanger, *New Developments in Heat Exchangers, ICHMT International Symposium, Lisbon.* (Available from Norwegian Institute of Technology, Trondheim.)

Austergard, A., Næss, E. and Sonju, O. K. (1994) Cross flow pressure drop in tube bundles, *Open Poster Forum at 10th International Heat Transfer Conference, Brighton, UK, August 1994.* (Available from Norwegian Institute of Technology, Trondheim.)

Bell. K. J. and Bergelin, O. P. (1957) Flow through annular orifices, *Transactions of the American Society of Mechanical Engineers*, **79**, 593–601.

Bergelin, O. P., Bell, K. J. and Leighton, M. D. (1958) Heat transfer and fluid friction during flow across banks of tubes – VI The effect of internal leakages within segmentally baffled exchangers, *Transactions of the American Society of Mechanical Engineers*, **80**, 53–60.

Butterworth, D., Guy, A. R. and Welkey, J. J. (1996) Design and application of twisted-tube heat exchangers, *Advances in Industrial Heat Transfer*, IChemE, 87–95.

Chen, N. H. (1979) An explicit equation for friction factor in pipe, *Industrial and Engineering Chemistry, Fundamentals*, **80**(19), 229–230.

Churchill, S. W. (1977) Comprehensive correlating equations for heat, mass and momentum transfer in fully developed flow in smooth tubes, *Industrial and Engineering Chemistry, Fundamentals*, **16**, 109–115.

Churchill, S. W. (1988) The role of mathematics in heat transfer, *AIChE Symposium Series, Heat Transfer*, **84**(263), 1–13.
Churchill, S. W. (1992) The role of analysis in the rate process, *Industrial and Engineering Chemistry, Research*, **31**(3), 643–658.
Danilov, Yu, I. et al. (1986) *Teploobmen i gidrodinamika v kamalakh slozhonoy formy. (Heat transfer and hydrodynamics in complex geometry channels)* Moscow; Mashinostroyeniye.
Dzyubenko, B. V., Dreitser, G. A. and Ashmantas, L. -V. A. (1990) *Unsteady Heat and Mass Transfer in Helical Tube Bundles*, Washington: Hemisphere.
Daugherty, R. L., Franzini, J. B. and Finnemore, E. J. (1985) *Fluid Mechanics with Engineering Applications*, New York: McGraw-Hill, pp. 317–319.
Duncan, W. J., Thom, A. S. and Young, A. D. (1975) *Mechanics of Fluids*, London: Arnold, pp. 204–205.
Gentry, C. C. (1990) RODbaffle heat exchanger technology, *Chemical Engineering Progress*, **86**(7), 48–57.
Gentry, C. C. (1995) Experimental databases for RODbaffle geometries, Private communication, Phillips Petroleum Company, 3 April 1995.
Gentry, C. C. and Small, W. M. (1981) The RODbaffle heat exchanger, *2nd Symposium on Shell and Tube Heat Exchangers, Houston, 14–18 September 1981*, pp. 389–409. (Data for ARG configuration.)
Gentry, C. C. and Small, W. M. (1985) RODbaffle exchanger thermal-hydraulic predictive models over expanded baffle spacing and Reynolds number ranges, *23rd National Heat Transfer Conference, Denver, AIChE Symposium Series* **81**(245).
Gentry, C. C., Young, R. K. and Small, W. M. (1982) RODbaffle heat exchanger thermal-hydraulic predictive methods, *7th International Heat Transfer Conference, Munich, 6–10 September 1982, Vol. 6*, pp. 197–202. (Data for ARA configuration.)
Gentry, C. C., Young, R. K. and Small, W. M. (1984) RODbaffled heat exchanger thermal-hydraulic predictive methods for bare and low-finned tubes, *AIChE Symposium Series, Heat Transfer*, **80**(236), 104–109.
Gnielinski, V. (1990) Forced convection in ducts, In: *Heat Exchanger Design Handbook*, Washington: Hemisphere, Chapter 2. 5. 1.
Hesselgreaves, J. E. (1988) A mechanistic model for heat transfer and pressure drop in rod-baffled heat exchangers, *2nd UK National Conference on Heat Transfer, Univ. of Strathclyde, 14–16 September 1988, Paper C197/88*, pp. 787–800.
Hewitt, G. F., Shires, G. L. and Bott, T. R. (1993) *Process Heat Transfer*, Boca Raton, FL: CRC Press, Section 3, pp. 271–292.
Ievlev, V. M. Danilov, Yu. N. Dzyubenko, B. V. Dreitser, G. A., Ashmantas, L. A. (1990) *Analysis and Design of Swirl-augmented Heat Exchangers*, Washington: Hemisphere.
Kays, W. M. and London, A. L. (1964) *Compact Heat Exchangers*, New York: McGraw-Hill.
Kern, D. Q. (1986) *Process Heat Transfer*, New York: McGraw-Hill.
Manglic, R. M. and Bergles, A. E. (1993) Heat transfer and pressure drop correlations for twisted-tape inserts in isothermal tubes: I. Laminar flows, II. Transition and turbulent flows, *Transactions ASME, Journal of Heat Transfer*, **115**, 881–889, 890–896.
Næss, E., Knusden, H. and Sonju, O. K. (1990) Technical note on a novel high performance shell-and-tube heat exchanger, *Open Poster Forum at the 9th International Heat Transfer Conference, Jerusalem, 20–24 August 1990*. (Available from Norwegian Institute of Technology, Trondheim.)
Phadke, P. S. (1984) Determining tube counts for shell-and-tube exchangers, *Chemical Engineering*, September, **91**(18), 65–68.

Rehme, K. (1992) The structure of turbulence in rod bundles and the implications on natural mixing between the sub-channels, *International Journal of Heat and Mass Transfer*, **35**(2), 567–581.

Smith, E. M. (1986) Design of helical-tube multi-start coil heat exchangers, *Advances in Heat Exchanger Design, ASME Winter Annual Meeting, Anaheim, CA, ASME Publication HTD* **66**, 95–104. *7–12 December 1986*, (Revised 3 December 1994)

Smith, E. M. (1994) Direct thermal sizing of plate-fin heat exchangers, *The Industrial Sessions Papers, 10th International Heat Transfer Conference, Brighton, Institution of Chemical Engiuneers, 14–18 August 1994*, p. 12.

Tong, L. S. (1968) Pressure drop performance of a rod bundle, In: *Heat Transfer in Rod Bundles, ASME Nucleonics Heat Transfer Committee, K-13 Symposium at Winter Annual Meeting of ASME*.

White, F. M. (1986) In: *Fluid Mechanics*, New York: McGraw-Hill, pp. 262–267.

FURTHER READING

Barrington E. A. (1973) Acoustic vibration in tubular exchangers, *Chemical Engineering Progress*, **69**(7), 62–68.

Eilers, J. F. and Small, W. M. (1973) Tube vibration in a thermosiphon reboiler, *Chemical Engineering Progress*, **69**(7), 57–61.

Gentry, C. C., Young, R. K. and Small, W. M. (1980) The RODbaffle heat exchanger thermal-hydraulic performance, *20th National IMIQ/AIChE Conference, Acapulco, October 1980*.

Gentry, C. C., Young, R. K. and Small, W. M. (1982) Conceptual RODbaffle nuclear steam generator design, *IEEE/ASME/ASCE 1982 Joint Power Generation Conference, 17–21 October 1982, EI Conf. no. 02416*.

Jegede, F. O. and Polley, G. T. (1992) Optimum heat exchanger design, *Transactions of the Institution of Chemical Engineers*, **70A**, 133–141.

Miller, P., Byrnes, J. J. and Benforado, D. M. (1956) Heat transfer to water flowing parallel to a rod bundle, *Journal of the American Institute of Chemical Engineers*, **2**(2), 226–234. (Triangular pitch only – nuclear fuel elements.)

Peters, M. S. and Timmerhaus, K. D. (1958) *Plant Design and Economics for Chemical Engineers, 3rd Edn*, New York: McGraw-Hill 678–710. (Chapters on heat transfer equipment design and costs and on shell-and-tube optimisation.)

Polley, G. T., Panjeh Shahi, M. H. and Picon Nunez, M. (1991) Rapid design algorithms for shell-and-tube and compact heat exchangers, *Transactions of the Institution of Chemical Engineers*, **69A**, 435–444.

Roetzel, W. and Lee, D. (1993) Experimental investigation of leakage in shell-and-tube heat exchangers with segmental baffles, *International Journal of Heat and Mass Transfer*, **36**(15), 3765–3771.

Sangster, W. A. (1968) Calculation of rod-bundle pressure loss, *ASME Paper no. 68-WA/HT-35*.

Small, W. M. and Young, R. K. (1979) The RODbaffle heat exchanger (design and operations), *Heat Transfer Engineering*, **1**(2), 21–27.

Smyth, R. (1981) Comparative assessment of RODbaffle shell-and-tube heat exchangers, *Heat Transfer Engineering*, 2 (3/4) 90–94.

Sun, S.-Y., Lu, Y.-D. and Yan, C.-Q. (1993) Optimisation in calculation of shell and tube exchanger, *International Communications in Heat and Mass Transfer*, **20**(5), 675–686.

8

TRANSIENTS IN CONTRAFLOW EXCHANGERS

8.1 SOLUTION METHODS

The computation of transients is a growing area of research, and full solution of the problem of combined general transients in mass flow rate and temperature may not yet have been achieved. Three main approaches to solution of transients exist:

- method of characteristics
- Laplace transformation
- direct solution by finite differences

But Laplace transformation follows a hybrid approach. Laplace transforms are taken of the partial differential equations; the resulting ordinary differential equations are solved numerically and the solution is inverted numerically.

The mathematical effort involved in producing a solution to a real problem is always considerable, so it is possible only to outline the methods within the scope of this book.

Method of characteristics

This involves finding the characteristics of the problem, which are the 'natural co-ordinates' of the system, lines representing the trajectories of fluid particles as they pass through the exchanger. The method provides good physical insight into the problem, and transient discontinuities may pass through the exchanger travelling along the characteristics. This is a good approach, providing the characteristics can be found and appropriate mathematical schemes can be developed for numerical solution. Finding the unknown temperatures corresponds to integrating along the characteristics.

For constant physical properties and constant mass flow rates, the characteristics become straight lines and a numerical solution is straightforward, but it restricts the solution to *modest* temperature excursions. An example given later provides insight into what is happening.

But as soon as account is taken of significant temperature variations in physical properties at points throughout the exchanger, the heat transfer coefficients change, the density of the fluids change and the transit times change. All these factors lead to a change in direction of the characteristics, which then become curved.

The effect is more pronounced with variable mass flow rates. This is because the mass flow rate directly affects the transit times through the exchanger. The problem then grows to one of finding the curved characteristics and of establishing a numerical scheme for calculating points along the exchanger – not a simple task. Much effort has gone into solving problems in fluid dynamics using the method of characteristics, particularly for handling shock waves, but so far it has not found favour in dealing with transients in heat exchangers.

Laplace transforms with numerical inversion

The approach here is to transform the partial differential equations, plus initial conditions and boundary conditions, into ordinary differential equations containing the initial conditions plus transformed boundary conditions. Because the boundary conditions (i.e. transients) may not be recognisable mathematical functions, this may involve Fourier representation of the disturbances.

The ordinary differential equations equations are then solved by finite differences in the phase field, and the results are then inverted back into real time using fast Fourier transforms. The mathematical approach is clear and accommodates longitudinal conduction effects, but it requires a significant amount of computing power. A number of workers in heat transfer have gone down this route, as can be seen from papers listed in the References, and some important solutions have been obtained by Roetzel and co-workers, including the added feature of axial dispersion.

Direct solution by finite differences

Ames (1965) points out that 'one of the main advantages of the use of characteristics and a similar disadvantage in the use of finite differences is the fact that discontinuities in the prescribed initial values may be propagated along the characteristics.'

The third approach is to work with the *original* set of partial differential equations and the original initial conditions and boundary conditions. The numerical approach is to convert the set of partial differential equations into finite-difference form, employing the fully implicit but unconditionally stable Crank-Nicholson scheme for solution.

Sharifi *et al.* (1995) have presented results for such a computation. Their partial differential equations did not include the second-order terms for longitudinal conduction.

Combination of methods

Dzyubenko *et al.* (1990) completed a thorough study on the flattened and helically twisted tube design of exchanger, in which the full set of energy and momentum equations were first set out. The transient solution involved solving a one-dimensional energy equation by the method of characteristics and two Cauchy problems. Some simplifications were necessary to obtain a solution.

8.2 FUNDAMENTAL EQUATIONS

Governing equations

Transient equations for contraflow are developed in Appendix A as equations (A.1). For solution it is preferable to use θ for temperature, leaving T for dimensionless time, before transforming and normalising to canonical form.

$$\frac{\partial \theta_h}{\partial t} + u_h \frac{\partial \theta_h}{\partial x} = -\left(\frac{\alpha_h S}{\dot{m}_h C_h}\right) \frac{u_h}{L} (\theta_h - \theta_w)$$

$$\frac{\partial \theta_w}{\partial t} - \hat{\kappa}_w \frac{\partial^2 \theta_w}{\partial x^2} = +\left(\frac{\alpha_h S}{M_w C_w}\right)(\theta_h - \theta_w) - \left(\frac{\alpha_c S}{M_w C_w}\right)(\theta_w - \theta_c) \qquad (8.1)$$

$$\frac{\partial \theta_c}{\partial t} - u_c \frac{\partial \theta_c}{\partial x} = +\left(\frac{\alpha_c S}{\dot{m}_c C_c}\right) \frac{u_c}{L} (\theta_w - \theta_c)$$

To keep the analysis simple we shall consider temperature transients for constant physical properties and constant mass flow rates only, and compact the notation for this special case using fluid transit times ($\tau_h = L/u_h$) and $\tau_c = L/u_c$), and defining

$$n_h = \frac{\alpha_h S}{\dot{m}_h C_h} \quad \text{and} \quad n_c = \frac{\alpha_c S}{\dot{m}_c C_c} \quad \text{(local values of Ntu)}$$

$$R_h = \frac{M_w C_w}{\tilde{m}_h C_h} \quad \text{and} \quad R_c = \frac{M_w C_w}{\tilde{m}_h C_h} \quad \text{(ratios of thermal capacities)}$$

Equations (8.1) may be written

$$\frac{\partial \theta_h}{\partial t} + \frac{L}{T_h} \cdot \frac{\partial \theta_h}{\partial x} = -\frac{n_h}{\tau_h}(\theta_h - \theta_w)$$

$$\frac{\partial \theta_w}{\partial t} - \hat{\kappa}_w \frac{\partial^2 \theta_w}{\partial x^2} = +\frac{n_h}{R_h \tau_h}(\theta_h - \theta_w) - \frac{n_c}{R_c \tau_c}(\theta_w - \theta_c) \qquad (8.2)$$

$$\frac{\partial \theta_c}{\partial t} - \frac{L}{T_c} \cdot \frac{\partial \theta_c}{\partial x} = +\frac{n_c}{\tau_c}(\theta_w - \theta_c)$$

Scaling and normalisation

To non-dimensionalise, refer everything to the hot residence time and hot fluid flow.

With scaling $\quad T = \dfrac{t}{\tau_h} \quad$ gives $\quad \dfrac{\partial(\ )}{\partial t} = \dfrac{\partial(\ )}{\partial T} \cdot \dfrac{1}{\tau_h}$

With normalisation $\quad X = \dfrac{x}{L} \quad$ gives $\quad \dfrac{\partial(\ )}{\partial x} = \dfrac{\partial(\ )}{\partial X} \cdot \dfrac{1}{L}$

$$\text{and} \quad \dfrac{\partial^2(\ )}{\partial x^2} = \dfrac{\partial^2(\ )}{\partial X^2} \cdot \dfrac{1}{L^2} \quad \text{over} \ (0 \leq X \leq 1)$$

With $P = \dfrac{\tau_c}{\tau_h}$ (ratio of transit times) there results

$$\boxed{\begin{aligned}
\dfrac{\partial \theta_h}{\partial T} + \dfrac{\partial \theta_h}{\partial X} &= -n_h (\theta_h - \theta_w) \\
\dfrac{\partial \theta_w}{\partial T} - \text{Fo}\, \dfrac{\partial^2 \theta_w}{\partial X^2} &= +\dfrac{n_h}{R_h}(\theta_h - \theta_w) - \dfrac{1}{P} \cdot \dfrac{n_c}{R_c}(\theta_w - \theta_c) \\
P\dfrac{\partial \theta_c}{\partial T} - \dfrac{\partial \theta_c}{\partial X} &= \quad\quad\quad\quad\quad\quad\quad\quad +n_c (\theta_w - \theta_c)
\end{aligned}} \quad (8.3)$$

where $\text{Fo} = \dfrac{\hat{\kappa}\tau_h}{L^2}$ is a Fourier number

Perturbance equations

Steady-state and perturbance temperatures are now separated by assuming that transient temperatures (θ) are the sum of steady-state values $\bar{\theta}$ and perturbance values ($\hat{\theta}$), which is permissible for linear theory, then

$$\theta_h = \bar{\theta}_h + \hat{\theta}_h$$
$$\theta_w = \bar{\theta}_w + \hat{\theta}_w$$
$$\theta_c = \bar{\theta}_c + \hat{\theta}_c$$

The coupled equations separate into two sets of three equations.

Steady-state temperatures

$$\boxed{\begin{aligned}
+\dfrac{\partial \bar{\theta}_h}{\partial X} &= -n_h (\bar{\theta}_h - \bar{\theta}_w) \\
-\text{Fo}\cdot \dfrac{\partial^2 \bar{\theta}_w}{\partial X^2} &= +\dfrac{n_h}{R_h}(\bar{\theta}_h - \bar{\theta}_w) - \dfrac{1}{P} \cdot \dfrac{n_c}{R_c}(\bar{\theta}_w - \bar{\theta}_c) \\
-\dfrac{\partial \bar{\theta}_c}{\partial X} &= \quad\quad\quad\quad\quad\quad\quad\quad +n_c (\bar{\theta}_w - \bar{\theta}_c)
\end{aligned}} \quad (8.4)$$

Perturbance temperatures

$$\frac{\partial \hat{\theta}_h}{\partial T} + \frac{\partial \hat{\theta}_h}{\partial X} = -n_h(\hat{\theta}_h - \hat{\theta}_w)$$

$$\frac{\partial \hat{\theta}_w}{\partial T} - \text{Fo.}\frac{\partial^2 \hat{\theta}_w}{\partial X^2} = +\frac{n_h}{R_h}(\hat{\theta}_h - \hat{\theta}_w) - \frac{1}{P} \cdot \frac{n_c}{R_c}(\hat{\theta}_w - \hat{\theta}_c) \quad (8.5)$$

$$P\frac{\partial \hat{\theta}_c}{\partial T} - \frac{\partial \hat{\theta}_c}{\partial X} = +n_c(\hat{\theta}_w - \hat{\theta}_c)$$

The perturbance equations (8.5) are similar to the full transient equations, and it is preferable to work with *normalised disturbances* starting from zero initial values by defining

$$\theta = \frac{\hat{\theta} - \hat{\theta}_{c2}}{\hat{\theta}_{h1} - \hat{\theta}_{c2}} = \frac{\hat{\theta}}{\hat{\theta}_{h1}} \quad \text{giving the substitution} \quad \hat{\theta} = \theta\hat{\theta}_{h1} \quad \text{in equations (8.5).}$$

After the perturbance temperature distributions have been determined, the complete temperature transient can be assembled as the sum of *actual* steady-state values and *actual* perturbance values.

8.3 ANALYTICAL CONSIDERATIONS

Wall axial conduction and shell losses in the steady state

The effects of steady-state longitudinal conduction have been analysed by Kroeger (1967) assuming zero temperature gradients at the exchanger ends. This may be more nearly true for compact plate-fin exchangers than for exchangers with tube-plates. With shell-and-tube heat exchangers the effect of longitudinal conduction in the shell may also need consideration if the pressure-shell is thick. In special applications steady-state longitudinal conduction in the fluids themselves may be significant (liquid metals).

An analysis of losses from the exchanger shell surface has been made by Nesselman (1928), with further treatment by Hausen (1950). These effects were not considered in steady-state treatments because modern insulating materials can minimise the effect.

Axial conduction terms for the fluids

In generating the simultaneous partial differential equations for simple contra-flow in Appendix A, several assumptions were made to keep the analysis within reasonable bounds.

First, it was assumed that there was no temperature difference across the solid wall of the exchanger. It is simple to incorporate the thermal resistance of the wall in steady-state analysis, and it is possible to assess the reasonableness of the assumption at that time. Second, there was no allowance for the thermal capacity effect of the exchanger shell, or of heat leak from the insulated shell. Third, plug flow is assumed for both fluids, i.e. there was no attempt to distinguish between boundary layer and bulk flow.

It is possible to introduce longitudinal conduction terms for the fluids. Consider the energy balance equation for the hot fluid of Appendix A.1, and include a longitudinal conduction term (λ_h) and the associated cross-sectional area for flow (A_h). Then

$$-\lambda_h A_h \frac{\partial T_h}{\partial x} - \left[-\lambda_h A_h \frac{\partial}{\partial x}\left(T_h + \frac{\partial T_h}{\partial x}\delta x\right)\right] + \dot{m}_h C_h T_h$$

$$- \dot{m}_h C_h \left(T_h + \frac{\partial T_h}{\partial x}\delta x\right) - \alpha_h \left(S\frac{\delta x}{L}\right)(T_h - T_w) = \tilde{m}_h \left(\frac{\delta x}{L}\right) C_h \frac{\partial T_h}{\partial T}$$

Cleaning up and substituting ($\tilde{m}_h = \dot{m}_h \tau_h$), this becomes

$$\frac{\partial T_h}{\partial t} + u_h \frac{\partial T_h}{\partial x} - \left(\frac{\lambda_h A_h L}{\dot{m}_h \tau_h C_h}\right)\frac{\partial^2 T_h}{\partial x^2} = -\frac{n_h}{\tau_h}(T_h - T_w) \qquad (8.6)$$

Putting $X = \dfrac{x}{L}$

$T = \dfrac{t}{\tau_h}$, the same substitution being used for the cold fluid

$\theta = \dfrac{T - T_{c2}}{T_{h1} - T_{c2}}$ and with $T_{c2} = 0$ then $T = \theta T_{h1}$

These scalings and normalisation are substituted in equation (8.6) to produce

$$\frac{\partial \theta_h}{\partial T} + \frac{\partial \theta_h}{\partial X} - \left[\frac{\lambda_h A_h}{\dot{m}_h C_h L}\right]\frac{\partial^2 \theta_h}{\partial X^2} = -n_h(\theta_h - \theta_w)$$

The group in square brackets is dimensionless, and the full set of three equations will become

$$\boxed{\begin{aligned}
\frac{\partial \theta_h}{\partial T} + \frac{\partial \theta_h}{\partial X} - \left(\frac{\lambda_h}{\dot{m}_h C_h}\right)\left(\frac{A_h}{L}\right)\frac{\partial^2 \theta_h}{\partial X^2} &= -n_h(\theta_h - \theta_w) \\
\frac{\partial \theta_w}{\partial T} \qquad\qquad - \left(\frac{\kappa_w \tau_h}{V_w}\right)\left(\frac{A_w}{L}\right)\frac{\partial^2 \theta_w}{\partial X^2} &= +\frac{n_h}{R_h}(\theta_h - \theta_w) - \frac{1}{P}\frac{n_c}{R_c}(\theta_w - \theta_c) \quad (8.7) \\
P\frac{\partial \theta_c}{\partial T} - \frac{\partial \theta_c}{\partial X} - \left(\frac{\lambda_c}{\dot{m}_c C_c}\right)\left(\frac{A_c}{L}\right)\frac{\partial^2 \theta_c}{\partial X^2} &= \qquad\qquad\qquad +n_c(\theta_w - \theta_c)
\end{aligned}}$$

The cogency of the added terms seems to lie in the presence of the bracket $(\frac{A}{L})$, which may be interpreted as the slenderness ratio of flow channels in the exchanger. The smaller this value is kept, the smaller will be the parasitic losses caused by the second differential terms, which is something to consider when initially selecting the exchanger geometry. The set of thermal equations is simplified by neglecting longitudinal conduction in the fluids, as these terms are usually negligible.

Axial dispersion terms for the fluids

To allow for deviation from the assumed plug flow model for the fluids, a model with axial dispersion was introduced by Roetzel and co-workers. To understand the concept of dispersion it is helpful to examine the paper by Taylor (1954) describing a turbulent flow experiment in which a salt solution pulse is injected into a steady flow of pure water. The presence of the salt was determined by measuring electrical conductivity of the water and by displaying this on a recorder. Downstream the pulse is shown to spread significantly due to lateral turbulent diffusion. This is a momentum effect. The analogous thermal effect exists, and can be modelled in one dimension by replacing the conventional Fourier constitutive equation

$$\dot{q} = -\lambda \left(\frac{\partial T}{\partial x} + \frac{\partial T}{\partial y} + \frac{\partial T}{\partial z} \right) = -\lambda \nabla T \quad \text{for constant } \lambda$$

by its more exact representation, see for example Chester (1963) or Churchill (1988),

$$\tau \frac{\partial \dot{q}}{\partial t} + \dot{q} = -\lambda \nabla T$$

where τ is a relaxation time which allows for the build-up of the transient. This leads to inclusion of a second-order term similar to that for longitudinal conduction in the fluids.

Fortunately, most of the heat exchangers capable of being sized directly have small flow channels, i.e. small values of (A/L), and the assumption of plug flow is acceptable. Additionally, it is possible that the heat transfer and fluid flow correlations obtained by transient test methods may already take into account some of the effects of axial dispersion, although this might require separate correlations for heating and cooling. The model may be of greater value for flow through discontinuities like baffles in conventional shell-and-tube exchangers. For exchangers which can be directly sized, we shall neglect axial dispersion to keep the equations simple.

Thermal storage effects in transient solutions

In transient behaviour the solid wall equation must be introduced for realistic assessment of effects. Transient solutions that do not include thermal storage are of little practical value. The pressure shell of an exchanger may also contribute to thermal storage if it is thick, but the effect can be designed out by minimising heat flow from the exchanger core to the shell.

Transient disturbances

Two basic types of transient disturbance exist:

- temperature transients
- mass flow (velocity) transients

while in real plant there will be combinations of these two effects. Both transients involve temperature changes which affect physical properties such as absolute viscosity (η), thermal conductivity (λ), and hence the Reynolds and Prandtl numbers (Re, Pr) and the heat transfer coefficients (α).

The simpler of the basic effects to analyse is the temperature transient. By assuming *constant* physical properties and *constant* mass flow rates, a first illustrative treatment can be kept within reasonable bounds using the method of characteristics, which provides a clear picture of the role played by different terms in the equations. The method of characteristics requires an introduction to the theory of coupled hyperbolic equations which is provided by Fox and presented later in the chapter.

8.4 METHOD OF CHARACTERISTICS

Governing equations

The method of solution will be applied to a simplified set of equations (8.7) in which second differential terms are neglected. Also it will be assumed that steady-state and perturbance equations have already been separated, and operations will only be on the perturbance set with the circumflex on the temperature symbol omitted to simplify notation.

Computation of a modest temperature transient in a simple contraflow heat exchanger will be illustrated. A plate-fin exchanger is used as an example, but this is not a restriction on the approach for exchangers with consistent core geometry.

Characteristics (transformation)

Fox (1962) discusses the solution of simultaneous linear partial differential equations of the form

8.4 METHOD OF CHARACTERISTICS

$$\alpha \frac{\partial u}{\partial x} + \beta \frac{\partial u}{\partial y} + \gamma \frac{\partial v}{\partial x} + \delta \frac{\partial v}{\partial y} = 0$$

$$A \frac{\partial u}{\partial x} + B \frac{\partial u}{\partial y} + C \frac{\partial v}{\partial x} + D \frac{\partial v}{\partial y} = 0$$

and obtains the following quadratic equation for the characteristics

$$(\alpha C - A\gamma)\left(\frac{dy}{dx}\right)^2 - (\beta C - B\gamma + \alpha D - A\delta)\left(\frac{dy}{dx}\right) + (\beta D - B\delta) = 0$$

Taking the first and last of the perturbance equations of the set (8.9), each of which have two independent variables, substitute into the above scheme, dropping the circumflex, then

$$\frac{\partial \theta_h}{\partial X} + \frac{\partial \theta_h}{\partial T} + 0 + 0 = 0$$

$$0 + 0 - \frac{\partial \theta_c}{\partial X} + P \frac{\partial \theta_c}{\partial T} = 0$$

giving the characteristic equation

$$\left(\frac{dT}{dX}\right)^2 - (1 - P)\frac{dT}{dX} - P = 0$$

from which

$$\frac{dT}{dX} = -P \quad \text{and} \quad \frac{dT}{dX} = 1$$

Integrating

$$T = -PX + \text{const.} \qquad T = X + \text{const.}$$

Let these new constants be α and β they are the characteristics of the problem

$$\alpha = T + PX \qquad \beta = T - X$$

Thus

$$\frac{\partial \theta_h}{\partial \theta_h} = \left(\frac{\partial \theta_h}{\partial \alpha}\right)P + \left(\frac{\partial \theta_h}{\partial \beta}\right)(-1) \qquad \frac{\partial \theta_c}{\partial X} = \left(\frac{\partial \theta_c}{\partial \alpha}\right)P + \left(\frac{\partial \theta_c}{\partial \beta}\right)(-1)$$

$$\frac{\partial \theta_h}{\partial T} = \left(\frac{\partial \theta_h}{\partial \alpha}\right)1 + \left(\frac{\partial \theta_h}{\partial \beta}\right)1 \qquad \frac{\partial \theta_c}{\partial \alpha} = \left(\frac{\partial \theta_c}{\partial \alpha}\right)1 + \left(\frac{\partial \theta_c}{\partial \beta}\right)1$$

216 TRANSIENTS IN CONTRAFLOW EXCHANGERS

so that

$$\frac{\partial \theta_h}{\partial T} + \frac{\partial \theta_h}{\partial X} = (1+P)\frac{\partial \theta_h}{\partial \alpha} \qquad P\frac{\partial \theta_c}{\partial T} - \frac{\partial \theta_c}{\partial X} = (1+P)\frac{\partial \theta_c}{\partial \beta}$$

making it necessary to put

$$\alpha = \frac{T+PX}{1+P} \qquad \beta = \frac{T-X}{1+P}$$

to obtain

$$\frac{\partial \theta_h}{\partial T} + \frac{\partial \theta_h}{\partial X} = \frac{\partial \theta_h}{\partial \alpha} \qquad P\frac{\partial \theta_c}{\partial T} - \frac{\partial \theta_c}{\partial X} = \frac{\partial \theta_c}{\partial \beta}$$

The original transient equations for the fluids can now been transformed into an equivalent pair of equations along the characteristic directions or *natural co-ordinates* of the problem, which represent the trajectories of particles of fluid in the exchanger. Treating the wall equation similarly gives

$$\frac{\partial \theta_w}{\partial T} = \frac{\partial \theta_w}{\partial \alpha}\left(\frac{d\alpha}{dT}\right) + \frac{\partial \theta_w}{\partial \beta}\left(\frac{d\beta}{dT}\right) = \frac{\partial \theta_w}{\partial \alpha} + \frac{\partial \theta_w}{\partial \beta}$$

$$= \frac{\partial \theta_w}{\partial \gamma} \quad \text{say, where } \gamma \text{ is related to directions } \alpha \text{ and } \beta$$

The perturbance set of three equations, without longitudinal conduction, becomes

$$\left.\frac{d\theta_h}{d\alpha}\right|_\beta = -n_h(\theta_h - \theta_w)$$

$$\left.\frac{d\theta_w}{d\gamma}\right|_{\alpha,\beta} = +\frac{n_h}{R_h}(\theta_h - \theta_w) - \frac{1}{P}\cdot\frac{n_c}{R_c}(\theta_w - \theta_c) \qquad (8.8)$$

$$\left.\frac{d\theta_c}{d\beta}\right|_\alpha = +n_c(\theta_w - \theta_c)$$

Figure 8.1 illustrates the physical problem, from which temperatures and times will eventually be obtained for disturbances across the exchanger.

For constant thermal capacity rates, the grid is determined by

- choice of stations across the exchanger (0 to m)
- space increment, $\Delta X = 1/m$
- β-characteristic, $T = X + const$
- α-characteristic, $T = -PX + const$

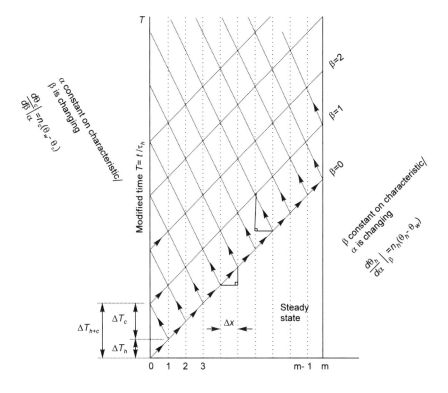

Fig. 8.1 Grid points of characteristics: single disturbance

and the time intervals $(\Delta T_h, \Delta T_c, \Delta T_{(h+c)})$ by

$$\left.\begin{array}{l}\Delta T_h = T_h/m \\ \Delta T_c = T_c/m \\ \Delta T_{(h+c)} = (T_h + T_c)/m\end{array}\right\} \quad \text{(dimensionless time intervals)}$$

The grid shown in Fig. 8.1 (Tan and Spinner 1984) has abscissa X (modified length along exchanger), ordinate T (modified time), and characteristics α and β. The value of L is now 1, the ordinate at the end of the first characteristic ($\beta = 0$) is $T_h = f(\text{time}/\tau_h)$, and the 'shape' of the grid is controlled by the parameter P.

Temperature calculation scheme

Before the disturbance propagation line $\beta = 0$, values of θ_h and θ_c are the (zero) initial values.

Along the disturbance propagation line $\beta = 0$, the expression

218 TRANSIENTS IN CONTRAFLOW EXCHANGERS

$$\left.\frac{d\theta_h}{d\alpha}\right|_\beta = -n_h(\theta_h - \theta_w)$$

is used to calculate new values of θ_h, starting at $X = 0$, and the triplet of values $(\theta_h, \theta_w, \theta_c)$ is now known along characteristic $\beta = 0$.

Along the solid wall characteristic, the expression

$$\left.\frac{d\theta_w}{d\beta}\right|_\alpha = +\frac{n_h}{R_h}(\theta_h - \theta_w) - \frac{n_c}{R_c}(\theta_w - \theta_c)$$

is used to calculate new values of θ_w at the characteristic line $\beta=10$

Along the disturbance propagation line $\alpha = 0$, the expression

$$\left.\frac{d\theta_c}{d\beta}\right|_\alpha = +n_c(\theta_w - \theta_c)$$

is used to calculate new values for θ_c.

At the hot inlet ($X = 0$), disturbance values of θ_h are known for all T; whereas at the cold inlet, the value of θ_c is always zero for no inlet disturbance. Thus the numerical process can be repeated indefinitely.

Grid construction

From the grid pattern in Fig. 8.1 it can be seen that any disturbance at $X = 0$ must be described by points spaced at $\Delta T_{(h+c)}$. Thus the number of grid points (m) along the modified X-axis, may have to be adjusted to allow an input disturbance along the modified T-axis to be sufficiently described.

Setting up the characteristics

Temperatures are now functions of the dimensionless independent variables α and β, indicating that the problem should be solved in (α, β) space. It remains to find values for the finite-difference spacing ($\Delta\alpha$, $\Delta\beta$, $\Delta\gamma$).

The expressions for α and β provide

$$\frac{\partial \alpha}{\partial T} = \frac{1}{1+P} \qquad \frac{\partial \beta}{\partial T} = \frac{1}{1+P}$$

$$\frac{\partial \alpha}{\partial X} = \frac{P}{1+P} \qquad \frac{\partial \beta}{\partial X} = \frac{-1}{1+P}$$

From the total differential $d\alpha = \dfrac{\partial \alpha}{\partial T} dT + \dfrac{\partial \alpha}{\partial X} dX$

8.4 METHOD OF CHARACTERISTICS

$$\Delta\alpha = \left(\frac{1}{1+P}\right)dT + \left(\frac{P}{1+P}\right)dX = \left(\frac{1}{1+P}\right)\left(\frac{1}{m}\right) + \left(\frac{P}{1+P}\right)\left(\frac{1}{m}\right) \quad (8.9)$$

From the total differential $d\beta = \dfrac{\partial \beta}{\partial T}dT + \dfrac{\partial \beta}{\partial X}dX$

$$\Delta\beta = \left(\frac{1}{1+P}\right)dT + \left(\frac{-1}{1+P}\right)dX = \left(\frac{1}{1+P}\right)\left(\frac{P}{m}\right) + \left(\frac{-1}{1+P}\right)\left(\frac{1}{m}\right) \quad (8.10)$$

Examine the equivalent solution diagram in (α, β) space (Fig. 8.2). Because α and β are independent, they must be at right angles. But equations (8.9) and (8.10) are to be interpreted as vector equations; thus scalar values of $\Delta\alpha$ and $\Delta\beta$ are to be regarded as sides of a right-angled triangle whose hypotenuse is calculated as follows:

$$\Delta\alpha = \sqrt{\left[\left(\frac{1}{1+P}\right)\left(\frac{1}{m}\right)\right]^2 + \left[\left(\frac{P}{1+P}\right)\left(\frac{1}{m}\right)\right]^2} = \frac{\sqrt{1+P^2}}{m(1+P)}$$

$$\Delta\beta = \sqrt{\left[\left(\frac{1}{1+P}\right)\left(\frac{P}{m}\right)\right]^2 + \left[\left(\frac{-1}{1+P}\right)\left(\frac{1}{m}\right)\right]^2} = \frac{\sqrt{1+P^2}}{m(1+P)}$$

Both finite-difference increments are equal, requiring Fig. 8.2 also to possess a square grid, (cf. the triangles in Fig. 8.1). The characteristic directions were not at right angles in Fig. 8.1, and it follows that superposition of Fig. 8.2 on Fig. 8.1 would never be correct. The algorithms can be written using Fig. 8.1 in this simple case.

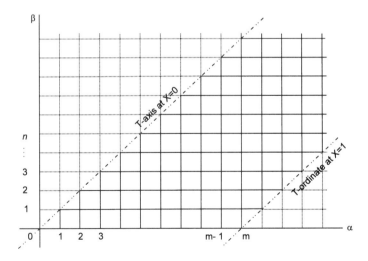

Fig. 8.2 Characteristic grid in (α, β) field

220 TRANSIENTS IN CONTRAFLOW EXCHANGERS

The original T directions on this new grid for $X = 0$ and $X = 1$ are shown on Fig. 8.2, and the solution must lie between these lines.

There remains the problem of finding a value for $\Delta\gamma$. For wall temperature terms of the form $(\partial\theta_w/\partial\gamma)$, substituting from the characteristic expressions for α and β

$$\frac{\partial\theta_w}{\partial\gamma} = \frac{\partial\theta_w}{\partial\alpha} + \frac{\partial\theta_w}{\partial\beta}$$

Finite-difference increments are to be combined vectorially, thus

$$\gamma = \sqrt{\Delta\alpha^2 + \Delta\beta^2}, \quad \text{giving} \quad \Delta\gamma = \frac{\sqrt{2(1+P^2)}}{m(1+P)} = \sqrt{2}.\Delta\alpha$$

Information necessary to produce a numerical solution for non-dimensional temperatures on the (α, β) field is now to hand.

Input data

Data for the partly optimised plate-fin contraflow exchanger described in Section 4.5 will be used to illustrate calculation of a simple temperature transient which does not affect the physical properties. In this example, the effects of longitudinal conduction have only been diminished by partly optimising surface geometry, but we shall neglect the term anyway.

Because overall values of Ntu are equal, steady-state temperature profiles must be linear and parallel. After any transient has settled down, the temperature profiles must again be linear and parallel, and also the *normalised* temperature ratios should be the same as for the steady state.

The data for this illustrative calculation is obtained from the numerical steady-state direct-sizing calculations and is given in Table 8.1. Whenever equal values of overall Ntu occur ($N_h = N_c$), it does not necessarily follow

Table 8.1 Data for transient response with wall equation

Hot fluid		Cold fluid		Transit times	
$N_h = \left(\dfrac{US}{\dot{m}_h C_h}\right)$	$= 9.1739$	$N_c = \left(\dfrac{US}{\dot{m}_c C_c}\right)$	$= 9.1739$	$P = \left(\dfrac{T_c}{T_h}\right)$	$= 6.5151$
$n_h = \left(\dfrac{\alpha_h S}{\dot{m}_h C_h}\right)$	$= 17.5818$	$n_c = \left(\dfrac{\alpha_c S}{\dot{m}_c C_c}\right)$	$= 19.2564$		
$R_h = \left(\dfrac{M_w C_w}{\tilde{m}_h C_h}\right)$	$= 1723.08$	$R_c = \left(\dfrac{M_w C_w}{\tilde{m}_c C_c}\right)$	$= 264.473$	$\begin{pmatrix}\tau_c = 0.093\,709\\ \tau_h = 0.014\,373\end{pmatrix}$	

that the local values will also be equal. The relationship between local and overall values of Ntu is easily shown to be

$$\frac{N_h}{n_h} + \frac{N_c}{n_c} = 1$$

Inlet disturbance (modest temperature change only)

The remaining data required is the shape of the imposed input disturbance. In this case, a cosine curve provides representation of a rapidly rising transient, followed by holding the new steady state at the maximum temperature. For the hot fluid, the disturbance is

$$\theta_h = \frac{1}{2}\left[1 - \cos\left(\frac{\pi k}{z+1}\right)\right] \quad \text{where} \quad z = \text{TRUNC}\left(\frac{\tau_h}{\Delta T_{(h+c)}}\right) + 1$$

with time constant $\tau_h = 0.2$, and integer $k = 0$ to $k = z$ where $z = 50$.

Computational results

The transient response of the exchanger is shown in Figs. 8.3, 8.4 and 8.5; solid lines represent the hot fluid temperature (θ_h), dashed lines the cold fluid temperature (θ_c) and dotted lines the solid wall temperature (θ_w)

In the modified plots of Figs. 8.4 and 8.5, the approach to expected steady-state profiles is evident. Substantial computer memory is essential for transient

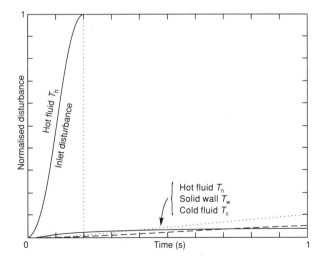

Fig. 8.3 Transient temperature histories in the exchanger for the case of full thermal storage in the wall

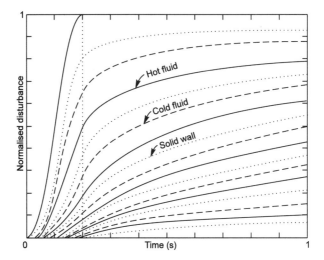

Fig. 8.4 Transient temperature histories in the exchanger for thermal storage in the wall, but with reduced thermal storage (mass/1000)

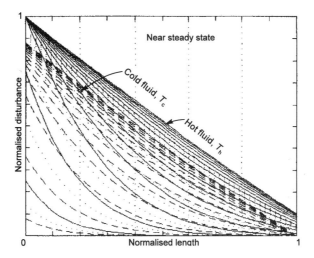

Fig. 8.5 Transient temperature profiles across exchanger for thermal storage in the wall, but with reduced thermal storage (mass/1000)

calculations. Initially there was very little rise in the temperature of either the solid wall or the cold fluid (Fig. 8.3). Because the author's modest computer has 1 Mbyte of RAM and no hard disk, it was necessary to reduce the mass of the exchanger 1000-fold to obtain reasonable temperature curves in the computable time span, which also confirms that it is *essential* to include the solid wall equation in analysis of transients.

A cautionary remark is, however, in order. When attempting to run the program with a mass reduction factor of 10 000-fold, the program failed. As the steady-state was approached, instability in computation was first encountered with a reduction factor around 2500. The difficulty is one of software accuracy, not mathematical error. The writer was using the simplest possible forward difference scheme, and there is scope for development of improved algorithms.

Disturbances in both fluids

When both hot fluid and cold fluid input disturbances begin within a time interval less than the transit time for the hot fluid, it may be necessary to return to Fig. 8.1 and revise the numerical solution. This possible case is illustrated in Fig. 8.6, where the cold fluid disturbance has to be evaluated for the triangular area at the bottom right before it meets the characteristic for the hot fluid coming from the left.

It is a straightforward matter to calculate the cold fluid transient as if the hot fluid transient were not present, then interpolate the results for missing data along the first hot fluid characteristic. Fortunately this case may not arise very often, for the transit times through the exchanger are very short compared with the time for fluids to pass through the rest of the system, and once continuous disturbances on both sides are present, the computational difficulty is reduced to finding sufficient storage space for the data.

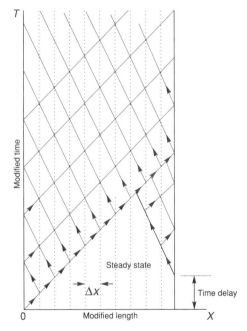

Fig. 8.6 Grid points of characteristics: double-inlet disturbance

With contraflow, a single disturbance will propagate forwards and backwards across the exchanger until it can no longer be seen, unlike crossflow, where the disturbance will pass through the exchanger without propagating further disturbances in the exchanger. This is perhaps a reason for considering the two-pass cross-flow design for use as a gas-turbine recuperator (see Sections 1.12 and 3.6). Specific power rating for a two-pass design may lie somewhere between for one-pass cross-flow, and for contraflow.

Evaporators and condensers

Although the problem of transients in an exchanger with one infinite capacitance rate fluid may be a contraflow problem, it will be considered in Chapter 9 under transients in cross-flow, as similar problems arise on the faces of a cross-flow exchanger.

Curved characteristics (variable mass flows and temperature transients)

The transient temperature problem solved above assumed constant physical properties, constant heat transfer coefficients and constant mass flow rates, allowing that temperature dependence of physical properties will curve the characteristics, i.e. the previously solved temperature transient must remain small.

When mass flow rates change continuously, it is essential to keep track of velocities at each station; the concepts of fixed transit times no longer apply and the approach illustrated earlier is inadequate because the characteristic grid lines have become curved. Reynolds numbers have to be recalculated at each station to obtain new heat transfer coefficients, and new temperature-dependent physical properties are similarly required.

When a step change in mass flow rate occurs, the effect can be imagined with reference to Fig. 8.1, e.g. for the hot fluid a new value of transit time would mean a change in slope of the characteristic. As this new characteristic propagates into the exchanger, it chases a characteristic of different slope. There will exist a step discontinuity immediately the flow front enters.

An introduction to methods of computation with curved characteristics can be found in Ames (1965). A considerable amount of work with curved characteristics has been done in fluid flow, see e.g. Jeffrey and Taniuti (1964).

Papers by Carver (1980) and by Herron and Rosenberg (1966) describe analytical solutions of the problem of a step change in velocity of one fluid in a contraflow exchanger. The wall equation is missing in these works, however, the results may provide a useful check on numerical methods developed using the method of characteristics. Oscillations in finite-difference solutions and the means for adding dissipation are discussed by Carver.

Engineering decisions

A quick estimate of transients may be possible using the method of characteristics with constant coefficients to provide sufficient indication of the thermal response of an exchanger. As a change in mass flow rate changes the slope of the characteristics, two calculations can be made: one with the initial flow rate and one with the final flow rate. This may provide sufficient information to allow non-critical engineering decisions to be made.

A complete treatment involving continuous changes in mass flow rates introduces non-linear characteristics, and such a treatment does not yet seem to have been obtained. Mass flow rate changes also introduce changes in temperature, velocity and heat transfer coefficients, requiring recalculation of all variables at each grid mathpoint for any solution method used.

8.5 LAPLACE TRANSFORMS WITH NUMERICAL INVERSION

Contraflow and parallel-flow exchangers

The fundamental paper is that of Roetzel and Xuan (1992a) which examined transient response of simple exchangers excluding axial dispersion terms. Since then the main thrust has included plate-and-frame exchangers, serpentine tube panels and conventional shell-and-tube exchangers, all of which may require consideration of axial dispersion for design-critical applications.

Designing out dispersion

The several papers by Roetzel and others include axial dispersion terms for the fluids, which use the Peclet number as a means of representing this fine detail in the flow. Roetzel and Lee (1993) confirmed that for values of Pe > 40 computed results are virtually identical with those for plug flow. By induction the term with a value of Fo < 0.025 could also possibly be neglected, but some computed results would be required for validation. Earlier it was seen that small values of (A/L) help to minimise the effects of the second-differential term. Compact plate-fin and porous matrix exchangers have small channels, thus axial dispersion effects and possibly longitudinal conduction terms can be effectively designed out. For RODbaffle designs, Pe may be a function of the spacing between baffles, and transient axial dispersion may have to be considered.

The Roetzel approach using Laplace transforms is probably the most comprehensive method of solution for transient problems currently available and able to handle second-differential terms. First normalise length $(X = x/L)$, scale time $(T = t/\tau_h)$ and separate steady-state and perturbance sets of equations; then normalise the perturbance temperatures. This ensures that the initial conditions will transform trivially. Retaining only the axial conduction term for the solid wall, the perturbance equations become

$$\frac{\partial \theta_h}{\partial T} + \frac{\partial \theta_h}{\partial X} = -n_h(\theta_h - \theta_w)$$
$$\frac{\partial \theta_w}{\partial T} - \text{Fo}\frac{\partial^2 \theta_w}{\partial X^2} = +\frac{n_h}{R_h}(\theta_h - \theta_w) - \frac{1}{P}\cdot\frac{n_c}{R_c}(\theta_w - \theta_c) \quad (8.11)$$
$$P\frac{\partial \theta_c}{\partial T} - \frac{\partial \theta_c}{\partial X} = +n_c(\theta_w - \theta_c)$$

Boundary conditions for this problem are

$$\left.\begin{array}{l}\dfrac{\partial \theta_g}{\partial X} = 0 \\ \dfrac{\partial \theta_w}{\partial X} = 0\end{array}\right\} \text{ at } X = 0 \qquad \left.\begin{array}{l}\dfrac{\partial \theta_c}{\partial X} = 0 \\ \dfrac{\partial \theta_w}{\partial X} = 0\end{array}\right\} \text{ at } X = 1$$

The wall boundary conditions have been encountered previously in discussing the effects of longitudinal conduction in Chapter 3. But both fluid boundary conditions are new, and they do not correspond to those fixed temperature values assumed when obtaining steady-state solutions. There seems no alternative to this assumption for transients, but it raises the interesting question as to whether the same constraints should be applied to steady-state problems.

After Laplace transformation, the transient equations will be reduced to finite-difference form. When considering the end-stations it is simpler to eliminate terms in the differential equations which are zero before taking finite differences. Not listed above are the boundary conditions for the second-order conduction term containing the Fourier number. This would also have to be zero for the end-conditions.

Finite-difference solution with numerical inversion

Laplace transforms are taken of the set of equations (8.11) plus the initial and boundary conditions, which reduces the set to ordinary differential equations plus boundary conditions in phase space. The ordinary differential equations plus boundary conditions are then converted to the finite-difference representation for numerical solution in phase space, which is carried out by matrix inversion.

For a solution in real time, inversion of the solution requires either the Gaver-Stehfest algorithm (Jacquot *et al.* 1983), or the Fourier series approximation (Ichikawa and Kishima 1972). Roetzel and co-workers found that the Gaver-Stehfest inversion took very little computational time, but was not suitable for handling disturbances containing rapid oscillatory components. The Fourier series approximation was preferred in handling oscillations, but convergence was slow. This may be speeded up by using the fast Fourier transform (Crump 1976, Press *et al.* 1992).

Boundary conditions are required as functions of dimensionless time. They can be expressed in terms of combinations of functions that can be transformed, e.g. step, ramp, exponential and sine. Care in selecting the appropriate solution method may be necessary. For example, summation of infinite Fourier series does not represent square waveforms accurately; the overshoot remains finite at 18% at each change of amplitude, the Gibbs phenomenon (Mathews and Walker 1970). However, real transients in heat exchangers tend to be mathematically smooth.

The method is mathematically complex, and does not yet handle arbitrary changes in mass flow rates.

Plate-fin exchangers

For plate-fin exchangers the solution scheme adopted by Das and Roetzel (1995) does not allow for the important cross-conduction effect analysed by Haseler (1983), which is discussed in Chapter 11. Apart from this, the approach is well developed and powerful, and it has been applied to several exchanger configurations.

With plate-fin exchangers, extra design flexibility is available by fitting half-height channels at each end of the block.

Plate-and-frame exchangers

This exchanger configuration was not included in the class for direct sizing because the single flow channel between plates is too wide without carefully designed flow distributors. However, the exchanger has some similarity to the plate-fin design.

The fundamental differential equations for plate-and-frame exchangers can be rewritten to distinguish between flow in the end channels, which only transfer heat to or receive heat from a single surface, and the internal flow channels, which transfer heat to or receive heat from two surfaces. Typical equations are presented by Sharifi *et al.* (1995), which are suitable for rewriting in the notation of this text.

An additional complication with plate-and-frame exchangers exists under transient conditions: a phase-lag effect occurs because the disturbed inlet fluid arrives at the entry to channels with different delay times. The delay is caused by the necessary headering arrangements for the plate-and-frame exchanger. For fast response, U-type headering gives quicker response than Z-type headering (Das and Roetzel 1995). Dispersion modelling may also be required.

Shell-and-tube exchangers with small tube inclinations

Serpentine tube-bundles and helical-coil tube-bundles fall into the class of exchangers that can be directly sized as parallel or contraflow exchangers having cross-flow heat transfer coefficients, providing there are sufficiently many tubes passes.

228 TRANSIENTS IN CONTRAFLOW EXCHANGERS

For a small number of tube passes (less than 10 say), Hausen (1950, 1983) derived expressions for steady-state temperature distributions. For such serpentine tube-bundles, Roetzel and Xuan found the dispersion model to be effective at modelling flow maldistributions in the steady state. Extension of the dispersion model to transient conditions in serpentine exchangers has also been reported by Roetzel and Xuan (1992b, c).

8.6 DIRECT SOLUTION BY FINITE DIFFERENCES

The ultimate solution method involves solving the three conservation equations of balance of mass, momentum and energy simultaneously with the constitutive equations for a perfect gas, using the Lax Wendroff schemes (Ames 1969).

Ames (1965) comments: 'There is some justification for operating with finite differences on the original equations without reducing them to characteristic form.' The problem with direct solution by finite differences is that it is more difficult to keep track of the progress of the transient, and alternating-difference implicit methods may be necessary.

Sharifi *et al.* (1995) solved the governing partial differential equations using a number of numerical techniques, using backward differencing for the convection term, and evaluating different time marching schemes. Only transient temperatures disturbances were investigated. For a step input disturbance, it was found that the *explicit* method did not reach the exchanger outlet until the fluid itself arrived. With an *implicit* method, the differencing smoothed out the response and would produce an effect at the outlet instantaneously.

The effect is similar to solving the problem of a step change in temperature at the surface of a semi-infinite plate. The classic paper by Churchill (1988) on the role of mathematics in heat transfer reviews the correct solution of the equation for thermo-acoustic convection. The approach due to Roetzel which includes Peclet number terms would seem to overcome the problem.

Lakshmann and Potter (1984, 1990) claimed that their 'cinematic' model requires the least amount of computation to provide fast and accurate results. They assumed that the effect of temperature on densities and heat capacities was negligible, and as they also assumed that the wall equation could be neglected, their solution is less representative than the elementary solution using the method of characteristics given earlier in this chapter.

Ontko and Harris (1990) presented a finite difference solution to the transient contraflow problem, neglecting longitudinal conduction, but including thermal storage in the wall. A full description of the method by Ontko (1989) includes computer source listings. Only results for step changes in temperature and fluid flow have been presented, but the approach is obviously capable of development, and the authors claim that the computational method is easily programmed.

8.7 ENGINEERING APPLICATIONS

The thermal storage terms retained in the solid wall equation are sufficient to delay outlet response to an input disturbance to the order of minutes (or tens of minutes) in the case of the contraflow example given by Campbell and Rohsenow (1992). This is because the terms R_h and R_c are so large, which may introduce a control constraint due to thermal storage in the exchanger.

A recuperative gas-turbine propulsion system applied to a ship can produce very significant fuel savings of the order of 30%, leading to cost savings of $1.5 million per annum (Cownie 1993, Crisalli and Parker 1993). Under high speed manoeuvring conditions, control of the gas turbine must take into account thermal storage effects in the heat exchanger. Multiple units are desirable for manufacturing, for maintenance and for damage control. A plate-and-frame design derivative operating in contraflow has been developed (Crisalli and Parker 1993, Valenti 1995). These units operate with an effectiveness of over 0.88 at full power, and over 0.95 at low power. U-type headering arrangements have been adopted for fast response; see the analysis by Das and Roetzel (1995) of plate-and-frame exchangers.

Water-cooled compact plate-fin heat exchangers are also being developed as intercoolers (Crisalli and Parker 1993, Bannister *et al.* 1994); they are manufactured in segments.

8.8 CONCLUSIONS

1. The wall thermal storage equation in the exchanger must not be ignored, and any solution which neglects this effect should be disregarded.
2. Longitudinal conduction effects are always likely to be much more important in the solid wall than longitudinal conduction in gases and liquids (except liquid metals)
3. Axial dispersion terms and longitudinal conduction terms can be minimised at design time by selecting channels in the exchanger with small aspect ratios, i.e. small values of (A/L). For a gas-gas exchanger, flow area on the low pressure side should be proportionately larger than the flow area on the high pressure side, thus dispersion effects may be first encountered there.
4. Second-order differential terms in the set of three perturbance equations can be neglected for Pe > 40 and Fo > 0.025.
5. Three principal methods of solution exist: the method of characteristics, Laplace transformation with numerical inversion and direct solution by finite differences.
6. An assessment of modest temperature transients is possible with the method of characteristics using constant coefficients. This may be sufficient for engineering decisions in non-critical applications.

7. A complete treatment involving continuous changes in mass flow rates introduces non-linear characteristics, and does not yet seem to have been obtained. Mass flow rate changes also introduces changes in temperature, velocity and heat transfer coefficients, requiring recalculation of all variables at each grid mathpoint for any solution method used.
8. The most highly developed transient solution method currently available in one-dimensional modelling is the Laplace transform method with numerical inversion due to Roetzel and co-workers. This approach has also handled step changes in mass flow rate.
9. Direct solution by the full finite-difference approach has been developed to a lesser extent, as the method is compuationally intensive.
10. None of the papers scanned to date handle continuously varying mass flow rates; only step changes in flow rate have been calculated.

REFERENCES

Acrivos, A. (1956) Method of characteristics technique, application to heat and mass transfer problems, *Industrial and Engineering Chemistry*, **48**(4), 703–710.

Ames, W. F. (1965) *Nonlinear Partial Differential Equations in Engineering*, New York: Academic Press, Section 7. 25 onwards.

Ames, W. F. (1969) *Numerical Methods for Partial Differential Equations* New York: Nelson, Section 4-14.

Bannister, R. L., Cheruvu, N. S., Little, D. A. and McQuiggan, G. (1994) Turbines for the turn of the century, *Mechanical Engineering*, **116**(6), 68–75.

Beckman, L., Law, V. J., Bailey, R. V. and von Rosenberg, D. U. (1990) Axial dispersion for turbulent flow with a large radial heat flux, *American Institute of Chemical Engineers Journal*, **36**(4), 598–604.

Campbell, J. F. and Rohsenow, W. M. (1992) Gas turbine regenerators: a method for selecting the optimum plate-finned surface pair for minimum core volume, *International Journal of Heat and Mass Transfer*, **35**(12), 3441–3450.

Carver, M. B. (1980a) Pseudo characteristic method of lines solution of the conservation equations, *Journal of Computational Physics*, **35**(1), 57–76.

Carver, M. B. (1980b) Pseudo characteristic method of lines solution of the first order hyperbolic equation systems, *Mathematics and Computers in Simulation*, **22**(1), 30–35.

Chester, M. (1963) Second sound in solids, *Physical Review*, **131**, 2013–2015.

Churchill, S. W. (1988) The role of mathematics in heat transfer, *AIChE Symposium Series, Heat Transfer*, **84**(263), 1–13

Collatz, L. (1960) *The Numerical Treatment of Differential Equations*, Berlin: Springer Verlag.

Cownie, J. (1993) Aerospace R & D 90 years on, *Professional Engineering*, **6**(11), 17–19.

Crisalli, A. J. and Parker, M. L. (1993) Overview of the WR-21 intercooled recuperated gas turbine engine system. A modern engine for a modern fleet, *International Gas Turbine and Aeroengine Congress and Exposition, Cincinnati, OH, 24–27 May 1993, ASME Paper 93-GT-231*.

Crump, K. S. (1976) Numerical inversion of Laplace transforms using a Fourier series approximation, *Journal of the Association of Computing Machinery*, **23**(1), 89–96.

Das, S. K. and Roetzel, W. (1995) Dynamic analysis of plate exchangers with dispersion in both fluids, *International Journal of Heat and Mass Transfer*, **38**(6), 1127–1140.

Dzyubenko, B. V., Dreitser, G. A. and Asmantas, L. -V. A. (1990) *Unsteady Heat and Mass Transfer in Helical Tube Bundles*, New York: Hemisphere. (helically twisted tubes in a shell-and-tube exchanger.)

Fox, L. (1962) *Numerical Solution of Ordinary and Partial Differential Equations*, Oxford: Pergamon Press, Chs. 17, 18, 28.

Gvozdenac, D. D. (1990) Transient response of the parallel flow heat exchanger with finite wall capacitance, *Ingenieur-Archive (Archive of Applied Mechanics)*, **60**(7), 481–490.

Haseler, L. E. (1983) Performance calculation methods for multi-stream plate-fin heat exchangers, In: *Heat Exchangers Theory and Practice*, Eds. J. Taborek, G. F. Hewitt and N. Afgan, New York: Hemisphere/McGraw-Hill, pp. 495–506.

Hausen, H. (1950) *Wärmeübertragung im Gegenstrom, Gleichstrom und Kreuzstrom*, Berlin: Springer Verlag, 1950.

Hausen, H. (1970) Berechnung der Wärmeübertragung in Regeneratoren bei zeitlische veranderlichem Mengenstrom, *International Journal of Heat and Mass Transfer*, **13**, 1753–1766.

Hausen, H. (1983) *Heat Transfer in Counterflow, Parallel Flow and Crossflow*, New York: McGraw-Hill, 2nd English Edn.

Herron, D. H. and Rosenberg, D. U. von (1966) An efficient numerical method for the solution of pure convective transport – problems with split boundary conditions, *Chemical Engineering Science*, **21**, 337–342

Huang, Y. M. (1989) Study of unsteady flow in the heat exchanger by the method of characteristics, *ASME Pressure Vessels and Piping Conference, Honolulu, 23–27 July 1989, ASME Pressure Vessels and Piping Division Publication PVP* **156**, 55–62.

Ichikawa, S. and Kishima, A. (1972) Application of Fourier series techniques to inverse Laplace transforms, *Kyoto University Memoires*, **34**, 53–67.

Jacquot, R. G., Steadman, J. W. and Rhodine, C. N. (1983) The Gaver-Stehfest algorithm for approximate inversion of Laplace transforms, *IEEE Circuits and Systems Magazine*, **5**(1), 4–8.

Jaswon, M. A. and Smith, W. (1954) Countercurrent transfer processes in the non-steady state, *Proceedings of the Royal Society*, **225A**, 226–244.

Jeffrey, A. and Taniuti, T. (1964) *Non-Linear Wave propagation with Applications to Physics and Magnetohydrodynamics*, New York: Academic Press.

Kabelac, S. (1988) Zur Berechnung des Zeitverhaltens von Wärmeübertragern (Calculation of transients in heat exchangers); *Wärme und Stoffübertragung (Heat and Mass Transfer)*, **23**(6), 365–370.

Kahn, A. R., Baker, N. S. and Wardle, A. P. (1988) The dynamic characteristics of a countercurrent plate heat exchanger, *International Journal of Heat and Mass Transfer*, **31**(6), 1269–1278.

Kroeger, P. G. (1966) Performance deterioration in high effectiveness heat exchangers due to axial conduction effects, *Proceedings of the 1966 Cryogenics Engineering Conference, Advances in Cryogenic Engineering*, **12**, Paper E-5, 363–372.

Lakshmann, C. C. and Potter, O. E. (1984) Dynamic simulation of heat exchangers and fluidised beds, *12th Australian Chemical Engineering Conference CHEMECA 84*, Vol. 2, pp. 871–878.

Lakshmann, C. C. and Potter, O. E. (1990) Dynamic simulation of plate heat exchangers, ('cinematic' method), *International Journal of Heat and Mass Transfer*, **35**(5), 995–1002.

Luo, X. and Roetzel, W. (1995) Extended axial dispersion model for transient analysis of heat exchangers, *Proceedings 4th UK National Heat Transfer Conference, IMechE, 1995, Paper C510/141/95*, pp. 411–416.

Massoud, M. (1990) Numerical analysis of transients in parallel and counter flow heat exchangers, In: *Thermal Hydraulics of Advanced Heat Exchangers, ASME Winter Annual Meeting, 1990, ASME Nuclear Engineering Division Publication NE* **5**, 99–104

Mathews, J. and Walker, R. L. (1970) *Mathematical Methods of Physics*, Reading, MA: Addison-Wesley, pp. 98–99.

Myers, G. E., Mitchell, J. W. and Lindeman, C. F. Jr. (1970) The transient response of heat exchangers having an infinite capacitance rate fluid, (condensers, evaporators), *Transactions ASME, Journal of Heat Transfer*, **92**, 269–275.

Nesselman, K. (1928) Der Einfluss der Warmeverluste auf Doppelrohrwarmeaustauscher, *Zeitschrift gestalte Kalte-Industrie*, **35**, 62

Ontko, J. S. (1989) A parametric study of counterflow heat exchanger transients, *Report IAR 89-10. Institute for Aviation Research*, Wichita, KS: Wichita State University.

Ontko, J. S. and Harris, J. A. (1990) Transients in the counterflow heat exchanger, In: *Compact Heat Exchangers – a Festschrift for A. L. London (Eds. R. K. Shah, A. D. Kraus and D. Metzger, Hemisphere, pp. 531–548.*

Press, W. H. Teukolsky, S. A., Vettering, W. T. and Flannery, B. P. (1992) *Numerical Recipies in FORTRAN* Cambridge: Cambridge University Press, 2nd Edn, p. 963.

Roetzel, W. Das, S. K. and Luo, X. (1994) Measurement of the heat transfer coefficient in plate heat exchangers using a temperature oscillation technique, *International Journal of Heat and Mass Transfer*, **37** (Suppl. 1), 325–331.

Roetzel, W. and Lee, D. (1993) Experimental investigation of leakage in shell-and-tube heat exchangers with segmental baffles, *International Journal of Heat and Mass Transfer*, **36**, (15), 3765–3771.

Roetzel, W. and Xuan, Y. (1991) Dispersion model for divided-flow heat exchanger, (serpentine tubes), In: *Design and Operation of Heat Exchangers*, Eds. W. Roetzel, P. J. Heggs and D. Butterworth, *Proceedings of the EUROTHERM Seminar no. 18 Hamburg, 27 February–1 March 1991*, Berlin: Springer Verlag, pp. 98–110.

Roetzel, W. and Xuan, Y. (1992a) Transient response of parallel and counterflow heat exchangers, *Transactions ASME, Journal of Heat Transfer*, **114**(2), 510–511.

Roetzel, W. and Xuan, Y. (1992b) Transient behaviour of multipass shell-and-tube heat exchangers, (serpentine tubes), *International Journal of Heat and Mass Transfer*, **35**(3), 703–710.

Roetzel, W. and Xuan, Y. (1992c) Analysis of transient behaviour of multipass shell and tube heat exchangers with the dispersion model, (serpentine tubes), *International Journal of Heat and Mass Transfer*, **35**(1), 2953–2962.

Roetzel, W. and Xuan, Y. (1993) The effect of core longitudinal heat conduction on the transient behaviour of multipass shell-and-tube heat exchangers, *Heat Transfer Engineering*, **14**(1), 52–61.

Roetzel, W. and Xuan, Y. (1993) Transient behaviour of shell and tube exchangers, *Chemical Engineering and Technology*, **16**(5), 296–302.

Roetzel, W. and Xuan, Y. (1993) The effect of heat conduction resistances of tubes and shells on transient behaviour of heat exchangers, *International Journal of Heat and Mass Transfer*, **36**(16), 3967–3973.

Roetzel, W. and Xuan, Y. (1993) Dynamics of shell-and-tube heat exchangers to arbitrary temperature and step flow variation, *American Institute of Chemical Engineers Journal*, **39**(3), 413–421

Romie, F. E. (1984) Transient response of the counterflow heat exchanger, *Transactions ASME, Journal of Heat Transfer*, **106**, 620–626.

Romie, F. E. (1985) Transient response of the parallel flow heat exchanger, *Transactions ASME, Journal of Heat Transfer*, **107**, 727–730.

Shah, R. K. (1981) The transient response of heat exchangers, In: *Heat Exchangers Thermal-Hydraulic Fundamentals and Design*, Eds. S. Kakac, A. E. Bergles, and F. Mayinger, Washington: Hemisphere, pp. 915–953.

Sharifi, F., Golkar Narandji, M. R. and Mehravaran, K. (1995) Dynamic simulation of plate heat exchangers, *International Communications in Heat and Mass Transfer*, **22**(2), 213–225.

Spang, B., Xuan, Y. and Roetzel, W. (1991) Thermal performance of split-flow heat exchangers, (serpentine tubes), *International Journal of Heat and Mass Transfer*, **34**(3), 863–874.

Stehfest, H. (1970) Numerical inversion of Laplace transforms, *Communications of the Association for Computing Machinery*, **13**, 47–49.

Tan, K. S. and Spinner, I. H. (1978) Dynamics of a shell-and-tube heat exchanger with finite tube-wall heat capacity and finite shell-side resistance, *Industrial and Engineering Chemistry, Fundamentals*, **17**(4), 353–358.

Tan, K. S. and Spinner, I. H. (1984) Numerical methods of solution for continuous countercurrent processes in the nonsteady state: Model equations and development of numerical methods and algorithms, Application of numerical methods, *American Institute of Chemical Engineers Journal*, **30**(5), 770–779, 780–786.

Tan, K. S. and Spinner, I. H. (1991) Approximate solutions for transient response of a shell and tube heat exchanger, *Industrial and Engineering Chemistry Research*, **30** 1639–1646.

Taylor, G. I. (1954) The dispersion of matter in turbulent flow through a pipe, *Proceedings of the Royal Society, Series A*, **223**, 446–468.

Valenti, M. (1995) A turbine for tomorrow's Navy, *Mechanical Engineering*, **117**(9), 70–73.

Wang, C. C. and Liao, N. S. (1989) Transient response of a double-pipe condenser to change in coolant flow rate, *International Communications in Heat and Mass Transfer*, **16**(3) 325–334.

Willmott, A. J. (1968a) Operation of Cowper stoves under conditions of variable flow, *Journal of the Iron and Steel Institute*, **206**, 33–38.

Willmott, A. J. (1968b) Simulation of a thermal regenerator under conditions of variable mass flow, *International Journal of Heat and Mass Transfer*, **11**, 1105–1116.

Willmott, A. J. (1969) The regenerative heat exchanger computer representation, *International Journal of Heat and Mass Transfer*, **12**, 997–1014.

Wolf, J. (1964) General solution of the equations for parallel flow multichannel heat exchangers, *International Journal of Heat and Mass Transfer*, **7**, 901–919

Xuan, Y. and Roetzel, W. (1993a) Dynamics of shell-and-tube heat exchangers to arbitrary temperature and step flow variations, *AIChE Journal*, **39**(3), 413–421.

Xuan, Y. and Roetzel, W. (1993b) Stationary and dynamic simulation of multipass shell and tube heat exchangers with the dispersion model for both fluids, (serpentine tubes), *International Journal of Heat and Mass Transfer*, **36**(7), 4221–4231.

Xuan, Y., Spang, B. and Roetzel, W. (1991) Thermal analysis of shell and tube exchangers with divided flow pattern, *International Journal of Heat and Mass Transfer*, **34**(3), 853–861.

9

TRANSIENTS IN CROSS-FLOW EXCHANGERS

9.1 SOLUTION METHODS

Fewer, and somewhat different, solutions are presently available for cross-flow compared with contraflow; they are reviewed in Section 9.4.

9.2 FUNDAMENTAL EQUATIONS

Transient equations for cross-flow are developed in Appendix A, neglecting longitudinal conduction in the fluids. For solution it is preferable to use θ for temperature, leaving T for dimensionless time, before transforming and normalising to canonical form.

Summarising the fundamental equations in tabular form

$$\frac{\partial T_h}{\partial t} + u_h \frac{\partial T_h}{\partial x} = -\left(\frac{\alpha_h S}{\dot{m}_h C_h}\right) \frac{u_h}{L_x} (T_h - T_w)$$

$$\frac{\partial T_w}{\partial t} - \hat{\kappa}_x \frac{\partial^2 T_w}{\partial x^2} - \hat{\kappa}_y \frac{\partial^2 T_w}{\partial y^2} = +\left(\frac{\alpha_h S}{M_w C_w}\right)(T_h - T_w) - \left(\frac{\alpha_c S}{M_w C_w}\right)(T_w - T_c)$$

$$\frac{\partial T_c}{\partial t} + u_c \frac{\partial T_c}{\partial y} = +\left(\frac{\alpha_c S}{\dot{m}_c C_c}\right) \frac{u_c}{L_y} (T_w - T_c)$$

(9.1)

For the special case of constant physical properties and constant mass flow rates, the equations can be similarly compacted as was done for contra-flow in Chapter 8.

Transient solutions are of no value when the wall thermal storage is neglected. It may be acceptable to neglect the second order conduction terms if a reduction factor has previously been applied to the mean temperature difference in cross-flow, or if the steady-state mean temperature difference has been calculated by allowing for the conduction terms.

Before examining possible approaches, it is appropriate to bring the equations into canonical form by scaling, normalisation, transformation and separation of steady-state and perturbance temperatures, as was done for the contraflow heat exchanger.

Scaling and normalisation

Referring everything to hot gas residence time τ_h,

with scaling $\quad T = \dfrac{t}{\tau_h} \quad$ then $\quad \dfrac{\partial(\ )}{\partial t} = \dfrac{\partial(\ )}{\partial T} \cdot \dfrac{1}{\tau_h}$

and normalisation $\quad X = \dfrac{x}{L_x} \quad$ giving $\quad \dfrac{\partial(\ )}{\partial x} = \dfrac{\partial(\ )}{\partial X} \cdot \dfrac{1}{L_x} \quad$ over $\ (0 \leq X \leq 1)$

$\quad Y = \dfrac{Y}{L_y} \quad$ giving $\quad \dfrac{\partial(\ )}{\partial y} = \dfrac{\partial(\ )}{\partial Y} \cdot \dfrac{1}{L_y} \quad$ over $\ (0 \leq Y \leq 1)$

Since $\quad u_h = \dfrac{L_x}{\tau_h} \quad$ then residence mass $\quad \tilde{m}_h = \dot{m}_h \tau_h$

$\quad u_c = \dfrac{L_y}{\tau_c} \quad$ then residence mass $\quad \tilde{m}_c = \dot{m}_c \tau_c$

Equations (9.1) now become

$$\dfrac{\partial \theta_h}{\partial T} + \dfrac{\partial \theta_h}{\partial X} = -n_h(\theta_h - \theta_w)$$

$$\dfrac{\partial \theta_w}{\partial T} - \mathrm{Fo}_x \cdot \dfrac{\partial^2 \theta}{\partial X^2} - \mathrm{Fo}_y \cdot \dfrac{\partial^2 \theta_w}{\partial Y^2} = +\dfrac{n_h}{R_h}(\theta_h - \theta_w) - \dfrac{1}{P} \cdot \dfrac{n_c}{R_c}(\theta_w - \theta_c)$$

$$P\dfrac{\partial \theta_c}{\partial T} + \dfrac{\partial \theta_c}{\partial Y} = +n_c(\theta_w - \theta_c)$$

(9.2)

where Fo_x, Fo_y are Fourier number, and $P = \dfrac{\tau_c}{\tau_h}$ is the ratio of transit times

Perturbance equations

Steady-state and perturbance temperatures are now separated by assuming that the transient temperatures θ are the sum of steady-state values $\bar{\theta}$ and perturbance values $\hat{\theta}$, (permissible for linear theory),

$$\theta_h = \bar{\theta}_h + \hat{\theta}_h$$
$$\theta_w = \bar{\theta}_w + \hat{\theta}_w$$
$$\theta_c = \bar{\theta}_c + \hat{\theta}_c$$

The coupled equations separate into two sets of three equations.

Steady-state temperatures

$$+\frac{\partial \bar{\theta}_h}{\partial X} = -n_h(\bar{\theta}_h - \bar{\theta}_w)$$

$$-\mathrm{Fo}_x \frac{\partial^2 \bar{\theta}_w}{\partial X^2} - \mathrm{Fo}_y \frac{\partial^2 \bar{\theta}_w}{\partial Y^2} = +\frac{n_h}{R_h}(\bar{\theta}_h - \bar{\theta}_w) - \frac{1}{P}\cdot\frac{c_c}{R_c}(\bar{\theta}_w - \bar{\theta}_c) \quad (9.3)$$

$$+\frac{\partial \bar{\theta}_c}{\partial Y} = +n_c(\bar{\theta}_w - \bar{\theta}_c)$$

Perturbance temperatures

$$\frac{\partial \hat{\theta}_h}{\partial T} + \frac{\partial \hat{\theta}_h}{\partial X} + = -n_h(\hat{\theta}_h - \hat{\theta}_w)$$

$$\frac{\partial \hat{\theta}_w}{\partial T} - \mathrm{Fo}_x\cdot\frac{\partial^2 \hat{\theta}_w}{\partial X^2} - \mathrm{Fo}_y\cdot\frac{\partial^2 \hat{\theta}_w}{\partial Y^2} = +\frac{n_h}{R_h}(\hat{\theta}_h - \hat{\theta}_w) - \frac{1}{P}\cdot\frac{n_c}{R_c}(\hat{\theta}_w - \hat{\theta}_c)$$

$$P\frac{\partial \hat{\theta}_c}{\partial T} + \frac{\partial \hat{\theta}_c}{\partial Y} = +n_c(\hat{\theta}_w - \hat{\theta}_c)$$

(9.4)

The perturbance equations (9.4) are similar to the full transient equations, but we can now work with *normalised disturbances* starting from zero initial values by redefining

$$0 = \frac{\hat{\theta} - \hat{\theta}_{c1}}{\hat{\theta}_{h1} - \hat{\theta}_{c1}} = \frac{\hat{\theta}}{\hat{\theta}_{h1}} \quad \text{giving the substitution } \hat{\theta} = \theta\hat{\theta}_{h1} \text{ in equations (9.4)}$$

238 TRANSIENTS IN CROSS-FLOW EXCHANGERS

After the perturbance temperature distributions have been determined, the complete temperature transient can be assembled as the sum of actual steady-state values and actual perturbance values.

9.3 METHOD OF CHARACTERISTICS

Characteristics (transformation)

Neglecting wall conduction terms and leaving the wall equation aside for the moment, as it depends on time alone, there obtains from the perturbance equations (9.4)

$$\frac{\partial \hat{\theta}_h}{\partial T} + \frac{\partial \hat{\theta}_h}{\partial X} = -n_h(\hat{\theta}_h - \hat{\theta}_w)$$

$$\frac{\partial \hat{\theta}_c}{\partial T} + \frac{\partial \hat{\theta}_c}{\partial Y} = +n_c(\hat{\theta}_w - \hat{\theta}_c)$$

(9.5)

Dropping the circumflex, the problem reduces to finding (α, β) such that

$$\frac{d\theta_h}{d\alpha} = -n_h(\theta_h - \theta_w)$$

$$\frac{d\theta_c}{d\beta} = +n_c(\theta_w - \theta_c)$$

There are three co-ordinates (X, Y, T) and only two equations available. With co-ordinate transformations

$$\alpha = AX + BY - T$$
$$\beta = CX + DY - T$$

then

$$\left(\frac{\partial \alpha}{\partial X} = A\right), \quad \left(\frac{\partial \alpha}{\partial Y} = B\right), \quad \left(\frac{\partial \alpha}{\partial T} = -1\right)$$

$$\left(\frac{\partial \beta}{\partial X} = C\right), \quad \left(\frac{\partial \beta}{\partial Y} = D\right), \quad \left(\frac{\partial \beta}{\partial T} = -1\right)$$

so that

$$\frac{\partial \theta_h}{\partial X} = \left(\frac{\partial \theta_h}{\partial \alpha}\right) A + \left(\frac{\partial \theta_h}{\partial \beta}\right) C \qquad \frac{\partial \theta_c}{\partial Y} = \left(\frac{\partial \theta_c}{\partial \alpha}\right) B + \left(\frac{\partial \theta_c}{\partial \beta}\right) D$$

$$\frac{\partial \theta_h}{\partial T} = \left(\frac{\partial \theta_h}{\partial \alpha}\right)(-1) + \left(\frac{\partial \theta_h}{\partial \beta}\right)(-1) \qquad \frac{\partial \theta_c}{\partial T} = \left(\frac{\partial \theta_c}{\partial \alpha}\right)(-1) + \left(\frac{\partial \theta_c}{\partial \beta}\right)(-1)$$

giving

$$\frac{\partial \theta_h}{\partial T} + \frac{\partial \theta_h}{\partial X} = \left(\frac{\partial \theta_h}{\partial \alpha}\right)(A-1) + \left(\frac{\partial \theta_h}{\partial \beta}\right)(C-1)$$

$$P \frac{\partial \theta_c}{\partial T} + \frac{\partial \theta_c}{\partial Y} = \left(\frac{\partial \theta_c}{\partial \alpha}\right)(B-P) + \left(\frac{\partial \theta_c}{\partial \beta}\right)(D-P)$$

To recover only $\left(\dfrac{d\theta_h}{d\alpha}\right)$ and $\left(\dfrac{d\theta_c}{d\beta}\right)$, we have to require that

$$(A-1) = 1, \quad (C-1) = 0, \quad (B-P) = 0, \quad (D-P) = 1$$

thus $\quad A = 2, \quad C = 1, \quad B = P, \quad D = 1 + P \quad$ and

$$\alpha = 2X + PY - T$$
$$\beta = X + (1+P)Y - T \qquad (9.6)$$

which are the set of simplifying transformations used in the classical analytical solution without the wall equation by Evans and Smith (1963). The wall equation becomes

$$\frac{\partial \theta_w}{\partial T} = \frac{\partial \theta_w}{\partial \alpha}\left(\frac{\partial \alpha}{\partial T}\right) + \frac{\partial \theta_w}{\partial \beta}\left(\frac{\partial \beta}{\partial T}\right) = \frac{\partial \theta_w}{\partial \alpha}(-1) + \frac{\partial \theta_w}{\partial \beta}(-1)$$

Thus

$$\frac{\partial \theta_w}{\partial T} = -\frac{\partial \theta_w}{\partial \alpha} - \frac{\partial \theta_w}{\partial \beta}$$

$$= \frac{\partial \theta_w}{\partial \gamma} \quad \text{say}$$

The original perturbed equations for the fluids (9.4) without the longitudinal conduction terms now become

$$\left. \frac{d\theta_h}{d\alpha} \right|_\beta = -n_h(\theta_h - \theta_w)$$

$$\left. \frac{d\theta_w}{d\gamma} \right|_{\alpha,\beta} = +\frac{n_h}{R_h}(\theta_h - \theta_w) - \frac{1}{P} \cdot \frac{n_c}{R_c}(\theta_w - \theta_c) \quad (9.7)$$

$$\left. \frac{d\theta_c}{d\beta} \right|_\alpha = +n_c(\theta_w - \theta_c)$$

Equations (9.7) can be drawn on a normalised grid representing the cross-section of the cross-flow exchanger. As time is a parameter of (α, β) such a grid exists only for the chosen value of T. At the next time interval, new equations (9.7) must exist on the grid.

Transients on the inlet faces

Boundary conditions along $X=0$ and $Y=0$ correspond to separate solution of the 'pseudo' transient evaporator and condenser problems. These may give a feel for the transient in much the same way as was done for steady-state longitudinal conduction in Section 3.8.

The solution is not exact physically, for it assumes that the plane material on each inlet face is not affected by the temperature of the inner material – a consequence of assuming zero longitudinal conduction. Notice that the concepts of $\dot{m}_c$ and C_c were retained, whereas for proper evaporation, the original solid wall equation would have to be rewritten as

$$\frac{\partial \theta_w}{\partial t} - \kappa_x \frac{\partial^2 \theta_w}{\partial x^2} = +\frac{\alpha_h S}{M_w C_w}(\theta_h - \theta_w) - \frac{\alpha_c S}{M_w C_w}(\theta_w - \theta_h)$$

$$= +\frac{n_h}{R_h \tau_h}(\theta_h - \theta_w) - \frac{\alpha_c}{\alpha_h} \cdot \frac{n_h}{R_h \tau_h}(\theta_w - \theta_c)$$

9.4 REVIEW OF EXISTING SOLUTIONS

A number of published solutions of the cross-flow problem exist, and it seemed appropriate to consider them together, collected under solution types.

Integral approach

Myers *et al.* (1967) develop an approximate integral solution for the associated problems of cross-flow, evaporation and condensation in which one fluid is mixed.

Method of characteristics

In a further paper, Myers *et al.* (1970) provide a solution for evaporation or condensation based on the method of characteristics, and they compare it with Rizika's (1956) exact solution.

Laplace transform methods

Using a single Laplace transform, Romie (1983) solved the transient response problem for gas-gas cross-flow exchangers with neither gas mixed. However, it was necessary to neglect fluid thermal capacity terms, and a complete solution for all fluids is thus not achieved.

Chen and Chen (1991) also neglect fluid thermal capacity terms, restricting their solution to gas-gas (unmixed/unmixed) heat exchangers.

Threefold Laplace transforms

Spiga and Spiga (1987, 1992) have published two solutions to the unmixed-unmixed problem for arbitrary inlet temperature disturbances only. The earlier paper is on gas-gas cross-flow, and fluid thermal capacity terms are neglected. The later paper includes fluid thermal capacity terms. In both cases the solution approach requires threefold Laplace transformation.

Finite-difference method

A interesting finite-difference solution is due to Yamashita *et al.* (1978) for one-pass cross-flow with both fluids unmixed. Only a step change in temperature is considered. Although computational time is heavy, the mathematical complication with analytical solutions may mean that computation is favoured when considering transients in two-pass cross-flow.

Laplace transformation with numerical solution

The approach developed by Roetzel and co-workers for contraflow exchangers does not yet appear to have been used on the cross-flow problem.

9.5 ENGINEERING APPLICATIONS

Transients in two-pass cross-flow

The problem in (unmixed/unmixed) cross-flow is made more difficult because the intermediate temperature profile between passes is not flat, as inspection of Fig. 3.16 will show. Much more computational effort would be required to calculate the transient response, as each half of the exchanger has to be iterated in turn at each time interval until matching intermediate temperatures are obtained. Nevertheless, this problem is of some importance, as examination of Fig. 1.10 may suggest.

9.6 CONCLUSIONS

The problem of transients in cross-flow involving mass flow rates has not been touched, and there is scope for extending existing transient temperature solutions for one-and two-pass cross-flow.

REFERENCES

Ames, W. F. *Numerical Methods for Partial Differential Equations*, London: Nelson.
Chen, H. -T. and Chen, K. -C. (1991) Simple method for transient response of gas-to-gas cross-flow heat exchangers with neither gas mixed, *International Journal of Heat and Mass Transfer*, **24**(11), 2891–2898.
Chen, H. -T. and Chen, K. -C. (1992) Transient response of cross-flow heat exchangers with finite wall capacitance – technical note, *Trans ASME, Journal of Heat Transfer*, **114**(3), 752–755.
Collatz, L. (1960) *The Numerical Treatment of Differential Equations*, Berlin: Springer Verlag, 3rd edn.
Courant, R. and Hilbert, D. (1953, 1962) *Methods of Mathematical Physics*, Vols. I, II, New York: Wiley.
Evans, F. and Smith, W. (1963) Cross-current transfer processes in the non-steady state, *Proceedings of the Royal Society*, **272A**, 241–269.
Fox, L. (1962) *Numerical Solution of Ordinary and Partial Differential Equations*, Oxford: Pergamon Press, Chs. 17, 18, 28.
Garabedian, P. (1964) *Partial Differential Equations*, New York: Wiley.
Gvodzenac, D. D. (1986) Analytical solution of the transient response of gas-to-gas crossflow heat exchanger with both fluids unmixed, *Transactions ASME, Journal of Heat Transfer*, **108**, 722–727.
Gvozdenac, D. D. (1991) Dynamic response of the crossflow heat exchanger with finite wall capacitance, *Warme-und Stoffubertragung*, **26**, 207–212.
Hausen, H. (1950) *Wämeübertragung im Gegenstrom, Gleichstrom und Kreuzstrom*, Berlin: Springer Verlag, pp. 213–229. (Second edition, 1976.)
Jacquot, R. G., Steadman, J. W. and Rhodine, C. N. (1983) The Gavir-Stehfest algorithm for approximate inversion of Laplace transforms, *IEEE Circuits System Management*, **5**(1), 4–8.
Jain, M. K. (1979) *Numerical Solution of Differential Equations*, New Delhi: Wiley Eastern.
McAdams, W. H. (1954) *Heat Transmission*, New York: McGraw-Hill, Table 10–6, p. 274.
Myers, G. E., Mitchell, J. W. and Norman, R. F. (1967) The transient response of crossflow heat exchangers, evaporators, and condensers, *Transactions ASME, Journal of Heat Transfer*, **89**, 75–80.

Myers, G. E., Mitchell, J. W. and Lindeman, C. F. (1970) The transient response of heat exchangers having an infinite capacitance rate fluid, *Transactions ASME, Journal of Heat Transfer*, **92**, 269–275.

Rizika, J. W. (1956) Thermal lags in flowing incompressible fluid systems containing heat capacitors, *Transactions ASME*, **78**, 1407–1413.

Romie, F. E. (1983) Transient response of gas-to-gas crossflow heat exchangers with neither gas mixed, *Transactions ASME, Journal of Heat Transfer*, **105**, 563–570.

Romie, F. E. (1994) Transient response of crossflow heat exchangers with zero core thermal capacitance, *Transactions ASME, Journal of Heat Transfer*, **116**, 775–777.

San, J. -Y. (1993) Heat and mass transfer in a two-dimensional cross-flow regenerator with a solid conduction effect, *International Journal of Heat and Mass Transfer*, **36**(3), 633–643.

Spiga, G. and Spiga, M. (1987) Two dimensional transient solutions for crossflow heat exchangers with neither gas mixed, *ASME, Journal of Heat Transfer*, **109**, 281–286.

Spiga, M. and Spiga, G. (1988) Transient temperature fields in crossflow heat exchangers with finite wall capacitance, *ASME, Journal of Heat Transfer*, **110**, 49–53.

Spiga, M. and Spiga, G. (1992) Step response of the crossflow heat exchanger with finite wall capacitance, *International Journal of Heat and Mass Transfer*, **35**(2), 559–565.

Yamashita, H., Izumi, R. and Yamaguchi, S. (1978) Analysis of the dynamic characteristics of cross-flow heat exchangers with both fluids unmixed, *Bulletin of the JSME*, **21**(153), 479–485. (On the transient response to a step change in the inlet temperature.)

10

SINGLE-BLOW TESTING AND REGENERATORS

10.1 ANALYTICAL AND EXPERIMENTAL BACKGROUND

The first part of this chapter contains solution of the regenerator transient equations for zero longitudinal conduction and outlines the approach for handling longitudinal conduction; the second part contains a short introduction to regenerators.

Since the pioneer work of Anzelius (1926), Nusselt (1927), Schumann (1929) and Hausen (1929), a variety of theoretical solutions for thermal response of an initially isothermal matrix subject to specific forms of inlet fluid temperature disturbances have been formulated. This paper provides a single general solution accommodating standard inlet temperature disturbances, which is suitable for determining the heat transfer performance of certain matrix geometries near ambient temperature.

The regenerator transient test technique may be used to determine heat transfer in crushed rock beds, plate-fin heat exchanger cores, tube-banks, etc., where a single surface heat transfer coefficient is to be determined.

In its simplest form, the experimental rig used in such applications comprises a duct containing the high thermal capacity test matrix, through which gas is arranged to pass at a steady mass flow rate. Fast response platinum resistance thermometers, placed immediately before and after the test matrix, record timewise variation in inlet and outlet temperatures. Upstream of the test matrix, a fast response electrical resistance heater is used to to impose a known temperature disturbance on the initially isothermal gas stream. Heat transfer performance of the test matrix is determined from the change in shape of outlet temperature response with respect to inlet temperature disturbance, through comparison with a mathematical model of the system (Smith and Coombs 1972, Smith and King 1978).

In precise testing it is essential that experimental conditions match the mathematical model in use. A different test matrix may require different mathematical models, e.g. a plate-fin exchanger core may require inclusion of a term for longitudinal conduction. The theory presented below is suitable for testing tube banks in cross-flow in which the longitudinal conduction term is absent.

10.2 PHYSICAL ASSUMPTIONS

1. Thermal and physical properties of the gas and matrix are independent of temperature (implying that the temperature change of inlet disturbance is small compared with the absolute temperature of the gas).
2. Thermal conductivity of the matrix material is infinitely large in the direction normal to the flow, and infititely small in the direction parallel to the flow (implying negligible heat loss from the test matrix casing, implying near-ambient test conditions, with a small axial conduction path within the matrix itself),
3. Thermal capacity of the gas in the matrix at any instant ($\tilde{m}_g$) is small compared with the thermal capacity of the matrix itself (M_b), perhaps implying air as test fluid, copper for test matrix.
4. Surface temperatures and bulk temperatures for the solid matrix during thermal transients are indistinguishable (implying Biot number Bi → 0, i.e. solid must be thin and/or have high thermal conductivity).
5. Initial test conditions should be isothermal. Circumstances may require departure from the above conditions, e.g. the requirement to test at much higher temperatures may introduce heat loss from the matrix surface and therefore transverse temperature gradients within the test matrix and the gas. Then the bulk temperature within the solid may have to be related to surface temperatures, and longitudinal diffusion within the gas may become significant. Additional terms in the equations will then be required.

For the physical assumptions specified, a variety of mathematical attacks on the transient test technique have been published for different input disturbances. It seems useful to collect them in a single general solution capable of accepting the range of input disturbances listed in Table 10.1. The analysis given is for isothermal initial conditions in the absence of longitudinal conduction.

There is, however, no *a priori* reason why the method of characteristics could not be developed for this application, instead of the approach via Laplace transforms which follows, see Section 8.1.

Table 10.1 Experimental papers

Input Disturbance	Representative Experimental papers	Year
Sine wave	Bell and Katz	1949
Sine wave	Meek	1962
Sine wave	Hart and Szomanski	1965
Step	Mondt	1961
Step	Pucci *et al.*	1967
Square wave	Close	1965
Exponential	Smith and Coombs	1972
Exponential	Liang and Yang	1975

10.3 THEORY

Coupled fluid and solid equations

Although the single-blow technique is for obtaining the heat transfer coeficient between fluid and solid, internal conduction in the solid also exists. Thus two subscripts are involved in describing the solid, b for bulk properties, and s for surface properties. The fluid is best chosen to be a perfect gas, and the symbol g is used for the fluid. For transient solutions $(\theta = T - T_{min})$ is used for temperature.

Without the longitudinal conduction term in the solid, heat transfer to a gas flowing through a porous prism is described by a simplified form of the equations given in Appendix A and involving one fluid only,

Fluid

$$\frac{\partial \theta_g}{\partial t} + u_g \frac{\partial \theta_g}{\partial x} = -\frac{n_g}{\tau_g}(\theta_g - \theta_s) = -\frac{n_g u_g}{L}(\theta_g - \theta_s) \qquad (10.1)$$

Solid

$$\frac{\partial \theta_b}{\partial t} = +\frac{n_g}{R_g \tau_g}(\theta_g - \theta_s) = +\frac{n_g}{R_g} \cdot \frac{u_g}{L}(\theta_g - \theta_s) \qquad (10.2)$$

The next step is non-dimensionalisation and scaling. In the overall notation scheme X and T would normally be used, but this would take the notation away from that normally favoured by workers in Laplace transforms and it was considered preferable to use ξ and τ. The residence time (τ_g) in equations (10.1) and (10.2) will not be used again in this analysis, we shall instead work with the right-hand expressions of these equations, and there should be no confusion. It is also true that the local value $(Ntu = n_g)$ is the only value of Ntu in this solution, from which the heat transfer correlation would eventually be constructed,

Non-dimensional scaling of length $\xi = \text{Ntu}\left(\frac{x}{L}\right)$

and non-dimensional modification and scaling of time

$$\tau = \left(\frac{\text{Ntu}}{R_g}\right)\left(\frac{ut-x}{L}\right)$$

then as $\text{Bi} \to 0$ we may put $\theta_s \to \theta_b$ with temperature excesses $B = (\theta_b - \theta_i)$ and $G = (\theta_g - \theta_i)$ over some initial value θ_i, and equations (10.1) and (10.2) become

$$\frac{\partial G}{\partial \xi} = (B - G) \quad \text{fluid} \quad (10.3)$$

$$\frac{\partial B}{\partial \tau} = (G - B) \quad \text{solid} \quad (10.4)$$

Solution of basic equations

Taking Laplace transforms

$$L\left\{\frac{\partial G}{\partial \xi}\right\} = \frac{d\hat{G}}{d\xi}$$

$$L\left\{\frac{\partial B}{\partial \xi}\right\} = s\hat{B} - B(\xi, 0)$$

Term $B(\xi, 0)$ is the initial temperature distribution in the matrix. For isothermal conditions at the start of blow $B(\xi, 0) = 0$, which keeps the solution simple, see e.g. Kohlmayr (1968a), then

Fluid

$$\frac{d\hat{G}}{d\xi} = \hat{B} - \hat{G} \quad (10.5)$$

Solid

$$s\hat{B} = \hat{G} - \hat{B} \quad (10.6)$$

Combining equations (10.5) and (10.6) to obtain fluid temperatures

$$\frac{d\hat{G}}{d\xi} + \left(1 - \frac{1}{s+1}\right)\hat{G} = 0$$

which has the solution

$$\hat{G} = A\exp\left[\left(\frac{1}{s+1} - 1\right)\xi\right]$$

where A is to be determined from the boundary conditions.

Boundary conditions

At inlet $\xi = 0$

$$\hat{G}_1(0, s) = A = \hat{g}(s)$$

where $\hat{g}(s)$ is defined as the Laplace transform of the inlet fluid disturbance. Thus

$$\hat{G} = \hat{g}(s)\exp\left[\left(\frac{1}{s+1} - 1\right)\xi\right] \qquad (10.7)$$

At outlet $\xi = \text{Ntu}$

$$\hat{G}_2(\text{Ntu}, s) = \hat{g}(s)\exp\left[\left(\frac{1}{s+1} - 1\right)\text{Ntu}\right]$$

Inverse transforms

Applying inverse Lapalace transforms to outlet fluid temperature response

$$G_2 = \exp(-\text{Ntu})L^{-1}\left\{\exp\left(\frac{\text{Ntu}}{s+1}\right)\hat{g}(s)\right\}$$

$$= \exp(-\text{Ntu})\int_0^\tau \left[\frac{\delta(\sigma) + \exp(-\sigma)\text{Ntu}I_1(2\sqrt{\text{Ntu}.\sigma})}{\sqrt{\text{Ntu}.\sigma}}\right]G_1(\tau - \sigma)d\sigma$$

$$= \exp(-\text{Ntu})\left[\int_0^\tau G_1(\tau - \sigma)\delta(\sigma)d\sigma + \int_0^\tau G_1(\tau - \sigma)\frac{e^{-\sigma}\text{Ntu}I_1(2\sqrt{\text{Ntu}.\sigma})}{\sqrt{\text{Ntu}.\sigma}}d\sigma\right]$$

where the Dirac δ-function has the property of 'sifting out' the value of another integrand at time zero, then

$$G_2 = \exp(-\text{Ntu})\left[G_1(\tau) + \int_0^\tau G_1(\tau-\sigma)R(\sigma)d\sigma\right]$$

$$\text{where} \quad R(\sigma) = e^{-\sigma}\frac{\text{Ntu}.I_1(2\sqrt{\text{Ntu}.\sigma})}{\sqrt{\text{Ntu}.\sigma}}$$

With non-dimensional inlet disturbances D given in Table 10.2, the general solution for outlet fluid temperature response becomes

$$G^{\#} = \exp(-\text{Ntu})\left[D(\tau) + \int_0^\tau D(\tau-\sigma)R(\sigma)d\sigma\right] \quad (10.8)$$

Table 10.2 Inlet disturbance

Inlet disturbance	Non-dimensional $D(\tau)$	At $x=0$
Step	1	$0 \to 1$
Exponential	$1 - k\exp(-\tau/\tau^*)$	$\tau/\tau^* = t/t^*$
First harmonic	$a_0 + a_1\cos(\varpi\tau) + b_1\sin(\varpi\tau)$	$\varpi\tau = \omega t$

When solid temperatures are required, combining equations (10.6) and (10.7)

$$\hat{B} = \hat{g}(s)\frac{1}{s+1}\exp\left[\left(\frac{1}{s+1}-1\right)\xi\right] \quad (10.9)$$

At outlet $\xi = \text{Ntu}$

$$\hat{B}_2(\text{Ntu}, s) = \hat{g}(s)\frac{1}{s+1}\exp\left[\left(\frac{1}{s+1}-1\right)\text{Ntu}\right]$$

The tables of Laplace transform inversions given in Appendix E include instances that were not found in the mathematical literature. Applying inverse Laplace transforms to the outlet matrix temperature response

$$B_2 = \exp(-\text{Ntu})L^{-1}\left\{\frac{1}{s+1}\exp\left(\frac{\text{Ntu}}{s+1}\right)\hat{g}(s)\right\}$$

$$= \exp(-\text{Ntu})\int_0^t e^{-\sigma}I_0\left(2\sqrt{\text{Ntu}.\sigma}\right)G_1(\tau-\sigma)d\sigma$$

$$= \exp(-\text{Ntu})\int_0^t G_t(\tau-\sigma)P(\sigma)d\sigma$$

where $P(\sigma) = e^{-\sigma} I_0(2\sqrt{\text{Ntu}.\sigma})$

With non-dimensional inlet disturbances D given in Table 10.2, the general solution for outlet matrix temperature response becomes

$$B^{\#} = \exp(-\text{Ntu}) \int_0^t D(\tau - \sigma) P(\sigma) d\sigma \qquad (10.10)$$

Temperatures elsewhere in the regenerator may be found by inserting other values for ξ in equations (10.8) and (10.10) or by using fictitious values for L.

For the step input disturbance it is easily shown that the temperature difference (gas-solid) at the outlet is

$$(G^{\#} - B^{\#})_{step} = \exp(\text{Ntu}.\tau) I_o(2\sqrt{\text{Ntu}.\tau})$$

and that the slope of the outlet response at any point is

$$\frac{dG^{\#}}{d\tau} = \exp(-\text{Ntu}) R(\tau)$$

and that in terms of an independent parameter (a), the locus of maximum slope

$$\text{Ntu}.P(\tau) = (1 + \tau) R(\tau)$$

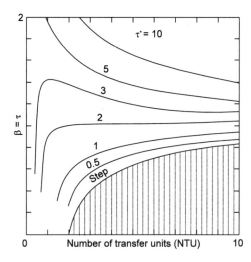

Fig. 10.1 Locus of maximum slope for exponential input

is given by

$$\tau = \frac{a\,I_0(a)}{2\,I_1(a)} - 1$$

subject to $2 \leq \text{Ntu} \leq \infty$ with $\text{Ntu} = (a^2/4\tau)$ and $\beta = (\tau/\text{Ntu})$. This last relationship was obtained in more complicated form by Kohlmayr (1968b), both expressions giving the identical curve shown shaded in Fig. 10.1.

Attempts to obtain a similar expression for the locus of maximum slope of the outlet response for exponential input disturbance leads to the condition

$$\int_0^t \exp\left(\frac{\sigma}{\tau^*}\right) R(\sigma)d\sigma = \frac{\tau^* R(\tau)}{\exp(-\tau/\tau^*)}$$

Evaluation of this expression was not carried out. Instead the position of maximum slope was determined numerically during evaluation of $G_{exp}^{\#}$ and the results are plotted in Fig. 10.1.

10.4 RELATIVE ACCURACY OF USING MATHEMATICAL OUTLET-RESPONSE CURVES IN EXPERIMENTATION

Comparison of methods

The Ntu value corresponding to a given experimental outlet response curve is determined through seeking the mathematical outlet response prediction which has the identical shape. Four techniques of comparison have been proposed:

- complete curve matching
- maximum slope
- initial rise
- phase angle and amplitude

The first three are appropriate to single-blow methods of testing. Complete curve matching for both step and exponential inlet disturbances may be by least squares fit or by using a direct optimisation simplex method; Spendley *et al.* (1962), Nelder and Meade (1965) and Parkinson and Hutchinson (1972). Maximum slope has been used by Locke (1950) and later by Howard (1964) with step inputs for the case of longitudinal conduction in the matrix.

For exponential input with zero longitudinal conduction, new Ntu versus locus of maximum slope curves are shown in Fig. 10.1 Practicable fast response heaters have exponential time constants around $\tau^* = 0.2$, giving a locus of maximum slope close to that for a step response. By choosing an inlet disturbance constant $\tau^* = 2.0$, an almost linear relationship between Ntu and $\tau_{max\ slope}$ may be obtained, and it also becomes possible to evaluate experimentally (with some resolution) values of Ntu down to about 1.0.

Figures 10.2 and 10.3 illustrate both step and exponential ($\tau^* = 0.2$) dimensionless response curves for zero longitudinal conduction calculated using equation (10.8). The initial rise technique proposed by Mondt and Siegla (1972) makes use of the fact that the intercept of the response to a step input at $\tau = 0$ has the value $\exp(-Ntu)$ from analytical solutions for both zero and infinite longitudinal conduction in the matrix (Mondt 1961); it is postulated that the same result will hold for intermediate values.

No heater has been devised which will produce a perfect step input (Kramers and Alberda 1953), and although output response curves for step and exponential input ($\tau^* = 0.2$) are virtually identical down to Ntu values of about 5.0, it is clear that the initial rise method should be avoided completely; the maximum slope method should be used only with knowledge of τ^*. The complete curve matching technique is the safest.

For first-harmonic responses with $a_0 = 1$, $a_1 = -1$, $b_1 = 0$ and $\varpi = 1.0$, Fig. 10.4 illustrates initial stages of steady-cyclic methods of testing, in which values of Ntu may be calculated either from the measurements of the ratio of amplitude of the varying fluid temperature at outlet to that of inlet, or alternatively from measurement of the phase lag between inlet and outlet fluid temperature variations.

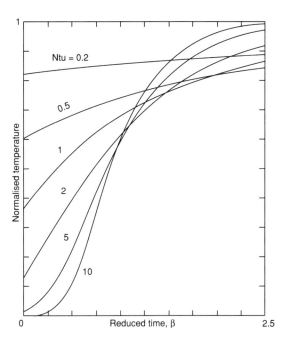

Fig. 10.2 Response from step input disturbances

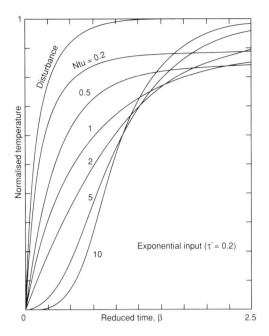

Fig. 10.3 Response from exponential input disturbances

A separate theoretical analysis may be used when steady-cyclic conditions have been attained, e.g. Bell and Katz (1949), Meek (1961) and Shearer (1962), who considered finite radial conductivity within the solid, and Stang and Bush (1972), who examined the case of longitudinal conduction within the matrix.

On precision of the cyclic method, Meek (1962) observed some apparent variation in measured heat transfer values against frequency, which he attributed to inaccuracies associated with very small downstream temperature amplitudes. Bell and Katz (1949) advise 10 heating cycles before taking measurements of amplitude and phase angle. For a given Ntu, Stang and Bush (1972) showed that one frequency exists at which best test results are produced for a given uncertainty in temperature measurement, and they recommend cyclic methods for values of Ntu for the difficult range $0.2 \leq$ Ntu < 5.0. However, from Fig. 10.4 it seems that there would be less difficulty in resolving exponential response curves using complete curve matching in the single-blow technique.

In all these methods involving use of Laplace transforms, the experimental input disturbance must follow, as near as is practicable, the form of the mathematical input disturbance. In single-blow testing this can be arranged using thyristor control of the fast-response electrical heater generating the input disturbance (Coombs 1970). This requirement reinforces the remarks made in Chapter 8 concerning the advantages of the method of characteristics in accommodating any input disturbance.

254 10 SINGLE-BLOW TESTING AND REGENERATORS

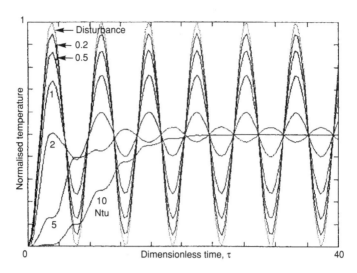

Fig. 10.4 Response from harmonic input disturbance

Generating theoretical response curves

In obtaining theoretical response curves, two methods are available to the investigator:

- direct application of numerical procedures to the physical problem
- mathematical derivation of integral expressions which are subsequently evaluated numerically

The direct method starting from the basic differential equations was favoured by Dusinberre, in a discussion to Coppage and London (1953). It has also been employed by Johnson (1948), by Howard (1964) and by other workers in an early form. With powerful computers, arbitrary shapes of input disturbance can now be handled, and presently this is the preferred approach.

A general solution for the second option is presented in this chapter in a more analytical form; this is to allow comparison of different experimental techniques, without which there would be no benchmark to compare the completely numerical method. Appendix E.2 provides a convenient method for dealing with the integration.

Additional effects which would complicate the canonical solution given in this chapter include

- longitudinal conduction in the solid   Stang and Bush (1972)
                                               Howard (1964)
- axial and longitudinal diffusion in the fluid   Amundson (1956)
- surface losses from the matrix exterior   Dabora (1957)
                                               Dabora et al. (1959)

- internal heat generation

- conduction into the solid interior

Clark *et al.* (1957, 1959, 1961)
Chiou and El-Walkil (1966)
Brinkley (1947)
Smith and Coombs (1972)
Meek (1961)

The most important effect is longitudinal conduction in the solid because high thermal conductivity in the solid matrix is desirable for accurate results. Conduction into the solid interior is less important because hollow sections can be used. In practice, Smith and Coombs (1972) found that little difference could be detected by replacing solid copper rods with hollow copper rods, providing the obvious adjustment is made in calculating the mass of the matrix.

10.5 CHOICE OF TEST METHOD

1. The initial rise method of determining Ntu is invalid for any practicable heater.
2. The maximum slope method requires accurate knowledge of heater *exponential* response time constant τ^*, and should be used with caution when determining values of Ntu<5.0.
3. Practical complications exist with the periodic method because Fourier analysis is required in order to extract the first harmonic from the inlet and outlet temperature waves.
4. Single-blow testing with complete curve matching by computer is recommended. Both inlet disturbance and outlet response curves have to be measured accurately for a fully numerical or an analytical/numerical method.
5. A fully numerical approach is desirable to eliminate 'tuning' of the electrical input heater to deliver exponential input disturbances. Both the direct finite-difference method, and the method of characteristics will accommodate arbitrary inlet disturbances close to the exponential form.

10.6 PRACTICAL CONSIDERATIONS

Earlier sections, directed at selection of the best method of testing, used only the simplified coupled equations (10.1) and (10.2).

Experimental testing requires some additional consideration of physical assumptions made in the analysis. In determining mean heat transfer coefficients in tube-bundles, the choice was of materials of construction (copper rods) for the matrix and Tufnol for the test section, to obtain physical properties best matching the mathematical assumptions made in the theory. The test fluid (ambient air) is close to a perfect gas, and the temperature rise introduced

256 10 SINGLE-BLOW TESTING AND REGENERATORS

into the flow during testing was kept small (e.g. 5–10 K) so as not to change bulk physical properties of the gas. For matrices which are connected in the longitudinal direction, the small temperature rise also helps minimise longitudinal conduction effects.

$$\frac{\partial \theta_g}{\partial t} + u_g \frac{\partial \theta_g}{\partial x} = -\frac{n_g}{\tau_g}(\theta_g - \theta_s) = -\frac{n_g u_g}{L}(\theta_g - \theta_s)$$

$$\frac{\partial \theta_b}{\partial t} = +\frac{n_g}{R_g \tau_g}(\theta_g - \theta_s) = +\frac{n_g}{R_g} \cdot \frac{u}{L}(\theta_g - \theta_s)$$

The fuller equations describing the single-blow process for a tube-bundle matrix are

$$\frac{\partial \theta_g}{\partial t} + u_g \frac{\partial \theta_g}{\partial x} = -\frac{n_g u_g}{L}(\theta_g - \theta_s) \qquad \text{(fluid)}$$

$$\frac{\partial \theta_b}{\partial t} = +\frac{n_g}{R_g} \cdot \frac{u_g}{L}(\theta_g - \theta_s) \qquad \text{(solid)}$$

$$\frac{\partial \theta_{r,x,t}}{\partial t} = \frac{\lambda_b}{\rho_b C_b}\left[\frac{\partial^2 \theta_{r,x,t}}{\partial r^2} + \frac{1}{r} \cdot \frac{\partial \theta_{r,x,t}}{\partial r}\right] \qquad \text{(transient conduction)}$$

$$\theta_{s(r,x,t)} = \theta_{a,x,t} \qquad \text{(surface temperature)}$$

$$\theta_{b(x,t)} = \frac{1}{\pi a^2}\int_0^a 2\pi r \theta_{r,x,t}\, dr \qquad \text{(bulk temperature)}$$

The solid equation relates the bulk temperature (θ_b) to the surface temperature (θ_s); the bulk temperature equation defines (θ_b) itself; and the surface temperature equation selects one solid temperature, surface temperature, from all possible solid temperatures ($\theta_{a,x,t}$).

The fuller set of equations is required when poor thermal conductivity in the solid causes bulk temperature to be different from surface temperature (i.e. slow thermal transients in the solid).

10.7 EQUATIONS WITH LONGITUDINAL CONDUCTION

The governing equations for the case including longitudinal conduction may be obtained by inspection of the equations for transients in contraflow (Appendix A). Here the usual subscripts for transient heat exchange are used. When the solid material has significant thermal gradients within its mass, additional equations are again required to represent bulk and surface temperatures.

$$\frac{\partial \theta_g}{\partial t} + u_g \frac{\partial \theta_g}{\partial x} = -\frac{n_g}{\tau_g}(\theta_g - \theta_s)$$

$$\frac{\partial \theta_b}{\partial t} - \kappa_x \frac{\partial^2 \theta_b}{\partial x^2} = +\frac{n_g}{R_g \tau_g}(\theta_g - \theta_s)$$

Remarks made above concerning effects within the solid still hold. When the bulk temperature (θ_b) can be taken as equal to surface temperature (θ_s), i.e. when the material thickness is small, these two temperatures can be replaced by the wall temperature (θ_w), which simplifes the solution considerably.

Further simplification is possible by normalising x, scaling t then normalising temperature (θ), as was done for transients in Chapters 8 and 9. Then

$$\frac{\partial \theta_g}{\partial T} + \frac{\partial \theta_g}{\partial X} = -n_g(\theta_g - \theta_{\hat{w}})$$

$$\frac{\partial \theta_w}{\partial T} - \mathrm{Fo}\frac{\partial^2 \theta_w}{\partial X^2} = +\frac{n_g}{R_g}(\theta_g - \theta_{\hat{w}})$$

where $\mathrm{Fo} = k(\frac{\tau_g}{L^2})$

Mathematical solution of the set of equations with longitudinal conduction may best be developed as a finite-difference method. Isothermal initial conditions simplify the mathematics, and adiabatic boundary conditions allow a test matrix for which only the inlet gas temperature disturbance and the outlet gas temperature response need be measured. The Laplace transform method with numerical solution and inversion offers an alternative approach.

10.8 REGENERATORS

Brief discussion of the regenerator problem is appropriate at this point, to indicate the work of others who have concentrated upon it.

A regenerator is a porous matrix through which hot and cold fluids flow alternately; the objective is to transfer thermal energy from one fluid to the other, the matrix acting as a temporary store for energy, or rather as a temporary store for exergy (Bejan 1988). The fluids occupy the same porous space in the matrix alternately, and may flow in opposite, directions, in the same directions or in cross-flow.

The matrix prism can be stationary, in which case the hot and cold fluid flows are intermittent, or two matrices can be provided, allowing the hot and cold fluids to flow continuously and to be directed by rotary valves. The matrix can be a slowly rotating disc, with hot fluid flowing steadily through one sector while the cold fluid flows steadily through the other sector. The sectors may not be equal.

Regenerator theory is related to single-blow theory, and to the cross-flow recuperator problem, but possesses its own attributes. Hausen's classical (1950) text contains analytical solutions, and the later finite-difference work by Willmott (1964) serves to confirm the accuracy of Hausen's original work.

Excellent reviews of the regenerator problem were prepared by Hausen (1979) and separately by Razelos (1979) in papers which appeared in the same publication. Schmidt and Willmott published a textbook on regenerators in 1981, and more recent papers include those of Evans and Probert (1987), Hill and Willmott (1989), Willmott and Duggan (1980), Romie (1990, 1991) and Foumeny and Pahlevanzadeh (1994). Van den Bulck (1991) and San (1993) have both considered optimal control and performance of cross-flow regenerators, whereas Shen and Worek (1993) have optimised the performance of a rotary regenerator considering both heat transfer and pressure loss. Interest in the Stirling cycle has produced quite a number of papers on regenerators under short-cycle conditions, of which Organ's (1992) book and other papers are representative.

The fundamental regenerator equations in canonical form are

$$\frac{\partial \theta_b}{\partial \eta} = -(\theta_b - \theta_g)$$
$$\frac{\partial \theta_g}{\partial \xi} = +(\theta_b - \theta_g)$$

where θ_g is dimensionless gas temperature and θ_b is dimensionless solid temperature, and both η and ξ are defined below.

$$\theta_g = \frac{T_g - T_{cl}}{T_{h1} - T_{cl}} \qquad \eta = \frac{\alpha St}{C_b V} \quad \text{reduced time variable}$$

$$\theta_b = \frac{T_b - T_{cl}}{T_{h1} - T_{cl}} \qquad \xi = \frac{\alpha St}{C_b L} \quad \text{reduced length variable}$$

The analytical problem may be complicated by a number of factors:

- allowing for residual fluid in the stationary matrix at the end of a blow
- allowing for carry-over in the case of a rotary matrix
- allowing for longitudinal conduction in a matrix
- allowing for the shape of real disturbances
- allowing for two-and three-dimensional effects in the matrix
- allowing for the disturbance not 'breaking-through' the matrix under short cycle times, or partially breaking through (Stirling cycle applications)

REFERENCES

Amundson, N. L. (1956) Solid-fluid interactions in fixed and moving beds: I. Fixed beds with small particles, II. Fixed beds with large particles, *Industrial and Engineering Chemistry, Industrial Edition*, **48**(1), 24–43.

Anzelius, A. (1926) Uber erwärmung vermittels durchströmender Medien, *Zeitschrift angewandte Mathematik und Mechanik*, **6**(4), 291–294.

Baclic, B. S. and Heggs, P. J. (1990) Unified regenerator theory and re-examination of the unidirectional regenerator performance, *Advances in Heat Transfer*, **20**, 133–179.

Bejan, A. (1988) *Advanced Engineering Thermodynamics*, New York: Wiley.

Bell, J. C. and Katz, E. F. (1949) A method for measuring surface heat transfer using cyclic temperature variations, *Proceedings of Heat Transfer and Fluid Mechanics*, 22–24 June 1949, p. 243–245.

Brinkley, S. R. (1947) Heat transfer between a fluid and a porous solid generating heat, *Journal of Applied Physics*, **18**, 582–585.

Brown, A. and Down, W. S. (1976) Melting and freezing processes as a means of storing heat, *6th Thermodynamics and Fluid Mechanics Conference, Durham, Institution of Mechanical Engineers, Thermodynamics and Fluid Mechanics Group, 6–8 April 1976, Paper 57/76*, pp. 157–163.

Chiou, J. P. and El-Wakil, M. M. (1966) Heat transfer and and flow characteristics of porous matrices with radiation as a heat source, *Transactions ASME, Journal of Heat Transfer*, **88**, 69–76.

Clark, J. A. Arpaci, V. S., and Treadwell, K. M. (1957, 1959, 1961) Dynamic response of heat exchangers having internal heat sources: I. *Trans ASME*, **57–SA–14**, 612; II. *Trans ASME*, **57–HT–6**, 625; III. *Trans ASME*, **58–SA–39**, 253; IV. *Trans ASME*, **60–WA–127**, 321.

Clenshaw, C. W. (1962) Chebyshev series for mathematical functions, *National Physical Laboratory, Mathematical Tables*, Vol. 5, London: HMSO.

Close, D. J. (1965) Rock pile thermal storage for comfort air conditioning, *Mechanical and Chemical Engineering Transactions, Inst, Engrs, Australia*, 1 May, **MC1**(1), 11–22.

Coombs, B. P. (1970) A transient test technique for evaluating the thermal performance of cross-inclined tube bundles, PhD Thesis, University of Newcastle upon Tyne.

Coppage, J. E. and London, A. L. (1953) The periodic flow regenerator – a summary of design theory, *Transactions of the American Society of Mechanical Engineers*, **75**, 779–787.

Dabora, E. K. (1957) Regenerative heat exchanger with heat-loss consideration, *AFOSR Tech. Note 57–613, 1957*.

Dabora, E. K., Moyle, M. P., Phillips, R., Nicholls, J. A. and Jackson, P. (1959) Description and experimental results of two regenerative heat exchangers, *AIChE Chemical Engineering Progress Symposium Series—Heat Transfer*, **55**(29), 21–28.

DeGregoria, A. J. (1991) Modelling the active magnetic regenerator, *Advances in Cryogenic Engineering*, **37B**, 867–873.

Evans, R. and Probert, S. D. (1987) Thermal performance of counterflow regenerators: a non-iterative method of prediction, *Applied Energy*, **26**(1), 9–46.

Foumeny, E. A. and Pahlevanzadeh, H. (1994) Performance evaluation of thermal regenerators, *Heat Recovery Systems and CHP*, **14**(1), 79–84.

Furnas, C. C. (1930) Heat transfer from a gas stream to a bed of broken solids – II, *Industrial and Engineering Chemistry, Industrial Edition*, **22**(7), 721–731.

Hamming, R. W, (1962) *Numerical Methods for Scientists and Engineers*, New York: McGraw-Hill, Ch. 26, pp. 445–458.

Handley, D. and Heggs, P. J. (1969) Effect of thermal conductivity of the material on transient heat transfer in a fixed bed, *International Journal of Heat and Mass Transfer*, **12**, 549–570.

Hart, J. A. and Szomanski, E. (1965) Development of the cyclic method of heat transfer measurement at Lucas Heights, *Mechanical and Chemical Engineering Transactions, Inst. Engrs. Australia*, May 1, **MC1**(1), 1–10.

Hausen, H. (1929) Uber die Theorie des Wärmeaustauchers in Regeneratoren, *Zeitschrift angewandte Mathematik und Mechanik*, **9**(3), 173–200.

Hausen, H. (1937) Feuchitgkeitsablagerung in Regenatoren, *Zeitschrift des Verein Deutscher Ingenieures, Beiheft 'Verfahrenstechnik'* **1937**(2), 62–67.

Hausen, H. (1979) Developments of theories on heat transfer in regenerators, In: *Compact Heat Exchangers – History, Technological and Mechanical Design Problems*, Eds. R. K. Shah, C. F. McDonald and C. P. Howard, *ASME Publication HTD* **10**, 79–89.

Heggs, P. and Burns, D. (1986) Single-blow experimental prediction of heat transfer coefficients: a comparison of four commonly used techniques, *Experimental Thermal and Fluid Science*, **1**(3), 243–252.

Hill, A. and Willmott, A. J. (1989) Accurate and rapid thermal regenerator calculations, *International Journal of Heat and Mass Transfer*, **32**(3), 465–476.

Howard, C. P. (1964) The single-blow problem including the effects of longitudinal conduction, *American Society of Mechanical Engineers*, Paper 64-GTP-11.

Johnson, J. E. (1948) Regenerator heat exchangers for gas turbines, *Aeronautical Research Council Reports and Memoranda (UK)*, R&M No. 2630, May, pp. 1025–1094.

Kohlmayr, G. F. (1966) Exact maximum slopes for transient matrix heat transfer testing, *International Journal of Heat and Mass Transfer*, **9**, 671–680.

Kohlmayr, G. F. (1968a) Analytical solution of the single-blow problem by a double Laplace transform method, *ASME, Journal of Heat Transfer*, **20**,(1), 176–178.

Kohlmayr, G. F. (1968b) Properties of the transient heat transfer (single-blow) temperature response function, *AIChE Journal*, **14**(3), 499-501.

Kohlmayr, G. F. (1971) Implementation of direct curve matching methods for transient matrix heat transfer testing, *Applied Scientfic Research*, **24**, 127–148.

Kramers, H. and Alberda, G. (1953) Frequency response analysis of continuous flow systems, *Chemical Engineering Science*, **2**, 173–181.

Liang, C. Y. and Yang, W. J. (1975) Modified single-blow technique for performance evaluation on heat transfer surfaces, *Transactions ASME, Journal of Heat Transfer*, **97**, 16–21.

Locke, G. L. (1950) Heat transfer and flow friction characteristics of porous solids, *Stanford University Technical Report No. 10*, Office of Naval Research NR-035-104.

Lowan, A. N., Davids, N. and Levinson, A. (1954) Table of the zeros of the Legendre polynomials of order 1–16 and the weight coefficients for Gauss mechanical quadrature formula, *Tables of Functions and Zeros of Functions, NBS Applied Mathematics Series, No. 37*, pp. 185–189.

Meek, R. M. G. (1961) The measurement of heat transfer coefficients in packed beds by the cyclic method, *Proceedings International Heat Transfer Conference, Boulder, CO, 1961–62, Paper 93*, pp. 770–780.

Meek, R. M. G. (1962) Measurement of heat transfer coefficients in randomly packed beds by the cyclic method, *NEL Report no. 54, 1962*, East Kilbride, UK: National Engineering Laboratory.

Mondt, J. R. (1961) Effects of longitudnal thermal conduction in the solid on apparent convective behaviour with data on plate-fin surfaces, *Proceedings of the International Heat Transfer Conference, Boulder, CO, ASME Paper 73*.

Mondt, J. R. and Siegla, D. C. (1972) Performance of perforated heat exchanger surfaces, *ASME Paper 72-WA-52*.

Nelder, J. A. and Meade, R. (1965) A simplex method for function minimisation, *Computer Journal*, 7, 308–313.

Nusselt, W. (1927) Die Theorie des Winderhitzers, *Zeitschrift Vereines deutscher Ingenieure*, 71(3), 85–91.

Organ, A. J. (1992) *Thermodynamics and Gas Dynamics of the Stirling Cycle Machine*, Cambridge CUP.

Organ, A. J. (1993) Flow in the Stirling regenerator characterised in terms of complex conditions, Part 1 – theoretical developments, *Proceedings of the Institution of Mechanical Engineers, Part C*, 207(2), 117–124.

Organ, A. J. (1994) The wire mesh regenerator of the Stirling cycle machine, *International Journal of Heat and Mass Transfer*, 37(16), 2525–2534.

Organ, A. J. and Rix, D. H. (1993) Flow in the Stirling regenerator characterised in terms of complex conditions, Part 2 – experimental investigation, *Proceedings of the Institution of Mechanical Engineers, Part C*, 207(2), 127–139.

Parkingson, J. M. and Hutchinson, D. (1972) An investigation into the efficiency of variants on the simplex method, In: *Numerical Methods for Non-Linear Optimisation*, Ed. F. A. Lootsma, New York: Academic Press.

Pfeiffer, S. and Huebner, H. (1987) Untersuchung zum einfrieren von Regeneratur-Wärmeübertragern (Investigation of freezing-up in regenerative heat exchangers), *Ki Klima Kälte Heizung*, 15(10), 449–452.

Pucci, P. F., Howard, C. P. and Piersall, C. H. (1967) The single-blow transient test technique for compact heat exchanger surfaces, *Transactions of the American Society of Mechanical Engineers, Series A*, 89(1), 29–40.

Razelos, P. (1979) History and advancement of regenerator thermal design theory, In: *Compact Heat Exchangers – History, Technological and Mechanical Design Problems* Eds. R. K. Shah, C. F. McDonald and C. P. Howard, *ASME Publication HTD* 10, 91–100.

Romie, F. E. (1990) Response of rotary regenerators to step changes in mass rates, *Transactions ASME, Journal of Heat Transfer*, 112(1), 43–48.

Romie, F. E. (1991) Treatment of transverse and longitudinal heat conduction in regenerators, *Transactions ASME, Journal of Heat Transfer*, 113, 247–249.

San, J. -Y. (1993) Heat and mass transfer in a two-dimensional cross-flow regenerator with a solid conduction effect, *International Journal of Heat and Mass Transfer*, 36(3), 633–643.

Schmidt, F. W. and Willmott, A. T. (1981) *Thermal Energy Storage and Regeneration*, Washington: Hemisphere, 1981.

Schumann, T. E. W. (1929) Heat transfer. A liquid flowing through a porous prism, *Journal of the Franklin Institute*, September, 405–416.

Shearer, C. J. (1962) Measurement of heat transfer coefficients in low conductivity packed beds by the cyclic method, *NEL Report No. 55, 1962*, East Kilbride, UK: National Engineering Laboratory.

Shen, C. M. and Worek, W. M. (1993) Second-law optimisation of regenerative heat exchangers including the effect of matrix heat conduction, *Energy*, **18**(4), 355–363.

Smith, E. M. (1979) General integral solution of the regenerator transient test equations for zero longitudinal conduction, *International Journal of Heat and Fluid Flow*, **1**(2), 71–75.

Smith, E. M. and Coombs, B. P. (1972) Thermal performance of cross-inclined tube bundles measured by a transient method, *Journal of Mechanical Engineering Science*, **14**(3), 205–220.

Smith, E. M. and King, J. L. (1978) Thermal performance of further cross-inclined in-line and staggered tube banks, *6th International Heat Transfer Conference, Toronto*, Vol. 4, paper HX-14.

Spendley, W., Hext, G. R. and Hinsworth, F. R. (1962) Sequential application of simplex design in optimisation and evolutionary optimisation, *Technometrics*, **4**, 441–461.

Stang, J. H. and Bush, J. E. (1972) The periodic method for testing heat exchanger surfaces, *Transactions of the American Society of Mechanical Engineers*, Paper 72-WA/HT-57.

Van den Bulck, E. (1991) Optimal thermal control of regenerator heat exchangers, In: *Design and Operation of Heat Exchangers*, Eds. W. Roetzel, P. J. Heggs and D. Butterworth, *Proceedings of the EUROTHERM Seminar no. 18, Hamburg, 27 February–1 March 1991*, pp. 295–304.

Willmott, A. J. (1964) Digital computer simulation of a thermal regenerator, *International Journal of Heat and Mass Transfer*, **7**, 1291–1302.

Willmott, A. J. and Duggan, R. C. (1980) Refined closed methods for the contra-flow thermal regenerator problem, *International Journal of Heat and Mass Transfer*, **23**, 655–662.

Willmott, A. J. and Hinchcliffe, C. (1976) The effect of heat storage upon the performance of the thermal regenerator, *International Journal of Heat and Mass Transfer*, **19**, 821–826.

Willmott, A. J. and Scott, D. M. (1993) Matrix formulation of linear simulation of the operation of thermal regenerators, *Numerical Heat Transfer, Part B – Fundamentals*, **23**(1), 43–65.

11

CRYOGENIC HEAT EXCHANGERS AND STEPWISE RATING

11.1 BACKGROUND

Before discussing stepwise rating of cryogenic heat exchangers, it is desirable to understand the procedure for laying out a liquefaction plant. Although several textbooks exist on the subject of cryogenics (e.g. Scott 1959, Haselden 1971 and Barron 1985), and although many examples of complete plants occur in textbooks and elsewhere, it is hard to find specific instructions on designing liquefaction plant from scratch.

The author proceeded to investigate the thermodynamics of the process on his own account, and what follows is a distillation of some of the results of his investigations. Early sections in this chapter will discuss a number of basic considerations as they affect plant design before examining the design of the heat exchangers themselves. In cryogenic plant, emphasis is placed on feasibility, simplicity and performance, and the difference between desirable and practical approaches will be discussed where appropriate.

Begin by considering Fig. 11.1, a representation of Carnot efficiency above and below the *dead-state temperature* at which all heat may be rejected without any possibility of generating further work. Although this temperature will vary from place to place on the earth's surface, as it is related to local ambient temperature, we will assume the dead-state temperature to be 300 K.

The *engine region* exists above 300 K; it is where thermal energy may be partially converted to work. The Carnot efficiency tends to the asymptote of 1.0 as temperature increases.

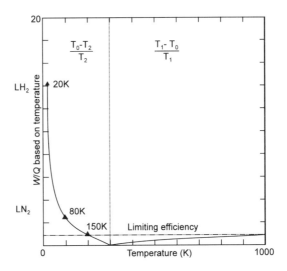

Fig. 11.1 Carnot efficiency above and below and dead state

The *heat-pump region* exists between 300 K and 150 K; here it is possible to take energy from one temperature level and reject it at a higher temperature level while doing less work than the energy being shifted. The limit of 150 K is where exactly the same amount of energy is shifted as work is done. If a slightly different dead-state temperature is chosen, then the lower temperature limit for the heat-pump region changes accordingly.

Below 150 K we have the true *cryogenic region* where more work is required to shift energy than the energy itself. At 80 K, just above liquid nitrogen (LN_2) saturation temperature at 1 bar, the Carnot work required is 2.75 times the cooling produced. At 20 K, close to liquid hydrogen (LH_2) saturation temperature at 1 bar, the Carnot work required is 14 times the cooling produced. There is thus every reason to seek the most efficient thermodynamic means for liquefying gases.

In the region 300 K–150 K it is appropriate to consider conventional refrigeration plant using evaporators and condensers because this is the most work-efficient method of cooling available.

Below 15 K there exist two principal means for cooling a refrigerating gas. The first involves expansion of high pressure refrigerating gas in a cryoturbine with very low frictional losses. With this apparatus, compression work is known to be a principal barrier to improvement in liquefaction performance. The second method involves the use of thermomagnetic regenerators whose matrix temperature may be changed by application and removal of strong magnetic fields, thus cooling the refrigerating gas at constant pressure. Thermomagnetic methods can be used with effect at and below liquid helium temperatures, and there have been attempts to extend the method to regions of higher temperature.

Most commercial plants employ the cryoturbine method, and this is what will be considered.

11.2 LIQUEFACTION CONCEPTS AND COMPONENTS

Liquefaction involves cooling a gas below its critical point and in large plant; this implies using gas as the refrigerant flowing in contraflow to the product stream. Table 11.1 lists primary cryogens of interest as possible cooling streams.

Mixtures of gases with high Joule-Thompson coefficients (e.g. nitrogen-methane-ethane) have produced significant improvement in cooling (Alfeev *et al*. 1971). In laboratory-scale testing, Little (1984) confirmed Russian claims that cooldown times were reduced from 18 to 2 minutes and that lower temperatures could be attained with mixtures than by using nitrogen alone. Further work is underway at Stanford University (Paugh 1990).

Any gas to be liquefied (sometimes a hydrocarbon) will be called the *product* stream, and the fluid doing the cooling will be called the *refrigerating* stream. While establishing the design procedure, we shall restrict ourselves to the gases in Table 11.1.

Table 11.1 Candidate refrigerant fluids

Fluid	Critical pressure (bar)	Critical temp., (K)	Sat. temp, at 1.0 bar, (K)	Latent heat (kJ/kg)	Gas, constant, (kJ/kg K)	Ratio, (C_p/C_v) (300 K)
Oxygen	50.9	154.77	90.18	212.3	0.2598	1.396
Argon	50.0	150.86	87.29	159.6	0.2082	1.670
Nitrogen	33.96	126.25	77.35	197.6	0.2968	1.404
Neon	26.54	44.40	27.09	86.1	0.4117	1.640
Hydrogen	12.76	32.98	20.27	434.0	4.157	1.410
Helium	2.3	5.25	4.2	21.0	2.075	1.662

Forms of hydrogen

Hydrogen has two forms, *ortho*-hydrogen and *para*-hydrogen, which differ in the spins of their protons (Fig. 11.2). These two forms are not isotopes. Above 300 K the *ortho:para* concentration ratio remains constant at 75:25 and this is known as *normal* hydrogen. Below 30 K each temperature level has an equilibrium concentration ratio as shown in Fig. 11.3. The desired final liquefaction state is 100% *para*-hydrogen, so the objective is to achieve the greatest *para* concentration at each temperature level throughout cooling of the process stream. This corresponds to removing the maximum amount of heat at the highest possible temperature levels, i.e. to achieving an equilibrium concentration ratio at each temperature level.

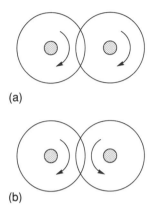

(a)

(b)

Fig. 11.2 Hydrogen molecule configurations: (a) *ortho* and (b) *para*

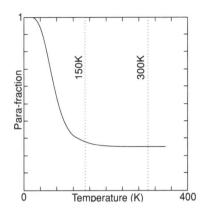

Fig. 11.3 *Para*-content vs. temperature (K)

The reader may therefore appreciate that the two forms of hydrogen have different thermodynamic properties. When consulting databooks it might be anticipated that properties would be listed for both *ortho*-and *para*-forms. It may come as a mild surprise to find that only normal hydrogen and *para*-hydrogen properties are listed. This means that some calculation is required to obtain the properties of equilibrium hydrogen at any temperature level.

The enthalpy of normal hydrogen is given by

$$_n h = 0.25 \, _p h + 0.75 \, _o h \tag{11.1}$$

Let x be the concentration of *para*-hydrogen at the desired temperature; the corresponding equilibrium enthalpy is then given by

$$_e h = x \, _p h + (1 - x) \, _o h \tag{11.2}$$

Substituting for the enthalpy of *ortho*-hydrogen from equation (11.1), the enthalpy of equilibrium hydrogen is obtained as

$$_e h = y \, _p h + (1 - y) \, _n h \text{ where } y = (4x - 1)/3 \tag{11.3}$$

Minimum work of liquefaction

This will be illustrated with reference to hydrogen, a more complicated case than will be encountered with other gases, but the principles remain the same.

The minimum work of liquefaction of equilibrium hydrogen from 300 K will be compared at pressure levels of 1, 15, 35 and 50 bar. This requires values of

11.2 LIQUEFACTION CONCEPTS AND COMPONENTS

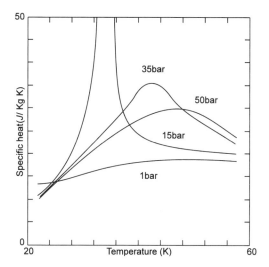

Fig. 11.4 Specific heat of equilibrium hydrogen at 1, 15, 35 and 50 bar

specific heat at constant pressure, obtained by cubic spline-fitting enthalpy data then differentiating once to obtain specific heat

$$C = \left(\frac{\partial h}{\partial T}\right)_p$$

The results for pressure levels of 1, 15, 35 and 50 bar are shown in Fig. 11.4. The minimum work of liquefation is evaluated as follows. In isobaric cooling through δT, the amount of heat removed is $\delta Q = C\delta T$, where C is the specific heat at constant pressure.

The Carnot efficiency is $\quad \eta = -\dfrac{W}{Q} = -\left(\dfrac{T_0 - T}{T}\right)$

Minimum work is then given by

$$\delta W_{min} = -\int \left(\frac{T_0 - T}{T}\right) C dT = -\int \left(\frac{CT_0}{T} - C\right) dT \quad (11.4)$$

When specific heat (C) is known as a simple mathematical function of temperature (T), then direct integration of equation (11.4) becomes possible. But clearly it is difficult fitting polynomials to the separate curves of Fig. 11.5, so an alternative method was developed.

If C is constant over a small range $(T_a - T_b)$ where the subscripts refer to initial and final states, then

$$\Delta W_{min} = -\bar{C}\left[T_0 \int_a^b \frac{dT}{T} - \int_a^b dT\right] = -\left(\frac{C_b + C_a}{2}\right)[T_0 \ln(T_b/T_a) - (T_b - T_a)]$$

and it is possible to evaluate $W_{min} = \sum \Delta W_{min}$ over an extended cooling range (T_1 to T_2) where the mean values of specific heat are taken over

5 K intervals for 300–150 K
2 K intervals for 150–50 K
1 K intervals for 50–20 K.

For the 1 bar pressure level, a latent heat term is added to W_{min}:

$$-\left(\frac{T_0 - T_2}{T_2}\right)h_{fg}$$

For the 15, 35 and 50 bar pressure levels, an isothermal work term $-RT_1 ln(r)$ is added to W_{min}, where r is the compression ratio.

The results are shown in Fig. 11.5. Figure 11.5 is a T-W diagram; it resembles a T-S diagram and the two vertical lines shown at the bottom left of the figure represent the maximum and minimum work requirements to liquefy.

Although least energy expenditure is achieved by cooling at 1 bar, there is very little difference in work expenditure if the hydrogen is first isothermally compressed to 35 bar and then cooled. A quick look at Fig. 11.1 confirms that any inefficiency in lifting a large amount of latent heat at 1 bar will completely

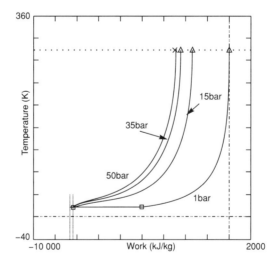

Fig. 11.5 Minimum work of liquefaction of hydrogen

negate the advantage of cooling at 1 bar. Thus in most liquefaction arrangements the product stream is first compressed to supercritical pressure before cooling commences.

Hydrogen tends to maintain its initial *ortho:para* ratio during cooling, and the conversion ratio can be made rapid enough only by using a catalyst; see collected references in Smith (1989a). Ideally the catalyst should be placed inside the heat exchangers used in the cooling, but there it can become contaminated. Current practice is to provide separate catalyst pots so they can be changed if required one manufacturer has placed catalyst inside the final heat exchanger.

New thermodynamic properties have to be calculated for the constant *ortho:para* condition between catalyst pots, but this is straightforward once the appropriate scheme for calculating thermodynamic properties has been set up. For other gases, the complication of different molecular forms does not exist.

Compressors

The product stream must be compressed to supercritical pressure so that cooling may proceed towards the liquid side of the saturation line in the T-S diagram. Refrigerating streams have also to be compressed to suitable pressures.

There is no problem in compressing such gases as oxygen and hydrogen using relatively slow-moving reciprocating compressors. Rotary compressors with fast-moving parts may safely be used for inert gases, and may also be used for some hydrocarbons if sufficient care is taken to avoid a high temperature rub between impellers and casings.

For comparison of prospective compressor arrangements, it is practicable to employ an isentropic index of compression to compute the work. When actual machines are constructed, an isentropic efficiency expression ($\eta_s = W_s/W_{real}$) can be used to relate actual performance to the computed value. This avoids having to guess a value for the polytropic index of compression (n), and the isentropic index ($\gamma = C_p/C_v$) can be used in its place.

Assuming k stages of compression with suction at $-p_1$, T_1, and final delivery at p_{k+1} with intercooling to T_2, the expression for minimum work for k stages of compression can be found by standard methods.

$$W_{min} = -R\left(\frac{\gamma}{\gamma-1}\right)\left[T_1\left(\rho_1^{\gamma-1/\gamma} - 1\right) + (k-1)T_2\left(\rho_2^{\gamma-1/\gamma} - 1\right)\right]$$

$$\text{where } \rho_1 = \left(\frac{p_{k+1}}{p_1}\right)^{\frac{1}{k}}\left(\frac{T_1}{T_2}\right)^{-\left(\frac{\gamma}{\gamma-1}\right)\left(\frac{k-1}{k}\right)} \quad (11.5)$$

$$\text{and } \rho_2 = \left(\frac{p_{k+1}}{p_1}\right)^{\frac{1}{k}}\left(\frac{T_1}{T_2}\right)^{\left(\frac{\gamma}{\gamma-1}\right)\frac{1}{k}}$$

270   11 CRYOGENIC HEAT EXCHANGERS

Whenever possible, a single-stage compressor is to be preferred (implying restriction of the compression ratio), and plant design may be configured accordingly.

Cryoexpanders
It is not easy to arrange for multi-staging in a single expansion turbine, and the most suitable turbine is the single-stage inward radial flow machine. The limitation on expansion ratio has then to be explored. A relatively crude analysis permits evaluation of comparative pressure expansion ratios for different refrigerant gases, using Fig. 11.6. This suffices for feasibility study of the overall liquefaction system.

For perfect gases: $$R = (C_p - C_v) = C_P\left(\frac{\gamma - 1}{\gamma}\right)$$

With the following subscript notation:

0 = nozzle inlet
1 = nozzle throat
2 = rotor exhaust

sonic velocity at the throat of the nozzle e_1 may be expressed as

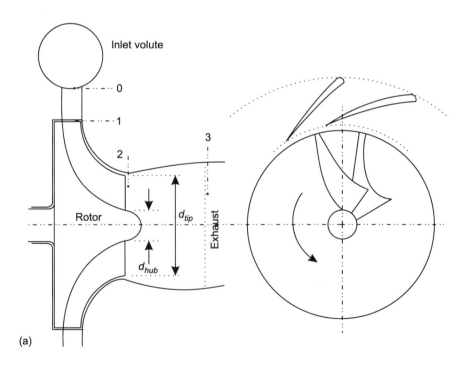

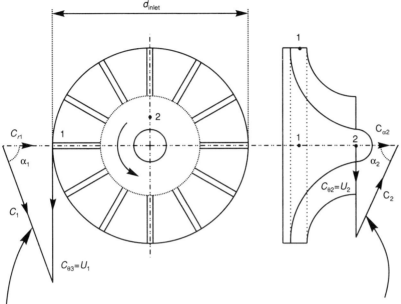

Fig. 11.6 Inward radial flow turbine

$$c_1^2 = \gamma RT_1 = C_p(\gamma - 1)T_1 = 2C_p\left(\frac{\gamma - 1}{\gamma + 1}\right)T_0$$

With an inward radial flow machine having a rotor tip speed U_1, an inlet angle (α_1), and equal inlet and outlet gas velocities V_1 and V_2 such that entering and leaving losses are the same, then

$$\cos\alpha_1 = \frac{V_1}{c_1}$$

For an isentropic efficiency η_s the pressure expansion ratio is given by

$$\frac{P_2}{P_0} = \left(\frac{1 - U_1^2}{\eta_s C_p T_0}\right)^{\frac{\gamma}{\gamma - 1}}$$

and the outlet temperature is given by

$$T_2 = T_0 - \frac{U_1^2}{C_p}$$

With substitution it is quickly shown that

$$\frac{P_2}{P_0} = \left[1 - \frac{2}{\eta_s}\left(\frac{\gamma-1}{\gamma+1}\right)\cos^2\alpha_1\right]^{\frac{\gamma}{\gamma-1}}$$
$$\frac{T_2}{T_0} = \left[1 - 2\left(\frac{\gamma-1}{\gamma+1}\right)\cos^2\alpha_1\right]$$

and both these relationships depend on γ and inlet angle α_1.

For the purpose of comparison, an isentropic efficiency of of 0.8 and an inlet angle of 10° will be assumed. Results for the expansion of five candidate refrigerant gases are presented in Table 11.2 in descending order of the ratio C_p/C_v.

Table 11.2 Cryoexpansion fluids

Gas	(C_p/C_v)	(P_0/P_2)	(T_2/T_0)
Argon	1.670	10.349	0.5133
Helium	1.662	10.145	0.5176
Neon	1.640	9.688	0.5298
Hydrogen	1.410	6.228	0.6700
Nitrogen	1.404	6.164	0.6740

Oxygen is not in Table 11.2 because of the very great risk of fire should a high speed turbine rotor come into contact with its casing, but oxygen is still a possible refrigerant gas as vapour return from the final stages of liquefaction.

The gases clearly fall into two groups: the monatomic group with expansion ratios of about 10/1, and the diatomic group with expansion ratios of about 6/1. The first group achieves the greatest amount of single-stage 'cooling'.

It is desirable to stay away from shock-wave losses whenever possible, and inward radial flow rotor design is eased when incompressible conditions are achieved at below approximately one-third of sonic velocity. Most practical plants try to keep expansion ratios below 3.0, but a better choice is 2.5 or less. In maintaining the expansion ratio constant, temperatures will fall in reducing geometric progression. For sequential expansion this gives optimum expansion ratios for minimisation of exergy loss found by Nesselman, reported briefly at the end of the paper by Grassmann and Kopp (1957).

Table 11.2 also indicates why there is current interest in mixed refrigerants. Cryoexpansion problems are eased, and some mixtures have been found capable of reaching lower temperatures than with a single component.

11.3 LIQUEFACTION OF NITROGEN

The design of radial inward flow turbines is discussed in the text by Whitfield and Baines (1990), but the later paper by Whitfield (1990) examines cryogenic turbines in more detail.

11.3 LIQUEFACTION OF NITROGEN

Nitrogen is almost always a first candidate for a refrigerating stream in liquefaction plant because of its abundance, inertness and low critical pressure. It does not have the properties of a monatomic molecule, but this disadvantage can be mitigated by mixing it with argon.

The present example of a liquefaction plant to produce LN_2 has been chosen so as to illustrate some features of a typical system. The example is untypical only in one respect: it is possible to mix product and refrigerating streams without affecting the product.

The reader should consider the T-S diagram for nitrogen. We have already decided to compress the product stream to supercritical pressure so that cooling can follow the liquid side of the saturation curve and get close to the final condition before throttling to produce liquid at almost ambient pressure. What pressure level should be chosen for the product stream? What governs its selection?

Low compression work is an important consideration, and thus we seek the lowest practicable pressure level. To do this, examine the H-T diagram for nitrogen (Fig. 11.7) and notice that it is possible to construct 'break-points' on the H-T curve such that straight lines joining these points provide a near approximation to the curve itself. The existence of these straight segments means that the temperature distributions in the heat exchangers will also be nearly linear.

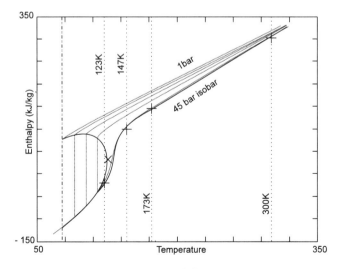

Fig. 11.7 Break-points on the nitrogen H-T diagram

274   11 CRYOGENIC HEAT EXCHANGERS

Only one supercritical curve is shown in Fig 11.7, but in practice many curves need to be examined, as the choice of pressure level changes the position and spacing of the break-points on the product-stream curve. Finding break-points which produce linear segments is a necessary but not sufficient condition for successful liquefaction; consideration must also be given to the expansion ratio and the temperature reduction achievable by cryoturbines feeding the refrigeration streams.

When different gases form the product and refrigeration streams, there is less incentive to match pressure levels elsewhere in the system. For nitrogen liquefaction using nitrogen as the refrigerant, it makes some sense to try to match pressure levels. This matching process is the art of engineering cryogenic plant.

Figures 11.8 and 11.9 show the T-S diagram for the plant and layout selected. All compressors shown are assumed to include aftercooling to 300 K.

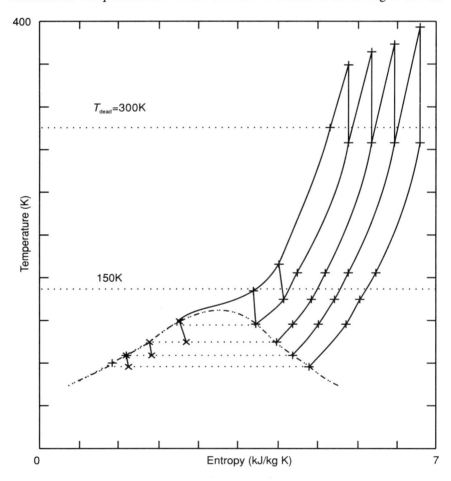

Fig. 11.8 T-S diagram for nitrogen liquefaction plant

11.3 LIQUEFACTION OF NITROGEN

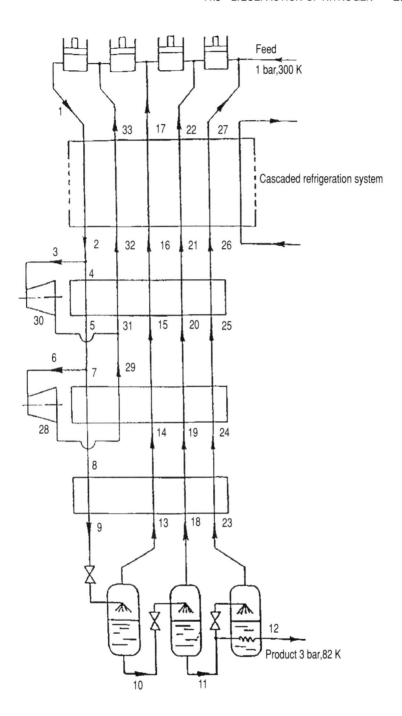

Fig. 11.9 Configuration of liquefaction plant to produce LN$_2$

The plant is not without scope for thermodynamic improvement, e.g. an additional cryoturbine could be fitted at point 14, discharging to point 18 and thus eliminating the return stream 14-15-16-17 in the multi-stream heat exchangers. The consequences of this change would need careful evaluation.

Essentially the product stream is compressed from 1 bar at around 300 K to 35 bar at 300 K; it is then cooled isobarically to 35 bar, 123 K. Compression was required to get the product stream to the left-hand side of the saturation line, but the work input has not yet produced any cooling in its own right. To remedy this, a succession of throttling steps are introduced which allows most of the product stream to 'walk down' the liquid saturation line (Marshall and Oakey 1985). Any cold vapour produced by throttling the product stream is made to return in parallel with the refrigerating streams. Here careful selection of pressure levels has allowed these return streams to be combined with the refrigerating streams, thus reducing the number of independent streams in the multi-stream heat exchangers – an important simplification.

The product stream is not expanded all the way to 1 bar; it is better to maintain the product liquid slightly above atmospheric pressure so that any leakage is outwards. However, the liquid product is undercooled as far as is practicable, making use of the last expansion stage for that purpose. This helps to counteract 'heat leak' from the insulated storage tanks. The first refrigeration stage in cooling the product stream is not shown, and this can be a series of cascaded conventional refrigeration plants, using appropriate working fluids; see e.g. Barron (1985). Some refrigerating fluids would be inappropriate for oxygen as a product stream.

For common product and refrigerant streams, it makes sense to select compressor pressure levels for the refrigerating system which match those generated by the product system. But which pairs should they be, 1 bar/3 bar, 3 bar/8 bar, 8 bar/20 bar or 20 bar/45 bar?

Returning to the *H-T* diagram, the 20 bar isobar has a small curvature which matches the 45 bar product stream slightly better that the isobars at 8, 3 or 1 bar. There is no *a priori* reason why suction should not be at 20 bar, providing the system is pressure-tight, and canned compressors can be used if necessary. The primary refrigeration compressor is smaller as a consequence of higher gas densities. Product return streams are not considered at this stage because they make lesser contributions to the cooling required.

Figure 11.9 shows the final plant configuration with four compressors, a cascaded refrigeration system, two cryoturbines and three throttling stages. The function of the lowest heat exchanger is simply to equalise the temperatures of the returning product vapour streams before serious cooling begins. There have been attempts in the industry to develop liquid expansion machines as a replacement for throttles, but absence of moving parts at cryogenic temperatures leads to plant reliability. It is possible to introduce another throttling stage at 20 bar, but whether this is worthwhile is a matter of economics. In the system shown, there is little opportunity for refrigerating fluid to flow in the wrong direction.

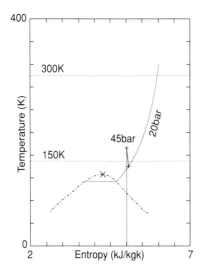
Fig. 11.10 Cryoturbine performance on a nitrogen T-S diagram

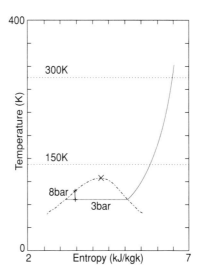
Fig. 11.11 Throttle performance on a nitrogen T-S diagram

In arriving at *temperatures* for the Fig. 11.9 configuration, a top-down procedure is used to find break-points that produce linear segments in the H-T curve, matching cryoturbine expansion ratios with the required break-points, matching temperature levels at entry and exit to the heat exchangers, and matching temperatures so that mixing losses do not occur, or are minimised.

Cryoturbine performance is determined by a simple calculation in which the isothermal efficiency is assumed to be 0.80, and a T-S diagram is used to check that the expansion is in the right position (Fig. 11.10). Throttle performance is assessed similarly (Fig. 11.11). In the multi-stream heat exchangers cooling performance of each return stream is assessed individually; each individual component performance allows mass flow *ratios* to be determined.

Assessment of heat exchanger performance at this stage is restricted to piecewise checking of the enthalpy balance along the exchanger, fixing appropriate cold inlet and outlet temperatures for the low pressure fluid, and appropriate warm inlet temperature for the high pressure fluid. Thermodynamic properties of both fluids are obtained from interpolating splinefits. An appropriate value of pinch point (temperature difference at point of closest approach) is chosen; the mass flow rate of the cold fluid is set to 1.0 kg/s; and the calculation iterated until the pinch point is achieved somewhere in the exchanger. This calculation provides five important items of design information along the exchanger (Figs. 11.12 and 11.13):

- shape of $\Delta T, T$ profiles
- shape of the H–T profiles
- high pressure, warm fluid outlet temperature

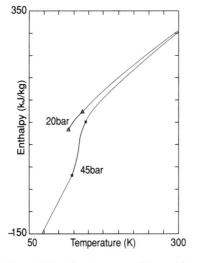 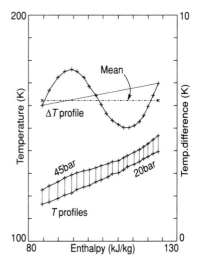

Fig. 11.12 Exchanger: (*H* vs. *T*) profiles

Fig. 11.13 Exchanger: (Δ*T*, *T* vs. *H*) profiles

- mass flow rate of warm fluid for 1.0 kg/s of cold fluid
- position of the temperature pinch point

If the outlet temperature of the high pressure, warm fluid is not the value desired (often a value corresponding to that of another 'mixing' stream, to avoid detrimental mixing losses), then the pinch-point temperature can be adjusted to achieve an outlet temperature match. If no suitable value of outlet temperature becomes available, it may be necessary to choose new temperature break-points (Fig. 11.7) or to reconfigure the plant. Where more than two fluids are present in an exchanger, several such calculations have to be made for each possible pair of fluids to determine the best possible combination of energy exchange balances.

Table 11.3 shows results of one such calculation for the two main fluids in the critical heat exchanger of the nitrogen liquefaction plant. That this is the critical exchanger can be confirmed by examining corresponding mass flow rates in Table 11.4. But generally the critical exchanger turns out to be the exchanger which straddles the critical temperature and most closely approaches the critical point of the fluid being cooled.

Once correct mass flow ratios for each exchanger have been determined, *true mass* flow rates for the whole plant system can be found starting from the bottom up (Fig. 11.9). This begins with free choice of the desired amount of undercooling for the product stream at 3 bar (noting that it is not possible to cool below the saturation temperature at 1 bar). Working back up a cryogenic system requires only arithmetic, except in a few cases when simple simultaneous algebraic equations may be needed to determine flow rates. Completion of Table 11.4 is necessary before the design of actual exchangers can proceed.

Table 11.3 Temperature profiles from enthalpy balance*

20 bar			45 bar			Tdiff, ΔT (K)
T (K)	h (kJ/kg)	Δh (kJ/kg)	$\Delta h \times R_m$ (kJ/kg)	h (kJ/kg)	T (K)	
140.0	123.6	–	–	101.4	147.0	7.00
		3.22	9.78			
137.7	120.4	–	–	92.1	143.7	5.958
		3.30	10.03			
135.4	117.1	–	–	81.7	140.7	5.286
		3.44	10.46			
133.1	113.6	–	–	71.1	138.1	5.030
		3.57	10.86			
130.8	110.1	–	–	60.2	136.0	5.160
		3.68	11.18			
128.5	106.4	–	–	48.7	134.2	5.678
		3.82	11.62			
126.2	102.6	–	–	37.2	132.7	6.459
		4.02	12.21			
123.9	98.6	–	–	25.4	131.1	7.230
		4.28	13.00			
121.6	94.3	–	–	12.2	129.2	7.606
		4.68	14.24			
119.3	89.6	–	–	−2.0	126.5	7.196
		5.12	15.57			
117.0	84.5	–	–	−17.5	123.0	5.996
Mean value of ΔT						6.236

*For 20 bar/45 bar section of multi-stream heat exchanger with mass flow ratio $R_m = (m_{20\text{bar}}/m_{45\text{bar}}) = 3.0407$.

11.4 HYDROGEN LIQUEFACTION PLANT

The same procedures are used in designing other liquefaction plant, except that in the case of hydrogen care has to be taken to use 'equilibrium' thermodynamic properties where appropriate.

In the very recent industrial-scale hydrogen liquefaction plant described by Bracha et al. (1994), liquid nitrogen is used to effect the first *ortho:para* hydrogen conversion, and the cold gaseous nitrogen is then used to refrigerate the incoming hydrogen streams. The refrigerating nitrogen stream in this plant is not recycled, but is continuously extracted from the air and discharged to the atmosphere.

Table 11.4 Thermodynamic and flow analysis of an LN$_2$ plant*

Station	Pressure (bar)	Temperature (K)	Mass flow (kg/s)	Enthalpy (kJ/kg)
1	45.0	300.0	7.2482	302.0
2	45.0	173.0	7.2482	148.5
3	45.0	173.0	2.9726	148.5
4	45.0	173.0	4.2756	148.5
5	45.0	147.0	4.2756	99.51
6	45.0	147.0	2.5357	99.51
7	45.0	147.0	1.7399	99.51
8	45.0	123.0	1.7399	−25.44
9	45.0	120.0	1.7399	−28.00
10	8.0	100.4 S	1.2480	−73.60 W
11	3.0	87.9 S	1.0697	−99.87 W
12	3.0	82.0 U	1.0000 PS	Lu
13	8.0	100.4 S	0.4919	87.70
14	8.0	117.0	0.4919	110.92
15	8.0	140.0	0.4919	137.3
16	8.0	165.0	0.4919	165.45
17	8.0	285.0	0.4919	293.88
18	3.0	87.9 S	0.1783	83.96 D
19	3.0	117.0	0.1783	118.3
20	3.0	140.0	0.1783	142.3
21	3.0	165.0	0.1783	169.0
22	3.0	285.0	0.1783	295.10
23	1.0	77.4 S	0.0697	76.80 D
24	1.0	117.0	0.0697	120.98
25	1.0	140.0	0.0697	144.2
26	1.0	165.0	0.0697	170.45
27	1.0	285.0	0.0697	295.60
28	20.0	117.0	2.5357	96.94
29	20.0	140.0	2.5357	123.6
30	20.0	140.0	2.9726	123.6
31	20.0	140.0	5.5083	123.6
32	20.0	165.0	5.5083	156.55
33	20.0	285.0	5.5083	291.05

Symbols: S=saturated, U=undercooled, Lu=liquid undercooled, W=wet, D=dry; PS=product stream.

*Data generated using Vargaftik (1975), except for a value of liquid specific heat which was obtained from Touloukian and Makita (1970).

11.5 PRELIMINARY DIRECT SIZING OF MULTI-STREAM HEAT EXCHANGERS

The paper by Bracha *et al.* provides an excellent example of a hydrogen liquefaction system, so there is no need to describe it here.

11.5 PRELIMINARY DIRECT SIZING OF MULTI-STREAM HEAT EXCHANGERS

Preliminary sizing of exchangers provides a best estimate for the exchanger cross-sections, e.g. edge length in plate-fin designs, and number of tubes and tube spacing in shell-and-tube exchangers. Stepwise rating becomes necessary when the asumption of constant thermophysical properties along the exchanger no longer holds. It is worth summarising the procedure for a two-stream exchanger, which is in several stages. The most awkward exchanger of the cryogenic plant in Fig. 11.9 is likely to be the multi-stream exchanger associated with the lowest cryoturbine because conditions for the 20 bar cooling stream and the 45 bar product stream are nearest to the critical point of nitrogen.

Stage 1
Table 11.3 is constructed to obtain a first estimate of mean temperature difference. Inlet and outlet temperatures of both fluids are chosen to meet the Grassman and Kopp requirement that $\Delta T = T/20$. A suitable number of intermediate stations is chosen along the temperature span of the cold fluid (usually 20, but 10 is used in Table 11.3 for compactness). Splinefitted temperature-enthalpy curves for both fluids are then used to calculate and match enthalpy increments on both sides of the exchanger, from which the corresponding temperature increments on the warm side can be found. The calculation requires knowledge of mass flow rates on both sides of the exchanger. It is convenient to set the cold mass flow rate to 1.00 kg/s and the warm fluid mass flow rate is iterated until the desired outlet temperature of the warm fluid is matched.

Usually a match is not obtained at the first trial, and the value of ΔT is then changed until the desired value of the warm fluid outlet temperature is obtained. The mean temperature difference for the exchanger is calculated as the average of the local temperature differences at stations along the exchanger, and this will be different from ΔT.

We now also have the *ratio* of the mass flow rates, and when this proceedure is followed for the whole cryogenic plant, then *actual* mass flow rates can be calculated.

Stage 2
The mean temperature difference of 6.236 K from Table 11.3 is used in direct sizing, together with actual mass flow rates, and the assumption of mean thermophysical properties. This proceedure is covered in Chapter 4 and includes an adjustment of the mean temperature difference to allow for longitudinal conduction. In this case the adjustment was 0.975, making the mean temperature difference 6.080K.

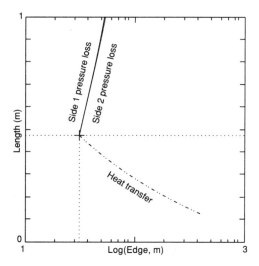

Fig. 11.14 Direct-sizing of the two-stream exchanger of Table 11.3 to determine maximum length

Direct sizing (Fig. 11.14) is carried out for three reasons: first to ensure that the pressure losses are as desired, second to optimise local surface geometries and approach the desired optimum exchanger, and third to obtain the edge length (E) and length (L) of the exchanger.

This is done for each combination pair of two flow streams in the multi-stream block. And for each combination pair, the design point is chosen at the upper left-hand end of the heat transfer curve corresponding to the maximum length (L) of that exchanger. If more than one stream is being cooled in the multi-stream block, then more than one selection of sets of combination pairs will prove possible, and an appropriate selection can be made at this stage.

Edge lengths (E) have to be whole multiples of block width in the multi-stream exchanger, and a suitable choice is made at this stage. The smallest value of (L) from the final set of combination pairs may also be selected as the block length of the multi-stream exchanger.

All selected combination pairs can now be recalculated to determine new pressure losses for the selected (E) and (L) values. It is desirable to use the same surface geometry for any stream which is split and serves more than one combination pair.

Stage 3
Stepwise rating of a single combination pair begins with assembling the necessary thermophysical data against temperature, either as tables or as interpolating splinefits. When a suitable number of stations are taken along the length of the exchanger in which the enthalpy balances are assured, then each small section of the exchanger can be dealt with as an individual exchanger.

11.6 STEPWISE RATING OF MULTI-STREAM HEAT EXCHANGERS

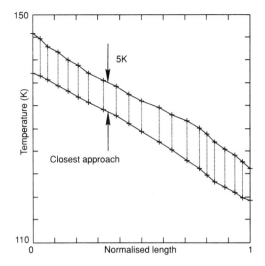

Fig. 11.15 Temperature profiles of the two-stream exchanger obtained using stepwise rating

The LMTD of individual sections can be calculated, together with the mean bulk temperatures of both fluids, and corresponding thermophysical properties can be found. Soyars (1992) did not find the ϵ-Ntu method accurate for this purpose.

Given the edge length (E) heat transfer and pressure losses may be determined for each section, and the required length and pressure loss for each section found. The calculation may be checked using the summed values of length and pressure loss for each section; they should be close to those obtained in the earlier direct-sizing step.

This produces the actual temperature-length profile for the exchanger of Table 11.3, which is shown in Fig. 11.15.

Obviously, each section of the multi-stream exchanger is likely to produce temperature profiles differing slightly from those shown in Fig. 11.15. This introduces further considerations:

- the desirability of using the same surface geometry in each section for the stream which is split
- the need to allow for cross-conduction effects
- appropriate choice of stacking pattern

11.6 STEPWISE RATING OF MULTI-STREAM HEAT EXCHANGERS

For multi-stream exchangers, temperatures of either the hot streams or the cold streams are not usually constant over each cross-section of the exchanger.

284 11 CRYOGENIC HEAT EXCHANGERS

Then *cross-conduction* effects between adjacent streams may significantly affect the performance. Haseler (1983) analysed this problem for the plate-fin design, using simple fin theory to evaluate cross-conduction effects, and he further showed how to incorporate an allowance for cross-conduction in the design process.

The algebra in Haseler's approach is compact and some assistance in getting quickly into his elegant solution seems appropriate. The differential equation governing heat conduction in a fin is

$$\frac{d^2 T_f}{dy^2} - \left(\frac{\alpha P}{\lambda A}\right)(T_f - T_c) = 0 \tag{11.6}$$

where T_f = in temperature
α = heat transfer coefficient
P = fin perimeter/unit length
λ = fin thermal conductivity
A = fin area for conduction

For a rectangular offset strip-fin equation (11.6) becomes

$$\frac{d^2 T_f}{dy^2} - \left(\frac{2\alpha}{\lambda tf}\right)(T_f - T_c) = 0 \quad \text{where } tf \text{ is fin thickness}$$

Putting $\theta_f = T_f - T_c$

$$\frac{d^2 \theta_f}{dy^2} - \left(\frac{2\alpha}{\lambda tf}\right)\theta_f = 0$$

for which the solution is

$$\theta_f = c_1 e^{my} + C_2 e^{-my} \quad \text{where} \quad m = \sqrt{\left(\frac{2\alpha}{\lambda tf}\right)}$$

Taking the origin at the centre of the fin, midway between two plates with spacing $b=2a$, the boundary conditions become

$$y = -b/2 = -a, \quad \text{say;} \quad \theta_f = \theta_{w1}$$
$$y = +b/2 = +a, \quad \text{say;} \quad \theta_f = \theta_{w2}$$

from which

$$C_1 = \frac{e^{ma}\theta_{w2} - e^{-ma}\theta_{w1}}{e^{2ma} - e^{-2ma}} \quad C_2 = \frac{e^{ma}\theta_{w1} - e^{ma}\theta_{w2}}{e^{2ma} - e^{-2ma}}$$

11.6 STEPWISE RATING OF MULTI-STREAM HEAT EXCHANGERS

and the solution for fin temperature becomes

$$\theta_f = \frac{\theta_{w2}(e^{ma}e^{my} - e^{-ma}e^{-my}) - \theta_{w1}(e^{-ma}e^{my} - e^{ma}e^{-my})}{(e^{2ma} - e^{-2ma})}$$

Digressing at this point, consider the expansion of

$$\frac{1}{2}(A+B)\frac{\cosh my}{\cosh ma} + \frac{1}{2}(A-B)\frac{\sinh my}{\sinh ma}$$

$$= A\left(\frac{e^{ma}e^{my} - e^{-ma}e^{-my}}{e^{2ma} - e^{-2ma}}\right) + B\left(\frac{e^{-ma}e^{my} - e^{ma}e^{-my}}{e^{2ma} - e^{-2ma}}\right)$$

which allows the solution for fin temperature to be written

$$\theta_f = \frac{1}{2}(\theta_{w2} + \theta_{w1})\frac{\cosh my}{\cosh ma} + \frac{1}{2}(\theta_{w2} - \theta_{w1})\frac{\sinh my}{\sinh ma} \tag{11.7}$$

Haseler writes heat transfer from the first wall in the y-direction as the sum of direct heat transfer and fin conduction

$$Q_L T = \alpha S_1 \theta_{w1} - N\lambda t f \left.\frac{dT}{dy}\right|_{y=-a} \tag{11.8}$$

where S_1 is primary surface per unit length along the exchanger
N is number of fins across the exchanger
$S_2 = 2aN$ is secondary surface per unit length along the exchanger

Differentiating equation (11.7) and substituting in (11.8)

$$Q_L T = \alpha S_1 \theta_{w1} + N\lambda t f m \left[\left(\frac{\theta_{w2} + \theta_{w1}}{2}\right)\tanh ma - \left(\frac{\theta_{w2} - \theta_{w2}}{2}\right)\coth ma\right]$$

$$= \alpha S_1 \theta_{w1} + \alpha\left(\frac{\tanh ma}{ma}\right)S_2\left(\frac{\theta_{w2} + \theta_{w1}}{2}\right) - \alpha\left(\frac{\coth ma}{ma}\right)S_2\left(\frac{\theta_{w2} - \theta_{w1}}{2}\right)$$

$$= \alpha S_1 \theta_{w1} + \alpha\left(\frac{\tanh ma}{ma}\right)S_2\left(\theta_{w1} + \frac{\theta_{w2} - \theta_{w1}}{2}\right) - \alpha\left(\frac{\coth ma}{ma}\right)S_2\left(\frac{\theta_{w2} - \theta_{w1}}{2}\right)$$

$$= \alpha S_1 \theta_{w1} + \alpha\left(\frac{\tanh ma}{ma}\right)S_2\theta_{w1} + \alpha\left(\frac{\tanh ma - \coth ma}{2ma}\right)S_2(\theta_{w2} - \theta_{w1})$$

The standard expression for fin efficiency is

$$\eta = \frac{\tanh ma}{ma}$$

Haseler defines fin 'bypass' efficiency as

$$\eta' = \frac{\coth ma - \tanh ma}{2ma} = \frac{\operatorname{cosech} 2ma}{ma}$$

then Haseler's equations become

$$\dot{Q}_{LT} = \alpha(S_1 + \eta S_2)\theta_{w1} + \alpha\eta' S_2(\theta_{w1} - \theta_{w2})$$
$$\dot{Q}_{LT} = \dot{Q}_L + \dot{Q}_B$$

where $\dot{Q}_{LT}$ = total heat flow from left-hand wall per unit length

$\dot{Q}_L$ = heat flow from left wall to or from fluid stream per unit length

$\dot{Q}_B$ = bypass heat flow per unit length

A similar set of equations exists for the right-hand wall

$$\dot{Q}_{RT} = \alpha(S_1 + \eta S_2)\theta_{w2} - \alpha\eta' S_2(\theta_{w1} - \theta_{w2})$$
$$\dot{Q}_{RT} = \dot{Q}_R + \dot{Q}_B$$

where $\dot{Q}_{RT}$ = total heat flow from left-hand wall per unit length

$\dot{Q}_R$ = heat flow from left wall to or from fluid stream per unit length

$\dot{Q}_B$ = bypass heat flow per unit length

The remainder of the analysis is straightforward, but it does require access to a large computer.

Instead of considering the number of separate streams, Haseler's analysis highlights the significance of *individual channel passages* when designing multi-stream exchangers, but more important, it focuses attention on the initial plant configuration stage, where there is opportunity to design out mixing losses and unacceptable temperature profiles, e.g. Fig. 11.9 and Table 11.4.

The full design process involves multipassage analysis (or perhaps multi-plate analysis), rather than multi-stream analysis, and solution of the simultaneous equations may then require a considerable amount of computational work. In an important paper, Suessman and Mansour (1979) provided a simple method for arriving at a good stacking pattern in the arrangement of individual flow passages. The stacking pattern is often repeated in an exchanger, and this can minimise the amount of computational work required by Haseler's method.

Mollekopf and Ringer (1987) indicate that Linde AG has developed an exact solution of the set of governing differential equations. This scheme assumes constant properties and is valid for incremental steps only, which is sufficient to allow computation of stacking patterns which deviate from the common wall temperature assumption.

Different philosophies are suggested by Haseler (1983) and by Prasad and Gurukul (1992) for carrying out the stepwise rating process in multi-stream exchangers. Prasad and Gurukul prefer to start the computation from the end where the temperature differential between hot and cold fluids is greatest. Haseler prefers to start the computation from the end at which only one stream temperature may be the true unknown; this is to avoid any instabilities that may otherwise appear in the calculation.

Papers by Prasad and Gurukul (1987), Paffenbarger (1990) and Prasad (1993) extend this work, and all papers cited in this section are recommended reading. Designing a multi-stream exchanger is not a fully explicit process.

Feasibility studies

A reasonable approximation to the final design can be achieved by adopting the stacking pattern of Suessman and Mansour, then working a stepwise rating design, assuming that hot fluid and cold fluid temperatures in any cross-section are reasonably constant. This should provide a first approximation on which to base cost estimates.

11.7 FUTURE COMMERCIAL APPLICATIONS

The thrust of the present work is towards new engineering developments, particularly in cryogenics. Electricity and hydrogen are the energy vectors of the future, and it is possible that new energy resource fields will exploit massive resources of hydropower and geothermal energy. It is already projected that hydrogen produced by electrolysis of water will be liquefied, and that bulk liquid hydrogen will then be shipped in sealed tanks mounted on skids to where the energy is required (Petersen *et al.* 1994).

Ceramic superconductors embedded in silver have been found which exhibit superconductivity up to 135 K. They have a current-carrying capacity greater than 10^6 A/cm^2, and are now being spun in lengths of 1000 m (Stansell 1994). Liquid nitrogen is a convenient, safe and inexpensive cryogen at 77 K. It is also the essential refrigerating cryogen in the technology of hydrogen liquefaction (Bracha *et al.* 1994). Interest in liquid nitrogen is set to increase as its applications grow in importance, including superconducting power generators and electricity storage in superconducting coils.

Liquid hydrogen is likely to find its first commercial application as a replacement fuel in aircraft populsion (Brewer 1993). The technical advantages of having a fuel with an energy content of 118.6 MJ/kg (which is 2.78 times that of conventional jet fuel) are considerable in the case of aircraft. The wings can be smaller, the landing gear lighter the engines less powerful engines are required, reducing gross take-off weight by 30%. These knock-on advantages do not exist for land-based or sea-based applications. And when the new technology arrives, advances in electronics, navigation and landing systems could mean that the aircraft will have no crew on the flight deck.

11.8 CONCLUSIONS

1. Considerations underlying the layout and design of a nitrogen liquefaction plant have been set out. This is an essential preliminary stage in obtaining parameters for heat exchanger design.
2. Factors affecting layout of a hydrogen liquefaction plant have been discussed, and an excellent example of such a plant is to be found in the paper by Bracha *et al.* (1994).
3. In the hydrogen liquefaction plant, coiled-tube heat exchangers have been used in sections where evaporation of a liquid is employed for *ortho:para* hydrogen conversion. This allows the evaporating shell-side fluid to equalise across the tube-bundle. Multi-stream plate-fin heat exchangers are preferred for heat exchange between single-phase gaseous fluids.
4. A method of arriving at a first estimate of the cross-section of multi-stream exchangers by direct-sizing has been outlined.
5. Papers on stepwise rating of multi-stream exchangers have been indicated, including important aspects of stacking pattern selection and allowance for cross-conduction effects.

REFERENCES

Alfeev, V. N., Brodyansky, V. M., Yagodin, V. M., Nikolsky, V. A. and Ivantsov, A. V. (1971) Refrigerant for a cryogenic throttling unit, *British patent 1336892*, 14 November 1973.
Barron, R. F. (1985) *Cryogenic Systems*, Oxford: OUP.
Bougard, J. and Afgan, N. (1987) *Heat and Mass Transfer in Refrigeration and Cryogenics*, New York: Hemisphere/Springer.
Bracha, M., Lorenz, G., Patzelt, A. and Wanner, M. (1994) Large-scale hydrogen liquefaction in Germany, *International Journal of Hydrogen Energy*, **19**(1), 53–59.
Brewer, R. D. (1993) *Hydrogen Aircraft Technology*, Boca Raton, FL: CRC Press.
Grassmann, P. and Kopp, J. (1957) Zur günstigen Wahl der temperaturdifferenz und der Wärmeuberganszahl in Wärmeaustauschern, *Kältetechnik*, **9**(10), 306–308.
Haselden, G. G. (1971) *Cryogenic Fundamentals*, New York: Academic Press.
Haseler, L. E. (1983) Performance calculation methods for multi-stream plate-fin heat exchangers, In: *Heat Exchangers: Theory and Practice*, Eds. J. Taborek,

G. F. Hewitt and N. Afgan, New York: Hemisphere/McGraw-Hill, pp. 495–506.

Little, W. A. (1984) Microminiature refrigeration, *Review of Scientific Instruments*, **55**(5) 661–680.

Marshall, J. and Oakey, J. D. (1985) Gas refrigeration method, *European patent application 85305248. 8*.

Mollekopf, N. and Ringer, D. U. (1987) Multistream heat exchangers—types, capabilities and limits of design, In: *Heat and Mass Transfer in Refrigeration and Cryogenics*, Eds. J. Bougard and N. Afgan, Hemisphere/Springer-Verlag, pp. 537–546.

Paffenbarger, J. (1990) General computer analysis of multi-stream plate-fin heat exchangers, In: *Compact Heat Exchangers – A Festschrift for A. L. London*, Eds. R. K. Shah, A. D. Kraus and D. Metzger, Washington: Hemisphere, pp. 727–746.

Paugh, R. L. (1990) New class of microminiature Joule-Thompson refrigerator and vacuum package, *Cryogenics*, **30**, 1079–1083.

Peschka, W. (1992) *Liquid Hydrogen: Fuel of the Future*, Berlin: Springer-Verlag.

Petersen, U. Wursig, G. and Krapp, R. (1994) Design and safety considerations for large-scale sea borne hydrogen transport, *International Journal of Hydrogen Energy*, **19**(7), 597–604.

Prasad, B. S. V. (1993) The performance prediction of multi-stream plate-fin heat exchangers based on stacking pattern, *Heat Transfer Engineering*, **12**(4), 58–70.

Prasad, B. S. V. and Gurukul, S. M. K. A. (1987) Differential methods for sizing multistream plate fin heat exchangers, *Cryogenics*, **27**, 257–262.

Prasad, B. S. V. and Gurukul, S. M. K. A. (1992) Differential methods for the performance prediction of multi-stream plate-fin heat exchangers, *Transactions ASME, Journal of Heat Transfer*, **114**, 41–49.

Scott, R. B. (1959) *Cryogenic Engineering*, New York: Van Nostrand.

Smith, E. M. (1984) A possible method for improving energy efficient para-LH_2 production, *International Journal of Hydrogen Energy*, **9**(11) 913–919.

Smith, E. M. (1989a) Slush hydrogen for aerospace applications, *International Journal of Hydrogen Energy*, **14**(3), 201–213.

Smith, E. M. (1989b) Liquid oxygen for aerospace applications, *International Journal of Hydrogen Energy*, **14**(11), 813–837.

Soyars, W. M. (1991) The applicability of constant property analyses in cryogenic helium heat exchangers, *Proceedings of the 1991 Cryogenics Engineering Conference, Advances in Cryogenic Engineering*, **37A**, 217–223.

Stansell J. (1994) *Sunday Times*, 3 April 1994, Business Section 3, p. 10.

Suessman, W. and Mansour, A. (1979) Passage arrangement in plate-fin exchangers, *Proceedings of XV International Congress of Refrigeration*, Vol. 1, pp. 421–429.

Touloukian, Y. S. and Makita, T. (1970) *Specific Heat – Nonmetallic Liquids and Gases*, New York: Plenum Press.

Vargaftik, N. B. (1975) *Tables on the Thermophysical Properties of Liquids and Gases*, New York: Hemisphere / Wiley.

Weimer, R. F. and Hartzog, D. G. (1972) Effect of maldistribution on the performance of multistream multipassage heat exchangers, *Proceedings of the 1972 Cryogenies Engineering Conference, Advances in Cryogenic Engineering*, **18**, Paper B-2, 52–64.

Whitfield, A. (1990) The preliminary design of radial inflow turbines, *Transactions ASME, Journal of Turbomachinery*, **112**, 50–57.

Whitfield, A. and Baines, N. C. (1990) *Design of Radial Turbomachines*, London: Longman.

12

VARIABLE HEAT TRANSFER COEFFICIENTS

12.1 WITH AND WITHOUT PHASE CHANGE

Real heat exchangers do not have constant heat transfer coefficients, even in single-phase designs; this is due to the temperature dependence of thermophysical properties. Some may have approximately constant properties, but others do not, e.g. the cryogenic exchanger discussed in Chapter 11.

In single-phase designs, the temperature dependence of physical properties is enough to change the values of the Reynolds number, the Prandtl number and therefore the Nusselt number. Ultimately it will change the overall heat transfer coefficient along the length of the heat exchanger.

Attempts have been made to adjust the expression for overall heat transfer coefficient U allowing for assumed mathematical variation of the overall coefficient along the exchanger (Schack 1965, Hausen 1950, 1983), but these analytical methods have less relevance now that computers are generally available. In fact it is necessary to design the exchanger first, in order to obtain the variation of U along its length.

For single-phase design, it is possible to size an exchanger incrementally by using splinefits to represent the physical properties involved and calculating each increment as if it were itself a small exchanger. The approach is no longer that of direct sizing, but direct sizing can still be used to obtain a good initial feel for the final size of the unit. Most of the earlier material in this text is relevant to designing single-phase heat exchangers by stepwise methods.

Once it has been accepted that stepwise design by computer is the most accurate way to go, it is straightforward to proceed to the more complicated design of heat exchangers involving two-phase flow. But first there has to be considered, the method of calculating pressure loss in two-phase flow. Second, there is a need to understand the several forms of two-phase flow which will exist in the design, so that appropriate correlations may be employed. Third, maldistribution and instability of flow in plate-fin and other exchangers has to be designed-out if possible.

The author has been highly selective in this chapter because the objective is simply to introduce the reader to the computer design approach. The wise reader will read more widely on two-phase flow and consult the several excellent texts now available before proceeding to his/her own design application. Good starting points are the texts by Collier (1972), Hewitt and Hall-Taylor (1970), Wallis (1969), Bergles *et al.* (1981), Chisholm (1983), Smith (1986), Carey (1992) and Hewitt *et al.* (1994). There are also good articles in the *Handbook of Heat Transfer*, (Rohsenow *et al.* 1985) and in the *Handbook of Multiphase Systems* (Hestroni 1982), and the reader is encouraged to use all journal and database sources to trace other authors and papers.

12.2 TWO-PHASE FLOW REGIMES

With extreme clarity, Rhee (1972) states: 'Knowing the flow patterns of a two-phase flow is as important as knowing whether the flow is laminar or turbulent in single-phase flow. ' Flow pattern maps for various tube geometries are to be found in Hewitt *et al.* (1994), among others. Here we shall be concerned only with forced-flow evaporation in a horizontal tube and we will use the relatively early work of Rhee (1972) simply to illustrate the computational approach to the problem.

Rhee found for refrigerant 12 that, with horizontal tubes, the general description of flow pattern, in order of increasing vapour quality, was nucleate (or bubbly) flow, stratified (sometimes slug or wavy) flow, annular flow (sometimes with mist, sometimes without) and mist flow itself.

Nucleate flow occurs for an extremely short length of tube when vapour bubbles appear as the liquid first reaches saturation temperature. It quickly changes to stratified flow as more vapour appears, so that separate streams of liquid and vapour flow in the tube. Sometimes liquid slugs form during this stage of evaporation, sometimes liquid waves appear.

At a later stage in the evaporation there is a transition from stratified flow to annular flow. This seems to occur when a higher heat transfer coefficient would result for annular flow than for stratified flow.

After a period of annular flow there is a sudden drop in heat transfer coefficient with a change to mist flow. This causes a jump in the temperature of the tube wall and is associated with the potentially dangerous condition of *burnout*, which will arise when the heat flux is being produced by external means (e.g. electrical or nuclear heating) and is not reduced immediately. In a normal heat exchanger this is simply a condition to evaluate, and the location of transition seems to be controlled principally by the Weber number. Mist flow then continues until all liquid has evaporated.

For his test fluid, refrigerant 12, Rhee found that there was one other condition to be noted, a condition related to mass velocity.
Above a critical mass velocity the flow pattern is

Nucleate flow ⇒ Stratified flow ⇒ Annular mist flow ⇒ Mist flow

Below the critical mass velocity the flow pattern is

Nucleate flow ⇒ Stratified flow ⇒ Annular (no mist) flow

The critical mass velocity needs to be evaluated before or during the computation, so that the correct flow pattern may be computed.

Other flow situations

Obviously there are many other possible two-phase flow design situations, e.g.
- internal forced-flow condensation in a tube
- external longidudinal forced-flow evaporation on a tube
- external transverse forced-flow evaporation on a tube
- external longitudinal forced-flow condensation on a tube
- annular forced-flow between tubes
- flow in plate-fin surfaces
- permanent dropwise condensation on a surface

The reader is encouraged to seek modern methods of design for these other flow situations in the references cited at the end of Section 12.1.

12.3 TWO-PHASE PRESSURE LOSS

It is necessary to know the saturation temperature at any point along a heat exchanger in order to calculate physical properties. As saturation temperature is dependent on saturation pressure, it follows that incremental pressure loss along the exchanger must be evaluated so that correct values of physical properties are obtained.

Several different models have been proposed for calculating pressure loss in two-phase flow; see e.g. Wallis (1969), Collier (1972), Friedel (1979), Bergles *et al.* (1981) and Chisholm (1983). According to Chisholm (1983), the Armand method is the most elegant and the Lockhart-Martinelli approach is the most easily applied because it does not explicitly consider flow pattern. Hewitt *et al.* (1994) recommend the method of Taitel and Dukler (1976) for predicting the flow pattern on horizontal flow, and they recommend the Friedel (1979) correlation for calculating pressure loss. These last two approaches are probably now to be preferred.

In his 1972 application, Rhee observed that pressure loss in two-phase flow was small, so it hardly seemed to matter which model was used; he therefore chose a simple *linear* fit of the Lockhart-Martinelli data. We shall stay with the Lockhart-Martinelli correlation so as not to depart from Rhee's calculations, but we shall use the better *curved* fit developed for this approach by Chisholm and Laird (1958), Chisholm (1967) and Collier (1972). It is not complicated; the analysis is given below.

Using the Lockhart-Martinelli model and defining quality of the vapour as x, then

$$x = \frac{\dot{m}_g}{\dot{m}_g + \dot{m}_f} \quad \text{from which} \quad \frac{\dot{m}_f}{\dot{m}_g} = \frac{1-x}{x}$$

Using frictional pressure loss only (neglecting acceleration loss, and with zero static head loss for a horizontal tube)

$$-\left(\frac{\Delta p}{\Delta \ell}\right) = \frac{4fG^2}{2\rho}\left(\frac{1}{d}\right)$$

and with

$$G = \frac{4\dot{m}}{\pi d^2}, \quad \text{Re} = \frac{dG}{\eta}, \quad f = a.\text{Re}^{-n}$$

$$-\left(\frac{\Delta p}{\Delta \ell}\right) = \left(\frac{4^{3-n}}{\pi^{2-n}} \cdot \frac{a}{2}\right)\left(\frac{\eta^n}{\rho}\right)\left(\frac{\dot{m}^{2-n}}{d^{5-n}}\right)$$

If the vapour fraction that is actually flowing were to alone occupy the pipe of diameter d, then

$$-\left(\frac{\Delta p}{\Delta \ell}\right)_g = (\text{const})\left(\frac{\eta_g^n}{\rho_g}\right)(\dot{m}_g^{2-n})$$

Similarly for the liquid fraction

$$-\left(\frac{\Delta p}{\Delta \ell}\right)_g = (\text{const})\left(\frac{\eta_f^n}{\rho_f}\right)(\dot{m}_f^{2-n})$$

Defining X where

$$X^2 = \frac{\left(\frac{\Delta p}{\Delta \ell}\right)_f}{\left(\frac{\Delta p}{\Delta \ell}\right)_g} = \left(\frac{\eta_f}{\eta_g}\right)^n \left(\frac{\rho_g}{\rho_f}\right)\left(\frac{\dot{m}_f}{\dot{m}_g}\right)^{2-n}$$

then X^2 provides a measure of the degree to which the two-phase mixture behaves like the liquid rather than like the gas.

Introducing the *two-phase multipliers* relating the pressure loss in each component flow to the same two-phase pressure loss

$$\left(\frac{\Delta p}{\Delta \ell}\right)_{tp} = \left(\frac{\Delta p}{\Delta \ell}\right)_{g} \times (\phi_g)^2$$

$$\left(\frac{\Delta p}{\Delta \ell}\right)_{tp} = \left(\frac{\Delta p}{\Delta \ell}\right)_{f} \times (\phi_f)^2$$

Thus

$$X^2 = \frac{(\phi_g)^2}{(\phi_f)^2} = \frac{\left(\frac{\Delta p}{\Delta \ell}\right)_f}{\left(\frac{\Delta p}{\Delta \ell}\right)_g} = \left(\frac{\eta_g}{\eta_f}\right)^n \left(\frac{\rho_g}{\rho_f}\right) \left(\frac{1-x}{x}\right)^{2-n}$$

Lockhart and Martinelli prepared empirical correlations from experimental data to relate ϕ_g, ϕ_f and X. Chisholm and Laird (1958), Collier (1972) and Chisholm (1983) report that these curves may be approximated graphically by the following expressions which are represented in Fig. 12.1:

$$(\phi_g)^2 = 1 + CX + X^2$$

$$(\phi_f)^2 = \frac{1}{X^2} + \frac{C}{X} + 1$$

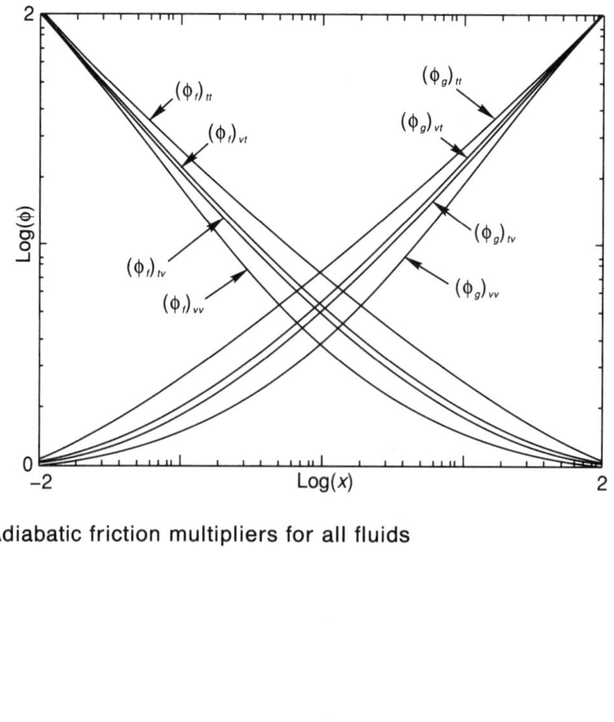

Fig. 12.1 Adiabatic friction multipliers for all fluids

The constant C is found from Table 12.1 for the two possible flow conditions in liquid and vapour streams. The turbulent-viscous case is not often encountered.

Table 12.1 Values of constant C (After Chisholm 1983)

Liquid-vapour	C	Re_f	Re_g
turbulent-turbulent (tt)	21*	> 2000	> 2000
viscous-turbulent (vt)	12	> 2000	< 1000
turbulent-viscous (tv)	10	< 1000	> 2000
viscous-viscous (vv)	< 5	< 1000	< 1000

*Some authors may use C = 20.

The above means may be used to determine two-phase pressure losses, approximately.

For the problem in question

The Lockhart and Martinelli approach provides

$$\left(\frac{\Delta p}{\Delta \ell}\right)_{tp} = (\phi_g)^2 \times \left(\frac{\Delta p}{\Delta \ell}\right)_g$$

$$X = \left(\frac{\eta_f}{\eta_g}\right)^{n/2} \left(\frac{\rho_g}{\rho_f}\right)^{1/2} \left(\frac{1-x}{x}\right)^{1-n/2}$$

For 100% vapour flowing turbulently in a tube

$$\left(\frac{\Delta p}{\Delta \ell}\right)_g = \frac{4fG^2}{2\rho}\left(\frac{1}{d}\right)$$

The Blasius correlation gives

$$f = 0.0791(Re)^{-0.25} \text{ from which } n = 0.25, \text{ thus}$$

$$X_{tt} = \left(\frac{\eta_g}{\eta_f}\right)^{0.125} \left(\frac{\rho_g}{\rho_f}\right) \left(\frac{1-x}{x}\right)^{0.875}$$

and the two-phase multiplier for turbulent-turbulent flow is

$$\phi_{tt}^2 = 1 + CX_{tt} + X_{tt}^2 \quad \text{where } C = 21$$
$$\text{for } 0.01 < X_{tt} < 10.0$$

Thus $\left(\frac{\Delta p}{\Delta \ell}\right)_{tp}$ can be found.

12.4 TWO-PHASE HEAT TRANSFER CORRELATIONS

Correlations presented in this section are those recommended by Rhee for calculating the performance of a double-tube exchanger with refrigerant 12 flowing in the central tube and in contraflow with water in the annulus. Hewitt *et al.* (1994) observe that the recommended Chen (1966) correlation was found to be seriously in error for refrigerant 12, and Shah (1976) considered that most correlations were not reliable beyond the range of data for which they were applicable (Smith 1986).

Rhee followed a systematic programme of experimentation on a purpose-built test rig which permitted adjustment of the two-phase inlet condition to any desired vapour quality, and he was thus able to explore each two-phase flow region with some precision and obtain extensive test results on which his correlations are based. The correlations themselves were developed after examining and assessing those of many other workers in the field of two-phase flow.

Rhee and Young are to be congratulated for seeking to find the solution to a real engineering problem, and for not being satisfied with exploring just one small part of the phenomenon of two-phase flow.

Nucleate boiling

This combines a Dittus-Boelter forced convection correlation with a McNelly pool boiling correlation, following a suggestion by Rohsenow that the two effects could be combined. The nucleate flow heat transfer coefficient is obtained from Nu.

$$\mathrm{Nu} = 0.023(\mathrm{Re}_f)^{0.8}(\mathrm{Pr}_f)^{0.4} + 0.30(\mathrm{Re}_{th})^{0.69}\left(\frac{\rho_f}{\rho_g} - 1\right)^{0.31}\left(\frac{p_i d}{\sigma}\right)^{0.31}$$

where $\mathrm{Re}_{th} = \dfrac{q_{flux}\, d}{\eta_f\, h_{fg}}$ is the thermal Reynolds number

σ is the surface tesion

These correlations are valid for very small values of vapour quality, say $x < 0.01$. There is a difficulty in evaluating the correlation because the thermal Reynolds number contains the same heat transfer coefficient as the Nusselt number on the left-hand side. An iterative approach is required, but this is not a serious impediment to solution as only one value of vapour quality is involved.

Stratified flow

The two-phase heat transfer coefficient (α_{tp}) is obtained from

$$\frac{\alpha_{tp}}{\alpha_f} = 313.76 \left(\frac{1+x}{1-x}\right)^{-0.542} (Fl)^{-0.448}$$

where the heat flux is $Fl = \dfrac{Gh_{fg}}{q_{flux}}$ and α_f is obtained from $Nu_f = 0.023 (Re)^{0.8} (Pr)^{0.4}$

There will be a numerical problem if attempts are made to evaluate the correlation at very high values of vapour quality (x). However, in design computation it is not necessary to approach the value $x = 1$ too closely, so the difficulty is avoided.

Annular-mist flow

The two-phase heat transfer coefficient (α_{tp}) is obtained from

$$\frac{\alpha_{tp}}{\alpha_f} = 59.03 \left(\frac{1+x}{1-x}\right)^{0.81} (Fl)^{-0.30}$$

where the heat flux is $Fl = \dfrac{Gh_{fg}}{q_{flux}}$ and α_f is obtained from $Nu_f = 0.023 (Re)^{0.8} (Pr)^{0.4}$

Annular (no mist) flow

The two-phase heat transfer coefficient (α_{tp}) is obtained from

$$\frac{\alpha_{tp}}{\alpha_f} = 140.8 \left(\frac{1+x}{1-x}\right)^{0.565} (Fl)^{-0.41}$$

where the heat flux is $Fl = \dfrac{Gh_{fg}}{q_{flux}}$ and α_f is obtained from $Nu_f = 0.023 (Re)^{0.8} (Pr)^{0.4}$.

Mist flow

$$Nu_f = 0.023 (Re_g)^{0.8} (Pr_g)^{0.4} (x)^{-1.23}$$
where x is the vapour quality

Transition from annular-mist flow to mist flow

This depends on the Weber number obtained from the following correlation:

$$We = 1.89 \times 10^{-7} (Re)_g \left(\frac{\rho_f}{\rho_g}\right)^{0.6} \left(\frac{d}{18.5318}\right)^{-0.48}$$

The form of the Weber number in the box was derived by Groothuis and Hendal (1959) for the case of heat addition to fluid flowing in a tube, when annular flow breaks down into mist flow. The Weber number is obtained using a velocity profile near the wall defined by the Prandtl universal profile, with the Blasius equation for wall friction,

$$We = \frac{x^2 G \eta_f}{\sigma \rho_g} \left(\frac{dG}{\eta_f}\right)^{0.125}$$

from which x can be determined

Demarcation mass velocity

Evaporation of refrigerant 12 in a horizontal tube proceeds in different ways depending on critical mass velocity G_{crit}.

Above G_{crit} the flow regimes being followed to 100% dry vapour are

Nucleate $\Rightarrow$ Stratified $\Rightarrow$ Annular mist $\Rightarrow$ Mist

Below G_{crit} the flow regimes being followed to 100% dry vapour are

Nucleate $\Rightarrow$ Stratified $\Rightarrow$ Annular no mist

Rhee found that the log-linear correlation $G_{crit} = B \exp(mT_{tp} + c)$ based on boiling temperature T_{tp} provided a good fit for refrigerant 12. For T_{tp} in K and G_{crit} in kg/(m² s) the constants take the following values:

$$B = 1.3563 \times 10^{-3} \quad m = 0.037\ 71 \quad C = 1.215\ 69$$

All the above correlations are for forced-flow evaporation of refrigerant 12 in a horizontal tube; they should not be used in any other circumstances without first checking their validity.

12.5 TWO-PHASE DESIGN OF A DOUBLE-TUBE EXCHANGER

The design exercise tackled was that of a double-tube heat exchanger with refrigerant 12 evaporating in the central tube, which was heated in contraflow by water flowing in the annulus. The undernoted parameters are for mass velocity $G = 221.40$ kg/(m²s).

Tube Parameters

Inner tube bore, m		d_i	= 0.011 887
Inner tube o.d., m		d	= 0.012 700
Inner tube wall thermal conductivity, J/(m s K)		λ_t	= 386.0
Outer tube bore, m		D	= 0.019 050

Refrigerant 12

Mass rate of flow of refrigerant, kg/s	$\dot{m}_r$	= 0.024 570
Inlet pressure of refrigerant, bar	p_1	= 4.102 21
Outlet pressure of refrigerant, bar	p_2	= 3.943 625

Water

Mass flow rate of water, kg/s	$\dot{m}_w$	= 0.653 94
Inlet temperature of water, K	T_1	= 287.37
Outlet temeperature of water, K	T_2	= 288.67

All physical properties were obtained from polynomial fits of data in a region close to the design conditions. In repeating the exercise the author converted all data to SI units before proceeding. In this exercise it was found that some of the datafits used by Rhee were not adequate, so new datafits were produced; thus the results presented here may differ somewhat from those of the original work by Rhee.

The first task is to determine the evaporative duty of the exchanger. A good approximation is obtained by using the latent heat of refrigerant 12 at the inlet condition. But this is not the correct duty because pressure loss due to friction and acceleration produces a different saturation condition at exit.

A good approximation to the other end conditions of both fluids is now available and the design can proceed. After a first design pass, the mass flow rate of water can be adjusted proportionately until the thermal duty on both sides becomes the same.

First design pass

The numerical procedure is by increments of vapor dryness fraction (x), and it is recommended that not less than 100 increments be used so that dryness is incremented in steps of 0.01.

The design proceeds by first evaluating G_{crit} to determine which correlations are to be used after stratified flow. It may be more accurate to evaluate G_{crit} at the end of stratified flow, but this is more easily done in a second design pass.

The correlation for nucleate flow is evaluated for only one very small increment of dryness, say 0.0001, as this two-phase flow region is extremely short.

12 VARIABLE HEAT TRANSFER COEFFICIENTS

All other correlations are to be evaluated separately for dryness increments of 0.01 over as much of the range as seems necessary, remembering that there is a numerical restriction in evaluating the term $(1+x)/(1-x)$ which appears in stratified flow, annular-mist flow and annular (no mist) flow. It is convenient to stop short of reaching 100% dryness as this does not affect the computation.

Each two-phase flow correlation and its associated Lockhart-Martinelli pressure loss correlation is placed inside a separate procedure body together with the heat transfer and pressure loss correlations for flow of water in the annulus. In each procedure body the dryness increment is used to calculate the following parameters, starting from the inlet end for refrigerant 12:

heat transferred in length increment $d\ell$
heat flux in length increment $d\ell$
overall heat transfer coefficent in length increment $d\ell$
pressure loss in length increment $d\ell$
pressure in refrigerant 12 at exit from length increment $d\ell$
water inlet temperature to length increment $d\ell$
cumulative length of tube $\Sigma d\ell$

Thus different curves can be produced over almost the whole length of the exchanger, showing how the two-phase heat transfer coefficient changes for each flow regime during evaporation (Figs. 12.2 and 12.3).

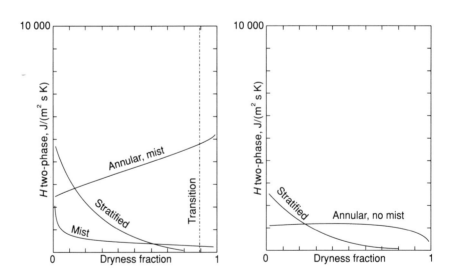

Fig. 12.2 Individual curves for two-phase flow above G_{crit} to base of dryness (x)

Fig. 12.3 Individual curves for two-phase flow below G_{crit} to base of dryness (x)

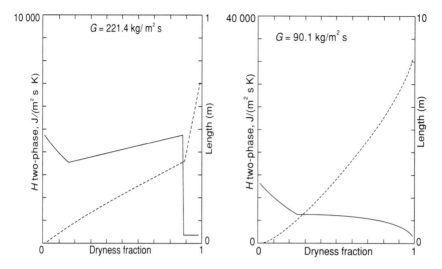

Fig. 12.4 Composite curve for two-phase flow above G_{crit} to base of dryness (x): heat transfer coefficient (solid), and exchanger length (dashed)

Fig. 12.5 Composite curve for two-phase flow below G_{crit} to base of dryness (x): heat transfer coefficient (solid) and exchanger length (dashed)

This information can be used to construct the actual behaviour of the evaporating fluid. Nucleate flow is the first point on the curve, at $x = 0$. Stratified flow proceeds until its heat transfer coefficient is exceeded either by

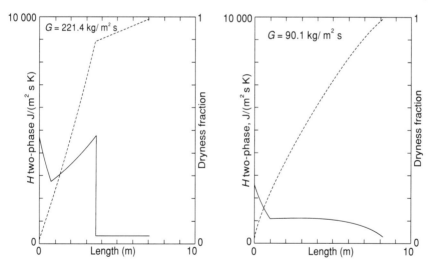

Fig. 12.6 Composite curve for two-phase flow above G_{crit} to base of length (ℓ): heat transfer coefficient (solid) and dryness (dashed)

Fig. 12.7 Composite curve for two-phase flow below G_{crit} to base of length (ℓ): heat transfer coefficient (solid) and dryness (dashed)

annular-mist flow or by annular (no mist) flow. If the refrigerant 12 mass velocity is below G_{crit} annular (no mist) flow continues to 100% dryness. If the refrigerant mass velocity is above G_{crit}, annular-mist flow continues to the dryness value determined by the Weber number then mist flow continues to 100% dryness (Figs. 12.4 to 12.7).

12.6 DISCUSSION

The software was written from scratch in SI units by the author, following guidelines provided by Rhee. The physical properties were not splinefitted, the recommended procedure, but were included as polynomial fits of data so they followed as closely as possible the method used by Rhee. Rhee's datafits were not used; instead the best available data was refitted by polynomials, and in the process, some serious discrepancies were found in the two representations of refrigerant 12.

Rhee used Du Pont data for Freon 12, whereas the author used ICI data for Arcton 12. Rhee admitted that there were disturbing inconsistencies with the Freon 12 data. In this light it cannot be certain that the computational predictions of Rhee are absolutely correct, consequently, the author's computations cannot be compared exactly with those of Rhee.

This is really not a serious problem as worldwide production of refrigerant 12 has now ceased because of damage to the ozone layer, and the above results are not likely to be used in anger.

However, the curves in Figs. 12.2 to 12.7 correspond very well in form to the test results obtained by Rhee (1972) and also to the independent experimental data of Chawla (1967) on refrigerant 11 boiling.

It will be noticed that the two-phase heat transfer coefficient in stratified flow always decreases as vapour dryness increases. Rhee reports that this effect was also experimentally noticed by other investigators in low mass flow rate studies (Chawla 1967, Zahn 1964), but that its explanation is straightforward:

> As the stratified flow develops, volume of the vapour on the top of the tube increases, lowering the value of the heat transfer coefficient in the upper part of the tube to that close to the heat transfer coefficient of the pure vapour. The more vapour generation, the larger the area covered by the vapour until it reaches the point where annular flow develops and the tube wall is again wetted with liquid.

It might be further remarked that, as annular-mist flow develops, liquid adheres to the tube wall because its higher viscosity allows a better match of slow-moving fluid to the stationary tube wall. In the core, the still higher speed vapour is happier to match its speed with the faster-moving liquid interface on its perimeter.

The 'dryout' transition in the two-phase heat transfer coefficient going from annular-mist flow to mist flow is not so sharp in the experimental data of

Chawla (1967), but this could be due to effects such as longitudinal thermal conduction in the tube wall.

In particular it is worth noting in Figs. 12.4 to 12.7 how the last small increment in dryness fraction requires a disproportionate length of the exchanger. It is clear that the much lower heat transfer coefficient on the water side is controlling this design during existence of the very high two-phase heat transfer coefficients, but when mist flow occurs there is closer correspondence with the heat transfer coefficients for water and refrigerant 12.

For those who may be despairing that no correlations yet exist for the two-phase fluid and horizontal surface geometry of their interest, initially it may be worth trying to establish the Weber number that provides the transition between high and low overall heat transfer coefficients. This will ease the task of designing because an inaccurate value for the two-phase flow heat transfer coefficient before transition will not greatly affect the design of an exchanger.

Where the problem may become more difficult is when both fluids in the exchanger change phase together. The computational problem becomes more complex, and the final heat exchanger will undoubtedly be short. This makes it easier for a slight change in operating conditions to perhaps move one fluid partially out of a short exchanger, as regards two-phase flow conditions. Caution is necessary because this situation is to be avoided.

In general, the work of Rhee and Young is a valuable contribution to design for two-phase flow, for it established a methodology of experimentation and also of heat exchanger design procedures on which future work may be based.

However, there is scope for reworking Rhee's data using the later paper of Friedel (1979), which provides the two-phase pressure loss correlations. There is little point in applying the Taitel and Dukler (1976) two-phase flow pattern map because Rhee's experimental technique has already identified each flow regime.

Friedel two-phase pressure loss correlation

This is based on evaluating friction factors for the pipe either totally filled with liquid (quality $x = 0$) or totally filled with dry vapour (quality $x = 1$).

Evaluate Reynolds numbers for flow with full mass velocity $G = \dot{m}/A$

$$\text{Reynolds number for liquid only } \text{Re}_f = \frac{dG}{\eta_f}$$

$$\text{Reynolds number for gas only } \text{Re}_g = \frac{dG}{\eta_g}$$

and determine the friction factors (f_f, f_g). Either the Fanning (16/Re), or the Moody (64/Re) definitions for friction factor will do, as subsequently the ratio of the two values will be taken.

The Friedel two-phase friction correlation is

$$(\phi_f)^2 = E + \frac{3.24FH}{(\text{Fr}_{tp})^{0.045}(\text{We}_{tp})^{0.035}}$$

where

$$E = (1-x)^2 + x^2 \left(\frac{\rho_f}{\rho_g}\right)\left(\frac{f_g}{f_f}\right)$$

$$F = x^{0.78}(1-x)^{0.24}$$

$$H = \left(\frac{\rho_f}{\rho_g}\right)^{0.91} \left(\frac{\eta_g}{\eta_f}\right)^{0.19} \left(1 - \frac{\eta_g}{\eta_f}\right)^{0.7}$$

Two-phase density

$$\rho_{tp} = \left[\frac{x}{\rho_g} + \frac{(1-x)}{\rho_f}\right]^{-1}$$

Froude number, where g is acceleration due to gravity (m/s²)

$$\text{Fr} = \frac{G^2}{gd(\rho_{tp})^2}$$

Weber number, where σ is surface tension (N/m)

$$\text{We} = \frac{G^2 d}{\sigma \rho_{tp}}$$

The two-phase pressure gradient is determined with the same expression as used in the Martinelli treatment

$$\left(\frac{\Delta p}{\Delta \ell}\right)_{tp} = \left(\frac{\Delta p}{\Delta \ell}\right)_f \times (\phi_f)^2$$

where the liquid-only pressure gradient (Fanning definition) is

$$-\left(\frac{\Delta p}{\Delta \ell}\right)_f = \frac{4 f_f G^2 (1-x)^2}{2 \rho_f} \left(\frac{1}{d}\right)$$

A full numerical example is to be found in Hewitt *et al.* (1994).

12.7 FLOW MALDISTRIBUTION

Cowans (1978) examined the problem of maldistribution of flow in single-phase heat exchangers, and has proposed the use of double-tapered flow passages (converging/diverging geometry for hot flows, and diverging/converging geometry for cold flows) to achieve a stable performance in a two-stream contraflow exchanger. He successfully demonstrated improved performance in a special purpose heat exchanger.

Recent two-phase work with plate-fin surfaces is to be found in the paper by Wadekar (1992), who considers vertical flow boiling of heptane, with earlier work on cyclohexane. Chen *et al.* (1981) have also studied boiling in plate-fin exchangers. Clarke and Robertson (1984) investigated convective boiling of liquid nitrogen in plate-fin heat exchanger passages; they found regions of superheated liquid in the exchanger, where boiling would have been expected, but the onset of boiling was delayed. This produced a considerable length of exchanger in which very low heat flux conditions existed and little heat transfer took place.

It is further remarked by Clarke and Robertson that the point of onset of evaporation appears to be affected by the method used to achieve the desired operating conditions, and they suggest that the onset conditions may be either stable or metastable.

The phenomenon of two-phase flow is not yet well understood in multi-stream plate-fin exchangers, and presently multi-start coil helical-tube heat exchangers may be preferred for evaporating service because the shell-side is fully interconnected. It seems obvious that compact plate-fin exchangers should also be configured to interconnect evaporating or condensing passages. This may involve the use of surface geometries like the rectangular offset strip-fin configuration, which is everywhere connected, plus transverse interconnection between all identical channels in the exchanger to equalise pressures in the evaporating or condensing stream.

This last concept will require reworking of the manufacturing process. Further experimental work is necessary on compact plate-fin exchangers to resolve the situation and to demonstrate stability in two-phase operation.

This effect may have similarities to roll-over in cryogenic tanks, where the temperature of liquid at the bottom of the tank may be higher than saturation at the evaporating surface. In vertical boiling in a channel, the column of liquid may exert sufficient pressure to suppress evaporation until explosive evaporation takes place. With vertical boiling it seems that the presence of gravitational forces may be anticipated in the theoretical correlations.

12.8 SOME FURTHER PROBLEMS

Air dehumidification using plate-fin and tube heat exchangers is discussed by Seshimo *et al.* (1989), with extension to frosting conditions by Ogawa and *et al* (1993). Related work is reported by Kondepudi and O'Neal (1989) and by Machielson and Kershbaumer (1989).

Readers with an interest in condensation should consult Chu and McNaught (1992), McNaught (1982, 1985), Bergles *et al.* (1981) and Collier (1972).

Ice harvesting during off-peak operation of gas-turbine plant has been used for cooling compressor inlet air during peak power operation as an effective way of increasing system capacity without increasing system operating costs. Heat transfer involving the growth of ice on refrigerated surfaces is a moving boundary problem or Stefan problem it is discussed in the text edited by Ockendon and Hodgkins (1975). A simple example of the non-linearity of the moving boundary problem is to be found in the paper by Perkeris and Schlichter (1933) about external ice formation on a tube, but the mathematics of that problem can be linearised without too much inaccuracy.

The problem becomes complex for airflow over iced surfaces, when simple geometries become modified by selective ablation, making the numerical problem more difficult (Date 1991).

One solution is to control the thickness of the ice sheet so that the assumed geometry is not lost. Another is to harvest the ice sheet by switching refrigerant fluid from liquid to gas inside the bayonet-tube or vertical hollow plate. Once the ice has fallen into a storage tank, a new cycle of ice production and harvesting can begin.

Mixtures of ice crystals and glycol are now of interest as secondary-circuit fluids in refrigeration systems.

Much remains to be done in multiphase systems. A programme of work on two-phase flow in compact heat exchangers is now underway at the laboratories of the Heat Transfer and Fluid Flow Service (HTFS), Harwell, UK, which should resolve some of the problems.

12.9 RATE PROCESSES

It would not be proper to close without brief mention of the many correlations used in heat transfer design. It was never the intention to cover this topic, yet it would be remiss not to mention the contributions of Churchill (1988, 1992) in explaining rate processes, and the asessment of his work by Kabel (1992).

REFERENCES

Bergles, A. E., Collier, J. G., Delhaye J. M., Hewitt, G. F. and Mayinger, F. (1981) *Two-Phase Flow and Heat Transfer in the Power and Process Industries*, Washington: Hemisphere.

Carey, V. P. (1992) *Liquid-Vapor Phase-Change Phenomena*, Washington: Hemisphere.

Chawla, J. M. (1967) Local heat transfer and pressure drop for refrigerants evaporating in horizontal tubes, *Kaltetechnik*, **19**, 246–252.

Chen, J. C. (1966) Correlation for boiling heat transfer to saturated fluids in convective flow, *Industrial and Engineering Chemistry, Process Design and Development*, **5**(3), 322–333.

Chen, C. C., Loh, J. U. and Westwater, J. W. (1981) Prediction of boiling heat transfer duty in a compact plate-fin heat exchanger using the improved local assumption, *International Journal of Heat and Mass Transfer*, **24**(12), 1907–1912.

Chisholm, D. (1967) A theoretical basis for the Lockhart-Martinelli correlation for two-phase flow, *International Journal of Heat and Mass Transfer*, **10**(12), 1767–1778.

Chisholm, D. (1983) *Two-Phase Flow in Pipelines and Heat Exchangers*, London: Longman.

Chisholm, D. and Laird, A. D. K. (1958) Two-phase flow in rough tubes, *Transactions ASME*, **80**(2), 276–286.

Chu, C. M. and McNaught, J. M. (1992) Condensation on bundles of plain and flow-finned tubes – effects of vapour shear and condensate inundation, *Institute of Chemical Engineers, Symposium Series*, **1**(129), 225–232.

Churchill, S. W. (1988) The role of mathematics in heat transfer, *AIChE Symposium Series, Heat Transfer*, **84**(263), 1–13.

Churchill, S. W. (1992) The role of analysis in the rate procesess, *Industrial and Engineering Chemistry, Research*, **31**, 643–658.

Clarke, R. H. (1992) Condensation heat transfer characteristics of liquid nitrogen in serrated plate-fin passages, *3rd UK National Conference Incorporating 1st European Conference on Thermal Sciences, Institution of Chemical Engineers, Symposium Series*, **2**(129), 1301–1309.

Clarke, R. H. and Robertson, J. M. (1984) Investigations into the onset of convective boiling with liquid nitrogen in plate-fin heat exchanger passages under constant wall temperature boundary conditions, *AIChE Symposium Series, Heat Transfer, Niagara Falls*, pp. 98–103.

Collier, J. G. (1972) *Convective Boiling and Condensation*, New York: McGraw-Hill.

Collier, J. G. and Thome, J. R. (1994) *Convective Boiling and Condensation*, New York: McGraw-Hill.

Cowans, K. W. (1978) A countercurrent heat exchanger that compensates automatially for maldistribution of flow in parallel channels, *Advances in Cryogenic Engineering*, **19**, 437–444.

Date, A. W. (1991) A strong enthalpy formulation for the Stefan problem, *International Journal of Heat and Mass Transfer*, **34**(9), 2231–2235.

Foumeny, E. A. and Heggs, P. J. (1991) *Heat Exchange Engineering*, Vos. 1, 2, New York: Ellis Horwood. (Vol. 2 covers compact heat exchangers: techniques of size reduction.)

Friedel, L. (1979) New friction pressure drop correlations for upward, horizontal and downward two-phase pipe flow, *HTFS Symposium, Oxford, 1979*. (Hoechst AG Reference 372217/24 698).

Groothuis, H. and Hendal, W. P. (1959) Heat transfer in two-phase flow, *Chemical Engineering Science*, **11**, 212–220.

Gungar, K. E. and Winterton, R. H. S. (1986) General correlation for flow boiling in tubes and annuli, *International Journal of Heat and Mass Transfer*, **29**(3), 351–358.

Hahne, E., Spindler, K. and Skok, N. (1993) New pressure drop correlations for subcooled flow boiling of refrigerants, *International Journal of Heat and Mass Transfer*, **36**(17), 4267–4274. (Refrigerants are R12 and R134a.)

Haseler, L. E. (1980) Condensation of nitrogen in brazed aluminium plate-fin heat exchangers, *ASME/AIChE National Heat Transfer Conference, Orlando 1980*, Paper 80-HT-57

Hausen, H. (1950) *Wärmeübertragung im Gegenstrom, Gleichstrom und Kreuzstrom*, Berlin: Springer-Verlag.

Hausen, H. (1983) *Heat Transfer in Counterflow, Parallel Flow and Cross Flow*, New York: McGraw-Hill.

Hestroni, G. (ed) (1982) *Handbook of Multiphase Systems*, New York: Hemisphere. (See Hewitt, G. F., Ch. 2, Gas-liquid flow.)

Hewitt, G. F. and Hall-Taylor, M. S. *Annular Two-Phase Flow*, Oxford: Pergamon.

Hewitt, G. F., Shires, G. L. and Bott, T. R. (1994) *Process Heat Transfer*, Boca Raton, FL: CRC Press, Ch. 10.

Kabel, R. L. (1992) Reflections on rates, *Industrial and Engineering Chemistry*, **31**, 641–643.

Kondepudi, S. N. and O'Neal, D. L. (1989) Effect of frost formation on the performance of louvred finned tube heat exchangers, *International Journal of Refrigeration*, **12**, 151–158.

Lockhart, R. W. and Martinelli, R. C. (1949) Proposed correlation of data for isothermal two-phase, two component flow in pipes, *Chemical Engineering Progress*, **45**(1), 39–48.

Machielson, C. H. M and Kerschbaumer, H. G. (1989) Influence of frost formation and defrosting on the performance of air coolers: standards and dimensionless coefficients for the system designer, *International Journal of Refrigeration*, **12**(5), 283–290.

McNaught, J. M. (1982) Two-phase forced convection heat transfer during condensation on a horizontal tube bundle, *7th International Heat Transfer Conference, Munich 1982*, Vol. 5, Ed. U. Grigull, Washington: Hemisphere, pp. 125–131.

McNaught, J. M. (1985) Condenser design, *Process Engineering*, **66**(1), 38–39, 42–43.

Ockendon, J. R. and Hodgkins, W. R. (eds) (1975) *Moving Boundary Problems in Heat Flow and Diffusion* Oxford: Clarendon Press.

Ogawa, K. Tanaka, N. and Takeshita, M. (1993) Performance improvement of plate-fin-and-tube heat exchangers under frosting conditions, *ASHRAE Transactions*, **99**(1), 762–771.

Paliwoda, A. (1989) Generalised method of pressure drop and tube length calculation with boiling and condensing refrigerants within the entire zone of saturation, *International Journal of Refrigeration*, **12**, 314–322.

Perkeris, C. L. and Schlichter, L. B. (1933) Problem of ice formation, *Journal of Applied Physics*, **10**, 135–137.

Rhee, B.-W. (1972) Heat transfer to boiling refrigerants R-12 and R-22 flowing inside a plain copper tube, PhD thesis, University of Michigan.

Rhee, B.-W. and Young, E. H. (1974) Heat transfer to boiling refrigerants flowing inside a plain copper tube, *Heat Transfer – Research and Design, AIChE Symposium Series*, **70**(138), 64–70.

Robertson, J. M. (1979) Boiling heat transfer with liquid nitrogen in brazed aluminium plate-fin heat exchangers, *AIChE Symposium Series*, **75**(189), 151–164.

Rohsenow, W. M. and Choi, H. Y. (1961) *Heat, Mass and Momentum Transfer*, Englewood Cliffs, NJ: Prentice-Hall.

Rohsenow, W. M., Hartnett, J. P. and Ganic, E. N. (1985) Two-phase flow, In: *Handbook of Heat Transfer Fundamentals*, New York: McGraw-Hill, Ch. 13.

Savkin, N., Mesarkishvili, Z. and Bartsch, G. (1993) Experimental study of void fraction and heat transfer under upward and downward flow boiling conditions, (water in an annulus), *International Communications in Heat and Mass Transfer*, **20**(6), 783–792.

Schack, A. (1965) *Industrial Heat Transfer*, London: Chapman & Hall.

Schlager, L. M., Pate, M. B. and Bergles, A. E. (1989) Heat transfer and pressure drop during evaporation and condensation of R22 in horizontal micro-fin tubes, *International Journal of Refrigeration*, **12**, 6–14.

Schlunder, E. U. (ed) (1990) *Heat Exchanger Design Handbook*, Washington: Hemisphere.

Seshimo, Y., Ogawa, K., Marumoto, K. and Fujii, M. (1989) Heat and mass transfer performances of plate-fin and tube heat exchangers with dehumidification, *Heat Transfer Japanese Research*, **18**(5), 79–94.

Shah, M. M. (1976) A new correlation for heat transfer during boiling flow through pipes, *ASHRAE Transactions*, **82**(2), 66–86.

Smith, R. A. (1986) *Vaporisers, Selection Design and Operation*, London: Longman, pp. 190–192.

Taitel, Y. and Dukler, A. E. (1976) A model for predicting flow regime transitions in horizontal and near horizontal gas-liquid flow, *Americal Institute of Chemical Engineers Journal*, **22**, 47–55.

Wadekar, V. V. (1991) Vertical slug flow heat transfer with nucleate boiling, *27th National Heat Transfer Conference, Minneapolis, July 1991 ASME HTD-159 Phase Change Heat Transfer*, pp. 157–161.

Wadekar, V. V. (1992) Flow boiling of heptane in a plate-fin heat exchanger passage, In: *Compact Heat Exchangers for Power and Process Industries, Proceedings 28th National Heat Transfer Conference, San Diego, CA, 1992, ASME HTD* **201**, 1–6.

Wallis, G. B. (1969) *One-Dimensional Two-Phase Flow*, New York: McGraw-Hill.

Weimar, R. F. and Hartzog, D. G. Effects of maldistribution on the performance of multistream multipassage heat exchangers, *Advances in Cryogenic Engineering*, **18**, 52–64.

Zahn, W. R. (1964) A visual study of two-phase flow while evaporating in a horizontal tube, *ASME Journal of Heat Transfer*, **86C**, 417.

APPENDICES

A

TRANSIENT EQUATIONS WITH LONGITUDINAL CONDUCTION AND WALL THERMAL STORAGE

A.1 TEMPERATURE TRANSIENTS IN CONTRAFLOW

The time taken for hot fluid to travel a differential length (δx) may be regarded as the *instantaneous residence time* or *transit time* (δt), and the relation between instantaneous residence mass ($\tilde{m}$) and instantaneous mass flow rate becomes ($\tilde{m} = \dot{m}\delta t$) where $u = \delta x/\delta t$ is the instantaneous velocity. Thus

$$\tilde{m}_h = \frac{\dot{m}_h \delta x}{u_h}$$

Similarly for cold fluid

$$\tilde{m}_c = \frac{\dot{m}_c \delta x}{u_c}$$

Striking energy balances for the hot fluid, solid wall and cold fluid, respectively.

Hot fluid

$$\left\{\begin{matrix}\text{energy entering}\\ \text{with hot fluid}\end{matrix}\right\} - \left\{\begin{matrix}\text{energy leaving}\\ \text{with hot fluid}\end{matrix}\right\} - \left\{\begin{matrix}\text{heat transferred}\\ \text{to solid wall}\end{matrix}\right\} = \left\{\begin{matrix}\text{energy stored}\\ \text{in hot fluid}\end{matrix}\right\}$$

$$\dot{m}_h C_h T_h - \dot{m}_h C_h \left(T_h + \frac{\partial T_h}{\partial x}\delta x\right) - \alpha_h \left(S\frac{\delta x}{L}\right)(T_h - T_w) = \tilde{m}_h C_h \frac{\partial T_h}{\partial t}$$

A TRANSIENT EQUATIONS WITH LONGITUDINAL CONDUCTION

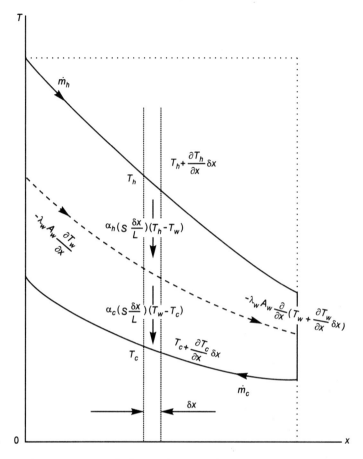

Fig. A.1 Differential energy balances in contraflow

$$-\dot{m}_h C_h \frac{\partial T_h}{\partial x} - \frac{\alpha_h S}{L}(T_h - T_w) = \frac{\dot{m}_h C_h}{u_h} \cdot \frac{\partial T_h}{\partial t}$$

$$\boxed{\frac{\partial T_h}{\partial t} + u_h \frac{\partial T_h}{\partial x} = -\frac{\alpha_h S}{\dot{m}_h C_h} \cdot \frac{u_h}{L}(T_h - T_w)}$$

Solid wall

$$\begin{Bmatrix}\text{energy entering}\\\text{by conduction}\end{Bmatrix} - \begin{Bmatrix}\text{energy leaving}\\\text{by conduction}\end{Bmatrix} + \begin{Bmatrix}\text{heat from}\\\text{hot fluid}\end{Bmatrix} - \begin{Bmatrix}\text{heat to}\\\text{cold fluid}\end{Bmatrix} = \begin{Bmatrix}\text{energy stored}\\\text{in solid wall}\end{Bmatrix}$$

$$-\lambda_w A_w \frac{\partial T_w}{\partial x} - \left[-\lambda_w A_w \frac{\partial}{\partial x}\left(T_w + \frac{\partial T_w}{\partial x}\delta x\right)\right] + \alpha_h\left(S\frac{\delta x}{L}\right)(T_h - T_w)$$

A.1 TEMPERATURE TRANSIENTS IN CONTRAFLOW

$$-\alpha_f \left(S\frac{\delta x}{L}\right)(T_w - T_g) = \rho_w(A_w \delta x)C_w \frac{\partial T_w}{\partial t}$$

With λ_w as constant

$$\lambda_w A_w \frac{\partial^2 T_w}{\partial x^2} + \frac{\alpha_h S}{L}(T_h - T_w) - \frac{\alpha_c S}{L}(T_w - T_c) = \rho_w A_w C_w \frac{\partial T_w}{\partial t}$$

Writing $M_w = \rho_w A_w L$ for solid wall, where necessary

$$\lambda_w A_w L \frac{\partial^2 T_w}{\partial x^2} + \alpha_h S(T_h - T_w) - \alpha_c S(T_w - T_c) = M_w C_w \frac{\partial T_w}{\partial t}$$

$$\frac{\partial T_w}{\partial t} - \left(\frac{A_w L}{V_w}\right)\kappa_x \frac{\partial^2 T_w}{\partial x^2} = \frac{\alpha_h S}{M_w C_w}(T_h - T_w) - \frac{\alpha_c S}{M_w C_w}(T_w - T_c)$$

Redefining thermal diffusivity as

$$\hat{\kappa} = \left(\frac{A_w L}{V_w}\right)\kappa_w \, (\mathrm{m^2/s})$$

$$\boxed{\frac{\partial T_w}{\partial t} - \hat{\kappa}_x \frac{\partial^2 T_w}{\partial x^2} = \left(\frac{\alpha_h S}{M_w C_w}\right)(T_h - T_w) - \left(\frac{\alpha_c S}{M_w C_w}\right)(T_w - T_c)}$$

Cold fluid

$$\left\{\begin{array}{c}\text{energy entering}\\ \text{with cold fluid}\end{array}\right\} - \left\{\begin{array}{c}\text{energy leaving}\\ \text{with cold fluid}\end{array}\right\} + \left\{\begin{array}{c}\text{heat transferred}\\ \text{from solid wall}\end{array}\right\} = \left\{\begin{array}{c}\text{energy stored}\\ \text{in cold fluid}\end{array}\right\}$$

$$\dot{m}_c C_c \left[T_c + \frac{\partial T_c}{\partial x}\delta x\right] - \dot{m}_c C_c T_c + \alpha_c \left(S\frac{\delta x}{L}\right)(T_w - T_c) = \tilde{m}_c C_c \frac{\partial T_c}{\partial t}$$

$$\dot{m}_c C_c \frac{\partial T_c}{\partial x} + \frac{\alpha_c S}{L}(T_w - T_c) = \frac{\dot{m}_c C_c}{u_c} \cdot \frac{\partial T_c}{\partial t}$$

$$\boxed{\frac{\partial T_c}{\partial t} - u_c \frac{\partial T_c}{\partial x} = +\frac{\alpha_c S}{\dot{m}_c C_c} \cdot \frac{u_c}{L}(T_w - T_c)}$$

A TRANSIENT EQUATIONS WITH LONGITUDINAL CONDUCTION

Collecting the three coupled partial differential equations for transients in contraflow

$$\frac{\partial T_h}{\partial t} + u_h \frac{\partial T_h}{\partial x} = -\left(\frac{\alpha_h S}{\dot{m}_h C_h}\right) \frac{u_h}{L} (T_h - T_w)$$

$$\frac{\partial T_w}{\partial t} - \hat{\kappa}_x \frac{\partial^2 T_w}{\partial x^2} = +\left(\frac{\alpha_h S}{M_w C_w}\right)(T_h - T_w) - \left(\frac{\alpha_c S}{M_w C_w}\right)(T_w - T_c)$$

$$\frac{\partial T_c}{\partial t} - u_c \frac{\partial T_c}{\partial x} = +\left(\frac{\alpha_c S}{\dot{m}_c C_c}\right) \frac{u_c}{L} (T_w - T_c)$$

(A.1)

When no longitudinal conduction is present, but wall thermal storage is important

$$\frac{\partial T_h}{\partial t} + u_h \frac{\partial T_h}{\partial x} = -\left(\frac{\alpha_h S}{\dot{m}_h C_h}\right) \frac{u_h}{L} (T_h - T_w)$$

$$\frac{\partial T_w}{\partial t} = +\left(\frac{\alpha_h S}{M_w C_w}\right)(T_h - T_w) - \left(\frac{\alpha_c S}{M_w C_w}\right)(T_w - T_c)$$

$$\frac{\partial T_c}{\partial t} - u_c \frac{\partial T_c}{\partial x} = +\left(\frac{\alpha_c S}{\dot{m}_c C_c}\right) \frac{u_c}{L} (T_w - T_c)$$

(A.2)

When no transients are present, but longitudinal conduction is important

$$+u_h \frac{\partial T_h}{\partial x} = -\left(\frac{\alpha_h S}{\dot{m}_h C_h}\right) \frac{u_h}{L} (T_h - T_w)$$

$$-\hat{\kappa}_x \frac{\partial^2 T_w}{\partial x^2} = +\left(\frac{\alpha_h S}{M_w C_w}\right)(T_h - T_w) - \left(\frac{\alpha_c S}{M_w C_w}\right)(T_w - T_c)$$

$$-u_c \frac{\partial T_c}{\partial x} = +\left(\frac{\alpha_c S}{\dot{m}_c C_c}\right) \frac{u_c}{L} (T_w - T_c)$$

(A.3)

Equations (A.3) are Kroegers equations, from which he obtained a closed-form solution for the case of equal water equivalents of the two fluids.

Note
It is demonstrated elsewhere that wall thermal storage virtually controls transient behaviour of the exchanger, thus solutions which ignore wall thermal storage have little practical value when normal liquids and gases are involved.

BIBLIOGRAPHY

Temperature transients
Acklin, L. and Laubli, F. (1960) Calculation of the dynamic behaviour of heat exchangers with the aid of analogue computers, *Sulzer Technical Review*, No. 4, 13–20.
Acrivos, A. (1956) Method of characteristics technique – application to heat and mass transfer problems, *Industrial and Engineering Chemistry*, **48**(4), 703–710.
Cima, R. M. and London, A. L. (1958) The transient response of a two-fluid counterflow heat exchanger, *Transactions ASME*, **80**, 1169–1179.
Cohen, W. C. and Johnson, E. F. (1956) Dynamic characteristics of double-pipe heat exchangers, *Industrial and Engineering Chemistry*, **48**, 1031–1034.
Gvozdenac, D. D. (1990) Transient response of the parallel flow heat exchanger with finite wall capacitance, *Ingenieure Archive*, **60**, 481–490.
Jaswon, M. A. and Smith, W. (1954) Countercurrent transfer processes in the non-steady state, *Proceedings of the Royal Society*, **225A**, 226–244.
Koppel, L. B. (1962) Dynamics of a forced-flow heat exchanger, *Industrial and Engineering Chemistry, Fundamentals*, **1**(2), 131–134.
London, A. L., Biancardi, F. R. and Mitchell, J. W. (1959) The transient response of gas-turbine plant heat exchangers – recuperators, intercoolers, pre-coolers, and ducting, *Transactions ASME. Journal of Engineering for Power*, **81**, 433–448.
Masubuchi, M. (1960) Dynamic response and control of multipass heat exchangers, *Transactions ASME. Journal of Basic Engineering*, **82D**, 51–55.
Mozley, J. M. (1956) Predicting dynamics of concentric pipe heat exchangers, *Industrial and Engineering Chemistry*, **48**, 1035–1041.
Paynter, H. M. and Takahashi, Y. (1956) A new method of evaluating dynamic response of counterflow and parallel flow heat exchangers, *Transactions ASME*, **78**, 749–758.
Tan, K. S. and Spinner, I. H. (1978) Dynamics of a shell-and-tube heat exchanger with finite tube-wall heat capacity and finite shell-side resistance, *Industrial and Engineering Chemistry, Fundamentals*, **17**(4), 353–358.
Tan, K. S. and Spinner, I. H. (1984) Numerical methods of solution for continuous countercurrent processes in the non-steady state: I. Model equations and development of numerical methods and algorithms, II. Application of numerical methods, *American Institute of Chemical Engineers Journal*, **30**(5), 770–779, 780–786.
Tan, K. S. and Spinner, I. H. (1991) Approximate solution for transient response of a shell and tube heat exchanger, *Industrial and Engineering Chemistry, Research*, **30**, 1639–1646.
Todo, I. (1976) Dynamic response of bayonet type heat exchangers, Part I – response to inlet temperature changes, *Bulletin of the Japanese Society of Mechanical Engineers*, **19**(136), 1135–1140.

Temperature transients with longitudinal conduction or interal heat sources
Clark, J. A. and Arpaci, V. S. (1958) Dynamic response of heat exchangers having internal heat sources – Part II, *Transactions ASME*, **80**, 623–634.

Clark, J. A. and Arpaci, V. S. (1959) Dynamic response of heat exchangers having internal heat sources – Part III, *Transactions ASME, Journal of Heat Transfer*, **81**, 233–266.

Clark, J. A., Arpaci, V. S. and Treadwell, K. M. (1958) Dynamic response of heat exchangers having internal heat sources – Part I, *Transactions ASME*, **80C**, 612–624.

Tan, K. S. and Spinner, I. H. (1978) Dynamics of a shell and tube heat exchanger with finite tube-wall heat capacity and finite shell-side resistance, *Industrial and Engineering Chemistry, Fundamentals*, **17**(4), 353–358.

Yang, W. J., Clark, J. A. and Arpaci, V. S. (1958) Dynamic response of heat exchangers having internal heat sources – Part IV, *Transactions ASME*, **83C**, 321–338.

Mass flow transients

Harmon, R. W. (1966) Forced-flow heat exchanger dynamics, *Industrial and Engineering Chemistry, Fundamentals*, **5**(1), 138–139.

Stermole, F. J. and Larson, M. A. (1963) Dynamic response of heat exchangers to flow rate changes, *Industrial and Engineering Chemistry, Fundamentals*, **2**(1), 62–67.

Todo, I. (1978) Dynamic response of bayonet-tube heat exchangers, Part II – response to flow-rate changes, *Bulletin of the Japanese Society of Mechanical Engineers*, **21**(154), 644–651.

Yang, W. J. (1964) Transient heat transfer in a vapour-heated heat exchanger with time-wise varient flow disturbance, *Transactions ASME, Journal of Heat Transfer*, **86C**, 133–142.

A.2 TEMPERATURE TRANSIENTS IN UNMIXED CROSS-FLOW

In direction x:

$$\text{instantaneous mass rate of flow } \dot{m}_h \text{ (kg/s)}$$
$$\text{instantaneous velocity of flow } u_h \text{ (m/s)}$$

Time taken for hot fluid to travel a differential length (δx) may regarded as *instantaneous residence time* or *transit time* (δt), and the relation between instantaneous residence mass ($\tilde{m}$) and instantaneous mass flow rate ($\dot{m}$) becomes ($\tilde{m} = \dot{m}\delta t$) where $u = \delta x/\delta t$ is the instantaneous velocity. Thus

$$\tilde{m}_h = \frac{\dot{m}_h \delta x}{u_h}$$

Similarly for cold fluid

$$\tilde{m}_c = \frac{\dot{m}_c \delta y}{u_c}$$

Surface area of exchanger $S = L_x L_y$

Striking energy balances for the hot fluid, solid wall and cold fluid, respectively.

A.2 TEMPERATURE TRANSIENTS IN UNMIXED CROSS-FLOW

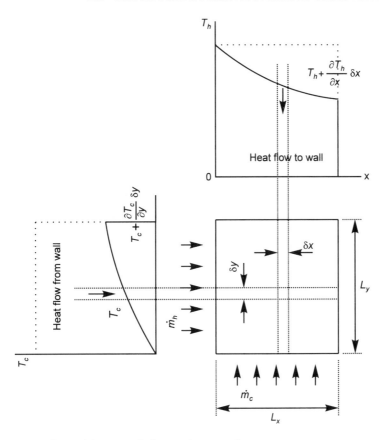

Fig. A.2 Differential energy balances in cross-flow

Hot fluid

$$\begin{Bmatrix}\text{energy entering}\\ \text{with hot fluid}\end{Bmatrix} - \begin{Bmatrix}\text{energy leaving}\\ \text{with hot fluid}\end{Bmatrix} - \begin{Bmatrix}\text{heat transferred}\\ \text{to solid wall}\end{Bmatrix} = \begin{Bmatrix}\text{energy stored}\\ \text{in hot fluid}\end{Bmatrix}$$

$$\left(\dot{m}_h \frac{\delta y}{L_y}\right) C_h T_h - \left(\dot{m}_h \frac{\delta y}{L_y}\right) C_h \left(T_h + \frac{\partial T_h}{\partial x}\delta x\right) - \alpha_h(\delta_x \delta_y)(T_h - T_w)$$

$$= \tilde{m}_h \left(\frac{\delta y}{L_y}\right) C_h \frac{\partial T_h}{\partial t}$$

$$-\left(\frac{\dot{m}_h \delta y}{L_y}\right) C_h \frac{\partial T_h}{\partial x}\delta x - \alpha_h(\delta x \delta y)(T_h - T_w) = \frac{\dot{m}_t \delta x}{u_h}\left(\frac{\delta y}{L_y}\right) C_h \frac{\partial T_h}{\partial t}$$

$$-\left(\frac{\dot{m}_h C_h}{L_y}\right)\frac{\partial T_h}{\partial x} - \alpha_h(T_h - T_w) = \left(\frac{\dot{m}_h C_h}{u_h L_y}\right)\frac{\partial T_h}{\partial x}$$

A TRANSIENT EQUATIONS WITH LONGITUDINAL CONDUCTION

$$\boxed{\frac{\partial T_h}{\partial t} + u_h \frac{\partial T_h}{\partial x} = -\left(\frac{\alpha_h S}{\dot{m}_h C_h}\right) \frac{u_h}{L} (T_h - T_w)}$$

Solid wall

$$\left\{\begin{array}{c}\text{heat from}\\\text{hot fluid}\end{array}\right\} - \left\{\begin{array}{c}\text{heat to}\\\text{cold fluid}\end{array}\right\} - \left\{\begin{array}{c}\text{heat conducted}\\\text{in x direction}\end{array}\right\} - \left\{\begin{array}{c}\text{heat conducted}\\\text{in y direction}\end{array}\right\} = \left\{\begin{array}{c}\text{energy stored}\\\text{in solid wall}\end{array}\right\}$$

$$\alpha_h(\delta x \delta y)(T_h - T_w) - \alpha_c(\delta x \delta y)(T_w - T_c) - \frac{\partial}{\partial x}\left[-\lambda_x\left(A_x \frac{\delta y}{L_y}\right)\frac{\partial T_w}{\partial x} \delta x\right]$$

$$-\frac{\partial}{\partial y}\left[-\lambda_y\left(A_y \frac{\delta x}{L_x}\right)\frac{\partial T_w}{\partial y} \delta y\right] = M_w\left(\frac{\delta x}{L_x}\cdot\frac{\delta y}{L_y}\right) C_w \frac{\partial T_w}{\partial t}$$

With λ_x and λ_y as constants

$$\alpha_h(T_h - T_w) - \alpha_c(T_w - T_c) + \lambda_x\left(\frac{A_x}{L_y}\right)\frac{\partial^2 T_w}{\partial x^2} + \lambda_y\left(\frac{A_y}{L_x}\right)\frac{\partial^2 T_w}{\partial y^2} = \left(\frac{M_w C_w}{L_x L_y}\right)\frac{\partial T_w}{\partial t}$$

Writing $S = L_x L_y$ for plate surface, and $M_w = \rho_w V_w$ where necessary

$$\frac{\partial T_w}{\partial t} - \left(\frac{A_x L_x}{V_w}\right)\kappa_x \frac{\partial^2 T_w}{\partial x^2} - \left(\frac{A_y L_y}{V_w}\right)\kappa_y \frac{\partial^2 T_w}{\partial y^2}$$

$$= \frac{\alpha_h S}{M_w C_w}(T_h - T_w) - \frac{\alpha_c S}{M_w C_w}(T_w - T_c)$$

Redefining thermal diffusivities as

$$\hat{\kappa}_x = \left(\frac{A_x L_x}{V_w}\right)\kappa_x \text{ (m}^2/\text{s)} \text{ and } \hat{\kappa}_y = \left(\frac{A_y L_y}{V_w}\right)\kappa_y \text{ (m}^2/\text{s)}$$

$$\boxed{\frac{\partial T_w}{\partial t} - \hat{\kappa}_x \frac{\partial^2 T_w}{\partial x^2} - \hat{\kappa}_y \frac{\partial^2 T_w}{\partial y^2} = \left(\frac{\alpha_h S}{M_h C_w}\right)(T_h - T_w) - \left(\frac{\alpha_c S}{M_w C_w}\right)(T_w - T_c)}$$

Cold fluid

$$\left\{\begin{array}{c}\text{energy entering}\\\text{with cold fluid}\end{array}\right\} - \left\{\begin{array}{c}\text{energy leaving}\\\text{with cold fluid}\end{array}\right\} - \left\{\begin{array}{c}\text{heat transferred}\\\text{from solid wall}\end{array}\right\} = \left\{\begin{array}{c}\text{energy stored}\\\text{in cold fluid}\end{array}\right\}$$

$$\left(\dot{m}_c \frac{\delta x}{L_x}\right) C_c T_c - \left(\dot{m}_c \frac{\delta x}{L_x}\right) C_c \left(T_c + \frac{\partial T_c}{\partial y}\delta y\right) - \alpha_c(\delta x \delta y)(T_w - T_c)$$

$$= \tilde{m}_c \left(\frac{\delta x}{L_x}\right) C_c \frac{\partial T_c}{\partial t}$$

A.2 TEMPERATURE TRANSIENTS IN UNMIXED CROSS-FLOW

$$-\left(\dot{m}_c \frac{\delta x}{L_x}\right) C_c \frac{\partial T_c}{\partial y} \delta y - \alpha_c (\delta x \delta y)(T_w - T_c) = \frac{\dot{m}_c \delta y}{u_c}\left(\frac{\delta x}{L_x}\right) C_c \frac{\partial T_c}{\partial t}$$

$$-\left(\frac{\dot{m}_c C_c}{L_x}\right) \frac{\partial T_c}{\partial y} - \alpha_c (T_w - T_c) = \left(\frac{\dot{m}_c C_c}{u_c L_x}\right) \frac{\partial T_c}{\partial t}$$

$$\boxed{\frac{\partial T_c}{\partial t} + u_c \frac{\partial T_c}{\partial y} = -\left(\frac{\alpha_c S}{\dot{m}_c C_c}\right) \frac{u_c}{L_y}(T_w - T_c)}$$

Collecting the three coupled partial differential equations for transients in cross-flow

$$\boxed{\begin{aligned}
\frac{\partial T_h}{\partial t} + u_h \frac{\partial T_h}{\partial x} &= -\left(\frac{\alpha_h S}{\dot{m}_h C_h}\right) \frac{u_h}{L_x}(T_h - T_w) \\
\frac{\partial T_w}{\partial t} - \hat{\kappa}_x \frac{\partial^2 T_w}{\partial x^2} - \hat{\kappa}_y \frac{\partial^2 T_w}{\partial y^2} &= +\left(\frac{\alpha_h S}{M_w C_w}\right)(T_h - T_w) - \left(\frac{\alpha_c S}{M_w C_w}\right)(T_w - T_c) \\
\frac{\partial T_c}{\partial t} + u_c \frac{\partial T_c}{\partial y} &= +\left(\frac{\alpha_c S}{\dot{m}_c C_c}\right) \frac{u_c}{L_y}(T_w - T_c)
\end{aligned}} \quad (A.4)$$

When no longitudinal conduction is present, but wall thermal storage is important

$$\boxed{\begin{aligned}
\frac{\partial T_h}{\partial t} + u_h \frac{\partial T_h}{\partial x} &= -\left(\frac{\alpha_h S}{\dot{m}_h C_h}\right) \frac{u_h}{L_x}(T_h - T_w) \\
\frac{\partial T_w}{\partial t} &= +\left(\frac{\alpha_h S}{M_w C_w}\right)(T_h - T_w) - \left(\frac{\alpha_c S}{M_w C_w}\right)(T_w - T_c) \\
\frac{\partial T_c}{\partial t} + u_c \frac{\partial T_c}{\partial y} &= +\left(\frac{\alpha_c S}{\dot{m}_c C_c}\right) \frac{u_c}{L_y}(T_w - T_c)
\end{aligned}} \quad (A.5)$$

When no transients are present, but longitudinal conduction is important

$$\boxed{\begin{aligned}
+ u_h \frac{\partial T_h}{\partial x} &= -\left(\frac{\alpha_h S}{\dot{m}_h C_h}\right) \frac{u_h}{L_x}(T_h - T_w) \\
- \hat{\kappa}_x \frac{\partial^2 T_w}{\partial x^2} - \hat{\kappa}_y \frac{\partial^2 T_w}{\partial y^2} &= +\left(\frac{\alpha_h S}{M_w C_w}\right)(T_h - T_w) - \left(\frac{\alpha_c S}{M_w C_w}\right)(T_w - T_c) \\
+ u_c \frac{\partial T_c}{\partial y} &= +\left(\frac{\alpha_c S}{\dot{m}_c C_c}\right) \frac{u_c}{L_y}(T_w - T_c)
\end{aligned}} \quad (A.6)$$

Note

It is demonstrated elsewhere that wall thermal storage virtually controls transient behaviour of the exchanger, thus transient solutions which ignore wall thermal storage have little practical value when normal liquids and gases are involved.

BIBLIOGRAPHY

Chen, H. -T. and Chen, K. -C. (1991) Simple method for transient response of gas-to-gas cross-flow heat exchangers with neither gas mixed, *International Journal of Heat and Mass Transfer*, **34**(11), 2891–2898.

Chen, H. -T. and Chen, K. -C. (1992) Transient response of cross-flow heat exchangers with finite wall capacitance – technical note, *Transactions ASME, Journal of Heat Transfer*, **114**, 752–755.

Dusinberre, G. M. (1959) Calculation of transients in a cross-flow heat exchanger, *Transactions ASME, Journal of Heat Transfer*, **81**, 61–67.

Evans, F. and Smith, W. (1963) Cross-current transfer processes in the non-steady state, *Proceedings of the Royal Society*, **272A**, 241–269.

Gartner, J. R. and Harrison, H.L. (1965) Dynamic characteristics of water-to-air cross-flow heat exchangers, *Transactions ASHRAE*, **71**(1), 212–224.

Gvozdenac, D. D. (1986) Analytical solution of the transient response of gas-to-gas crossflow heat exchanger with both fluids unmixed, *Transactions ASME, Journal of Heat Transfer*, **108**, 722–727.

Gvozdenac, D. D. (1991) Dynamic response of the cross-flow heat exchanger with finite wall capacitance, *Wärme-und Stoffübertragung*, **26**, 207–212.

Myers, G. E., Mitchell, J.W. and Norman, R. F. (1967) The transient response of crossflow heat exchanegrs, evaporators, and condensers, *Transactions ASME, Journal of Heat Transfer*, **89**(1), 75–80.

Myers, G. E., Mitchell, J. W. and Lindeman, C. F. Jr. (1970) The transient response of heat exchangers having an infinite capacity rate fluid, *Transactions ASME, Journal of Heat Transfer*, **92**, 269–275.

Romie, F. E. (1983) Transient response of gas-to-gas cross-flow heat exchangers with neither gas mixed, *Transactions ASME, Journal of Heat Transfer*, **105**, 563–570.

San, J. -Y. (1993) Heat and mass transfer in a two-dimensional cross-flow regenerator with a solid conduction effect, *International Journal of Heat and Mass Transfer*, **36**(3), 633–643.

Spiga, G. and Spiga, M. (1987) Two-dimensional transient solutions for crossflow heat exchangers with neither gas mixed, *Transactions ASME, Journal of Heat Transfer*, **109**, 281–286.

Spiga, M. and Spiga, G. (1988) Transient temperature fields in cross-flow heat exchangers with finite wall capacitance, *Transactions ASME, Journal of Heat Transfer*, **110**, 49–53.

Spiga, M. and Spiga, G. (1992) Step response of the cross-flow heat exchanger with finite wall capacitance, *International Journal of Heat and Mass Transfer*, **35**(2), 559–565.

Yamashita, H., Izumi, R. and Yamaguchi, S. (1978) Analysis of the dynamic characteristics of cross-flow heat exchangers with both fluids unmixed. (On the transient responses to a step change in the inlet temperature), *Bulletin of the JSME*, **21**(153), 479–485.

A.3 MASS FLOW AND TEMPERATURE TRANSIENTS IN CONTRA-FLOW

When both mass flow rates and temperatures vary together in transients, then three conservation equations of continuum mechanics are required, viz.

- balance of mass flow
- balance of linear momentum (+friction)
- balance of energy (+friction)

together with constitutive equations for each fluid, the simplest being for perfect gas.

In setting up the equations for temperature transients for constant mass flow in Sections A.1 and A.2 there was no mention of varying physical properties, or of varying heat transfer coefficients and pressure loss. For accuracy these need to be incorporated, and only a numerical approach is then capable of handling the problem.

Balance of energy

Energy of frictional dissipation is assumed negligible compared with other terms.

$$\frac{\partial T_h}{\partial t} + u_h \frac{\partial T_h}{\partial x} = -\left(\frac{\alpha_h S}{\dot{m}_h C_h}\right) \frac{u_h}{L} (T_h - T_w)$$

$$\frac{\partial T_w}{\partial t} - \kappa_x \frac{\partial^2 T_w}{\partial x^2} = +\left(\frac{\alpha_h S}{M_w C_w}\right)(T_h - T_w) - \left(\frac{\alpha_c S}{M_w C_w}\right)(T_w - T_c)$$

$$\frac{\partial T_c}{\partial t} - u_c \frac{\partial T_c}{\partial x} = +\left(\frac{\alpha_c S}{\dot{m}_c C_c}\right) \frac{u_c}{L} (T_w - T_c)$$

Equations are linear in (T_h, T_c) when (u_h, u_c) are known. (A.7)

Balance of mass

Standard equations from continuum mechanics.

$$\frac{\partial \rho_h}{\partial t} + \frac{\partial (\rho_h u_h)}{\partial x} = 0$$

$$\frac{\partial \rho_c}{\partial t} - \frac{\partial (\rho_c u_c)}{\partial x} = 0$$

(A.8)

Equations are linear in (u_h, u_c) when (ρ_h, ρ_c) are known.

A TRANSIENT EQUATIONS WITH LONGITUDINAL CONDUCTION

Balance of linear momentum

The two balance of linear momentum equations require evaluation of pressure loss gradient $(\partial p/\partial x)$, obtained using a simple force balance. The following pressure gradient equations are appropriate for incompressible conditions (Ma < 0.3).

Force balance $\qquad [p - (p - \delta p)]A = \left(\dfrac{f\rho u^2}{2}\right)\dfrac{A\delta x}{r_{hyd}}$

Pressure gradient due to friction $\qquad \dfrac{\partial p}{\partial x} = \dfrac{f\rho u^2}{2r_{hyd}}$

Then balance of linear momentum equations for hot and cold fluids may be written

$$\dfrac{\partial u_h}{\partial t} + u_h \dfrac{\partial u_h}{\partial x} = -\dfrac{1}{\rho_h}\dfrac{\partial p_h}{\partial x} - \dfrac{f_h u_h^2}{2r_h} \qquad (A.9)$$

$$\dfrac{\partial u_c}{\partial t} - u_c \dfrac{\partial u_c}{\partial x} = +\dfrac{1}{\rho_c}\dfrac{\partial p_c}{\partial x} - \dfrac{f_c u_c^2}{2r_c}$$

The equations are non-linear in (u_h, u_c) when (ρ_h, ρ_c) and $\left(\dfrac{\partial p_h}{\partial x}, \dfrac{\partial p_c}{\partial x}\right)$ are known.

Adding $u\partial\rho/\partial t$ to both sides of each equation and substituting for $u\partial\rho/\partial t$ from the mass flow equations gives

$$\left.\begin{array}{l}\dfrac{\partial(\rho_h u_h)}{\partial t} = -\dfrac{\partial(p_h + \rho_h u_h^2)}{\partial x} - \dfrac{f_h \rho_h u_h^2}{2r_h} \\[1em] \dfrac{\partial(\rho_c u_c)}{\partial t} = +\dfrac{\partial(p_c + \rho_c u_c^2)}{\partial x} - \dfrac{f_c \rho_c u_c^2}{2r_c}\end{array}\right\}$$

Without the friction terms these equations have been solved numerically by Longley (1960).

Constitutive equations

The equations for a perfect gas are to be interpreted locally.

$$\begin{array}{l} p_h = \rho_h R_h T_h \\ p_c = \rho_c R_c T_c \end{array} \qquad (A.10)$$

Numerical stability

With fiinite differences, a stable solution is possible when $(|u| + c)(\Delta t/\Delta x) \leq 1$ where c is the local speed of sound, (Ames 1965). Knowing $(c, \Delta x)$ this constraint provides the time interval (Δt) to reveal the desirable size and speed of computer required.

REFERENCES

Ames, W. F. (1965) *Nonlinear Partial Differential Equations In Engineering*, New York: Academic Press, pp. 453–454.

Dzyubenko, B. V., Dreitser, G. A. and Ashmantas, L. -V. A. (1990) *Unsteady Heat and Mass Transfer in Helical Tube Bundles*, New York: Hemisphere.

Longley, H. J. (1960) 'Methods of differencing in Eulerian hydrodynamics', *Los Alamos Scientific Laboratory*, New Mexico: Report LAMS-2379.

Mitchell, A. R. and Griffiths, D. F. (1980) *The Finite Difference Method in Partial Differential Equations*, Chichester: Wiley.

B

ALGORITHMS AND SCHEMATIC SOURCE LISTINGS

B.1 ALGORITHMS FOR MEAN TEMPERATURE DISTRIBUTION IN ONE-PASS UNMIXED CROSS-FLOW

The approach is explicit finite-difference, and accuracy is affected by error propagation, particularly from the steepest parts of the temperature fields. For equally spaced intervals, at least a (50 × 50) mesh should be used. Improvements can be expected using a smaller mesh near the starting corner, but it is simpler to increase the size of the full mesh.

Two algorithms are given below. The first is a simple algorithm designed to introduce the reader to the method of solution. The second algorithm employs the modified Euler-Cauchy method, which obtains an estimate of the slope at the new point, and takes the mean of this estimated slope and the known slope at the old point to obtain a better estimate for the slope in the interval.

FIRST ALGORITHM

```
{first origin square}
k := 0;
sl Tg : = -Tg [k, k + 1] + Tf [k, k + 1]        {initial slope}
sl Tf : = +Tg [k + 1, k] - Tf [k + 1, k]        {initial slope}
Tg[k + 1, k + 1] : = Tg[k, k + 1] + slTg * Ng/50;   {new Tg}
Tf[k + 1, k + 1] : = Tf[k + 1, k] + slTg * Nf/50;   {new Tf}
```

B.1 MEAN TEMPERATURE DISTRIBUTION IN ONE-PASS UNMIXED CROSS-FLOW

```
            FOR k = 1 TO 49 DO
BEGIN
        n : = k;
        FOR m : = n TO 49 DO                        {along X-dirn.}
BEGIN  slTg : = -Tg[m,n] + Tf[m,n];                 {initial slope}
        slTf : = +Tg[m+1, n-1] - Tf[m+1, n-1];      {initial slope}
        Tg[m+1,n] : = Tg[m,n] + slTg*Ng/50;         {new Tg}
        Tf[m+1,n] : = Tf[m+1,n-1] + slTf*Nf/50;     {new Tf}
END;
        m = k;
        FOR n :- = m TO 49 DO                       {along Y-dirn.}
BEGIN  slTg : = -Tg[m-1,n+1] + Tf[m-1,n+1];         {initial slope}
        slTf : = +Tg[m,n] - Tf[m,n];                {initial slope}
        Tg[m,n+1] : = Tg[m-1,n+1] + slTg*Ng/50;     {new Tg}
        Tf[m,n+1] : = Tf[m,n] + s1Tf*Nf/50;         {new Tf}
END;
        {next origin square}
        slTg : = -Tg[k,k+1] + Tf[k,k+1];            {initial slope}
        slTf : = +Tg[k+1,k] - Tf[k+1,k];            {initial slope}
        Tg[k+1,k+1] : = Tg[k,k+1] + slTg*Ng/50;     {new Tg}
        Tf[k+1,k+1] : = Tf[k+1,k] + slTf*Nf/50;     {new Tf}
END;    {FOR k-loop}
```

COMPUTE OTHER THERMAL PARAMETERS

```
        Sum : = 0.0;
        FOR m : = 1 TO 49 DO                        {sum over inside temps}
BEGIN                                               FOR n : = 1 TO 49 DO
        Sum : = Sum + (Tg[m,n] - Tf[m,n]);
END;

        FOR n : = 0 TO 50 DO                        {sum over edge temps}
        Sum : = Sum + ((Tg[0,n] - Tf[0,n]) + (Tg[50,n] - Tf[50,n]))/2;
        FOR m : = 0 TO 50 DO
        Sum : = Sum + ((Tg[m,0] - Tf[m,0]) + (Tg[m,50] - Tf[m,50]))/2;

        mean TDiff : = Sum/(50*50);                 {mean temp. difference}
        Tginn : = 1.0;                              {mixed warm inlet temp}
        Tfinn : = 0.0;                              {mixed cold inlet temp}

        Tspan : = Tginn-Tfinn;

        Sum : = 0.0;
        m : = 50;                                   {at warm gas outlet}
```

```
FOR n := 1 TO 49 DO                    {sum over inner temps}
   Sum := Sum+Tg[m,n];
Sum := Sum + (Tg[m,0] + Tg[m,50])/2;   {add mean outer temps}
Tgout := Sum/50;                       {mixed warm outlet}

Sum := 0.0;
n := 50;                               {at cold fluid outlet}
FOR m := 1 TO 49 DO
   Sum := Sum+Tf[m,n];                 sum over inner temps
Sum := Sum + (Tf[0,n] + Tf[50,n])/2;   {add mean outer temps}
Tfout := Sum/50;                       {mixed cold outlet}

TdiffG := Tginn-Tgout;
TdiffF := Tfout-Tfinn;
IF (TdiffG > TdiffF) THEN Tdiff := TdiffG
                     ELSE Tdiff := TdiftF;
                                       {max. temp. change
                                        ⇒ min. water equiv.}

Qg := TdiffG/Ng;                       {heat balance, Qg/US}
Qf := Tdiff/Nf;                        {heat balance, Qf/US}
Eff := Tdiff/Tspan;                    {exchanger effectiveness}
```

In using this algorithm it is to be recognised that the explicit type of solution always produces some error propagation. This affects both temperature sheets more or less equally. Consequently the value of 'mean TDiff' is more reliable than the computed outlet temperatures 'Tgout' and 'Tfout'.

The value of 'mean TDiff' is used in computation, but it is better to calculate mean outlet temperatures using expressions of the following type:

$$\text{Tout} = \text{Tinn} - Q/(m*Cp)$$

SECOND ALGORITHM

```
k := 0;                                {first origin square}
islTg := -Tg[k,k+1] + Tf[k,k+1];       {initial slope}
islTf := +Tg[k+1,k] - Tf[k+1,k];       {initial slope}
eTg := Tg[k,k+1] + islTg*Ng/50;        {estimated Tg}
eTf := Tf[k+1,k] + islTf*Nf/50;        {estimated Tf}
fslTg := eTg + eTf;                    {final slope}
fslTf := eTg + eTf;                    {final slope}
mslTg := (islTg + fslTg)/2;            {mean slope}
mslTf := (islTf + fslTf)/2;            {mean slope}
```

B.1 MEAN TEMPERATURE DISTRIBUTION IN ONE-PASS UNMIXED CROSS-FLOW

```
              Tg [k + 1, k + 1] : = Tg [k, k + 1] + msl Tg*Ng/50;  {new Tg}
              Tf [k + 1, k + 1] : = Tf [k + 1, k] + mslTf*Ng/50;   {new Tf}

              FOR k : = 1 TO 49 DO
BEGIN         n : = k;
              FOR m : = n TO 49 DO                                {along X-dirn.}
BEGIN         islTg : = -Tg[m, n] + Tf[m, n];                     {initial slope}
              islTf : = +Tf[m + 1, n - 1] - Tf[m + 1, n - 1];     {initial slope}
              eTg : = Tg[m, n] + islTg*Ng/50;                     {estimated Tg}
              eTf : = Tf[m + 1, n - 1] + islTf*Nf/50;             {estimated Tf}
              fslTg : = -eTg + eTf;                               {final slope}
              fslTf : = +eTg - eTf;                               {final slope}
              mslTg : = (islTg + fslTg)/2;                        {mean slope}
              mslTf : = (islTf + fslTf)/2;                        {mean slope}
              Tg[m + 1, n] : = Tg[m, n] + mslTg*Ng/50;            {new Tg}
              Tf[m + 1, n] : = Tf[m + 1, n - 1] + mslTf*Nf/50;    {new Tf}
END;
              m : = k;
              FOR n : = m TO 49 DO                                {along Y-dirn.}
BEGIN         islTg : = -Tg[m - 1, n + 1] + Tf[m - 1, n + 1];     {initial slope}
              islTf : = +Tg[m, n] - Tf[m, n];                     {initial slope}
              eTg : = Tg[m - 1, n + 1] + islTg*Ng/50;             {estimated Tg}
              eTf : = Tf[m, n] + islTf*Nf/50;                     {estimated Tf}
              fslTg : = -eTg + eTf;                               {final slope}
              fslTf : = +eTg - eTf;                               {final slope}
              mslTg : = (islTg + fslTg)/2;                        {mean slope}
              mslTf : = (islTf + fslTf)/2;                        {mean slope}
              Tg[m, n + 1] : = Tg[m - 1, n + 1] + mslTg*Ng/50;    {new Tg}
              Tf[m, n + 1] : = Tf[m, n] + mslTf*Nf/50;            {new Tf}
END;
              {next origin square}
              islTg : = -Tg[k, k + 1] + Tf[k, k + 1];             {initial slope}
              islTf : = +Tg[k + 1, k] - Tf[k + 1, k];             {initial slope}
              eTg : = Tg[k, k + 1] + islTg * Ng/50;               {estimated Tg}
              eTf : = Tf[k + 1, k] + islTf * Ng/50;               {estimated Tf}
              fslTg : = -eTg + eTf;                               {final slope}
              fslTf : = +eTg - eTf;                               {final slope}
              mslTg : = (islTg + fslTg)/2;                        {mean slope}
              mslTf : = (islTf + fslTf)/2;                        {mean slope}
              Tg[k + 1, k + 1] : = Tg[k, k + 1] + mslTg * Ng/50;  {new Tg}
              Tf[k + 1, k + 1] : = Tf[k + 1, k] + mslTf * Nf/50;  {new Tf}
END;          {FOR k-loop}
```

B.2 SCHEMATIC SOURCE LISTING FOR DIRECT SIZING OF COMPACT CROSS-FLOW EXCHANGER

MAIN PROGRAM

refMR=m1/m2	(desired mass flow ratio)
iterate Re1	(binary search)
f1=value from f-correlation	(splinefit of data)
G1=Re*mu1/D1	(mass velocity)
Lp1=dp1*2*rho1*D1/(4*f1*G1^2)	(length of channel)
E2=Lp1	(edge length, side-2)
iterate 'aspect'	(binary search)
Lp2=aspect*Lp1	(plate aspect=E1/E2)
E1=Lp2	(edge length, side-1)
Splate=E1*E2	(area of single plate)
Given dP2	(pressure loss, side-2)
iterate Re2	(binary search)
f2=value from f-correlation	(splinefit of data)
G2=Re2*mu2/D2	(mass velocity)
dp=4*f2*G2^2*L2/(2*rho2*D2)	(estimate of dP2)
until dp=dP2	(Re known on both sides)
Afront1=E1*(b1/2)	(Afront for 1/2 plate-spacing, both single & double-cells)
Aflow1=sigma1*Afront1	(Aflow for 1/2 plate-spacing)
mP1=G1*Aflow1	(Mflow for 1/2 plate-spacing)
Repeat last three lines for side-2	
newMR=mP1/mP2	(estimate of refMR)
until newMR=refMR	('aspect' for Re1)
Nw=TRUNC(m1/mP1)+1	(number of plates)
wide=Nw*(b1/2+tp+b2/2)	(exchanger width)
vol=E1*E2*wide	(exchanger core volume)
Sexchr=Nw*Splate	(total plate surface)
Stotal1=Sexchr*kappa1	(total surface, side-1)
Stotal2=Sexchr*kappa2	(total surface, side-2)
Pr1=Cp1*mu1/k1	(at mean bulk temperature)
St1=value from j-corelation	(splinefit of data)
uh1=St1*Cp1*G1	(heat trans. coeff, side-1)
Y1=b1/2	(approx. fin height)
mY1=Y1*SQRT(2*uh1/(kf1*tf1))	(fin parameter)
phi1=TANH(mY1)/mY1	(fin performance ratio)
eta=1-gamma1*(1-phi1)	(performance correction to total surface, side-1)
uh1t=uh1*eta1	(h.t. coeff @ total surface)
u1=uh1t*kappa1	(h.t. coeff @ plate surface)
u2=similarly for side-2	

```
u3=kp/tp                                 (plate coefficient)
U = 1/(1/u1 + 1/u2 + 1/u3)               (overall coeff. at plate)
Ntu1=U*Sexchr/(m1*Cp1)                   (Ntu, side-1)
Ntu2=U*Sexchr/(m2*Cp2)                   (Ntu, side-2)
determine mean TD                        (using T1, t1, Ntu1, Ntu2)
Qtem=U*Sexchr*MeanTD                     (exchanger duty at Re1)
until Qtemp=Q                            (Q is desired performance)
```

B.3 SCHEMATIC SOURCE LISTING FOR DIRECT SIZING OF COMPACT CONTRAFLOW EXCHANGER

MAIN PROGRAM

```
Re1 = 2500                               (mid-range value)
fric(Re1,loRe1F,hiRe1F,f1)                (Re1 fric correlation limits)
heat(Re1, loRe1H, hiRe1H, StPr^2/3)       (Re1 heat correlation limits)
(max loRe1)<Re1<(min hiRe1)               (valid range for Re1)
Re2 = 2500                                (mid-range value)
fric(Re2,loRe2F,hiRe2F,f2)                (Re2 fric correlation limits)
heat(Re2,loRe2H, hiRe2H, StPr^2/3)        (Re2 heat correlation limits)
(max loRe2)<Re2<(min hiRe2)               (valid range for Re2)
surf1(fins,b,c,x,D,tf,ts,beta,gamma)      (surface parameters)
surf2(fins,b,c,x,D,tf,ts,beta,gamma)      (surface parameters)
Scan over range of Re1                    (100 steps from loRe1→hiRe1)
   heatrans(Re1,forced Re2,Lh,Edge)       (procedure, find edge length)
   pdrop1(Re1,dp1,Lp1)                    (procedure, find Lp1 given dp1)
   pdrop2(Re2,dp2,Lp2)                    (procedure, find Lp2 given dp2)
until complete                            (full validity range)
plot curves, (Lh,Lp1,Lp2)vs Edge          (visual check)
iterate for RH intersection & L           (design point, Re1,Re2)
if Lp1 RH curve, calculate NEWdp2         (actual pressure loss)
if Lp2 RH curve, calculate NEWdp1         (actual pressure loss)
design                                    (procedure, exchr block
                                              parameters)

PROCEDURE heatrans(Re1,Re2,Lh,Edge)
G1=Re1 * mu1/D1                           (mass velocity)
Aflow1=m1/G1                              (total flow area, side-1)
Afront1=Aflow1/sigma1                     (total frontal area, side-1)
E=Afront1/(b1/2)                          (edge length E = z1*c1 = z2*c2
                                           both single and double cells)
z1=E/c1                                   (cells on side-1)
```

z2=E/c2   (cells on side-2)
Afront2=E * (b2/2)   (total frontal area, side-2)
Aflow2=sigma2*Afront2   (total flow area, side-2)
Re2=D2 * m2/(mu2 * Aflow2)   (forced Re2)
G2=Re2 * mu2/D2   (forced mass velocity)
PrX=EXP(2/3 * LN((Pr1))   (side-1, Pr^2/3)
heat(Re1, loRe1H, hiRe1H, StPr^2/3)   (Re1 heat transfer correlation)
St1 = (StPr^2/3)/PrX   (side-1, Stanton number)
uh1=St1 * Cp1 * G1   (side-1, mean h. trans coefficient)
Y1=b1/2   (side-1, approx. fin height)
mY1=Y1 * SQRT(2 * uh1/(kf1 * tf1))   (fin parameter)
phi1=TANH(mY1)/mY1   (fin performance ratio)
eta=1--gamma1 * (1-- phi1)   (performance correlation to total surface, side-1)

uh1t=uh1 * eta1   (h. trans. coeff. @ total surface)
u1=uh1t * kappa1   (h. trans. coeff. @ plate surface)
u2 similarly for side-2
u3=kp/tp   (plate coefficient)
U=1/(1/u1+1/u2+1/u3)   (overall h. trans. coeff. @ plate)
Splate=Q/U * LMTD)   (based on plate surface)
Lh=Splate/E   (length for Q)

PROCEDURE pdrop(Re1,dp1,Lp1)
fric(Re1, loRe1F, hiRe1F, f1)   (Re1 friction correlation)
G1=Re1 * mu1/D1   (mass velocity)
Lp1=dp1 * 2 * rho1 * D!/(4 * f1 * G12)   (length for dp1)

PROCEDURE pdrop(Re2,dp2,Lp2)   (as above for dp1)

PROCEDURE design
Splate = E * L   (total plate surface)
Ntu1 = U * Splate/(m1 * Cp1)   (whole exchanger)
Ntu2 = U * Splate/(m2 * Cp2)   (whole exchanger)
Stotal1 = Splate * kappa1   (total surface, side-1)}
Stotal2 = Splate * kappa2   (total surface, side-2)}
V = L * E * (b1/2 + tp + b2/2)   (volume of exchanger)}
zR1=TRUNC(z1)+1   (number of cells, side-1)
zR2=TRUNC(z2)+1   (number of cells, side-2)
Ac=X-section for longitudinal conduction   (dependent on surfaces, LMTD adjustment)

Am=X-section for mass evaluation   (dependent on surfaces)
Mblock = rhoM * Am * L   (mass of exchanger)
Py=1-MBlock/(rhoM * V)   (porosity of exchanger)

B.4 PARAMETERS FOR RECTANGULAR OFFSET STRIP-FINS

In running software for both rating and direct sizing, very careful attention must be paid to accurate definition of the surface geometry. Quite small deviations from correct values may cause significant change in exchanger performance or in final dimensions.

It was found that significant errors existed in some published data. For the rectangular offset strip-fin it is practicable to proceed from basic dimensions and compute consistent values. This procedure is recommended as the best way of avoiding data entry problems.

Definitions of parameters are provided in Table 4.4, but the reader may find the following algorithms helpful in the generation of useful and accurate values.

Single-cell geometries
(subscripts '1' for side-1 except where indicated for parameter 'alpha')

```
Cell : = (c - tf)*(b - tf)              {cell Aflow}
Cellx := Cell*x;
Per := 2*(c - tf) + 2*(b - tf);         {cell perimeter}
Perx : = Per*x + 2*(c/2)*f  +  2*(b - 2*tf)*tf;
                  {2 base ends}  {2 fin ends}
rh : = Cellx/Perx;                      {hydraulic radius}
D : = 4*rh                              {hydraulic diameter}
Per : = Perx/x;                         {effective perimeter}

{1-cell} {sides}   {plate} (--fin ends--}
Stotal : = 2*(b - tf) + 2*(c - tf) + 2*(b - 2*tf)*tf/x
                         + 2*(c/2)*tf/x;
                            {base end}

{1-cell}{sides}            {-fin ends--}
Sfins : = 2*(b - tf)       +2*(b - 2*tf)*tf/x
                           + 2*(c/2)*tf/x;
                            {base end}

Vtotal : = b*c                          {value at cell-level}
Sexchr : = 2*c;                         {value at cell-level}

beta : = Stotal/Vtotal;                 {Stotal/Vtotal}
alpha1 : = b1*beta1/(b1 + 2*tp + b2)    {Stotal/Vexchr, side-1}
alpha2 : = b2*beta2/(b1 + 2*tp + b2);   {Stotal/Vexchr, side-2}
gamma : = Sfins/Stotal;                 {Sfins/Stotal}
kappa : = Stotal/Sexchr;                {Stotal/Sexchr}
lambda : = Sfins/Sexchr;                {Sfins/Sexchr}
```

```
sigma := beta*rh;                      {Aflow/Afront}
omega := alpha/kappa;                  {Sexchr/Vexchr}
```

Double-cell geometries
(subscripts '1' for side-1 except where indicated for parameter 'alpha')

```
Cell  := (c - tf)*((b - ts)/2 - tf);   {cell Aflow}
Cellx := Cell*x;
Per   := 2*(c - tf) + 2*((b - ts)/2 - tf);  {cell perimeter}
Perx  := Per*x
         + 2*(c/2)*tf                  {two base ends}
         + 2*((b - ts)/2 - 2*tf))*tf;  {two fin ends}
rh    := Cellx/Perx;                   {hydraulic radius}
D     := 4*rh;                         {hydraulic diameter}
Per   := Perx/x;                       {effective perimeter}

{2-cells} {---sides---}    {splitter} {plate}
Stotal :=  4*((b - ts)/2 - tf)  +  2*(c - tf) + 2*(c - tf) +
           4*((b - ts)/2 - 2*tf)*tf/x  +  4*(c/2)*tf/x;
           {-----fin ends-----}   {base ends}

{2-cells} {---sides---}    {splitter}
Sfins  :=  4*((b - ts)/2 - tf)  +  2*(c - tf) +
           4*((b - ts)/2 - 2*tf)*tf/x  +  4*(c/2)*tf/x;
           {-----fin ends-----}   {base ends}

Vtotal := b*c;                         {value at cell-level}
Sexchr := 2*c;                         {value at cell-level}

beta   := Stotal/Vtotal;               {Stotal/Vtotal}
alpha1 := b1*beta1/(b1 + 2*tp + b2);   {Stotal/Vexchr, side-1}
alpha2 := b2*beta2/(b1 + 2*tp + b2);   {Stotal/Vexchr, side-2}
gamma  := Sfins/Stotal;                {Sfins/Stotal}
kappa  := Stotal/Sexchr;               {Stotal/Sexchr}
lambda := Sfins/Sexchr;                {Sfins/Sexchr}
sigma  := beta*rh;                     {Aflow/Afront}
omega  := alpha/kappa;                 {Sexchr/Vexchr}
```

B.5 TYPICAL INPUT/OUTPUT FOR DIRECT SIZING OF COMPACT CONTRAFLOW EXCHANGER

To distinguish between commentary and computer listings, the commentary is in italics. Typical input and output is given for direct sizing of a contraflow exchanger with plate-fin surfaces. This is an example of the computer runs involved in the optimisation discussed in Section 4.7.

B.5 INPUT/OUTPUT FOR DIRECT SIZING COMPACT CONTRAFLOW EXCHANGER

Input data

The first three lines, including the 'header', are common to all source listings written by the author. They are more valuable for Pascal source listings, when it helps to reserve the first eight spaces for commands like 'BEGIN' and 'END', and to use the next 40 spaces for coding, leaving the remaining spaces for{comments}. Source listings with a good set of comments are relatively easy to understand after they have been left for six months.

```
{-----+----------------------------------------+---------------------------------}
{Filename BERGDAT - input data for BergFin}
{-----+----------------------------------------+---------------------------------}
```
These three lines are STRINGS used to identify the computer run
Contraflo|<<<<<<<<<<<<<<<<<<<<<<<<'tagit', Data identifier-9 chars
1⇒Gas-1.1bar |<<<<<<<<<<<<<<<<<<<'fluid1'-23 characters
2⇒Air-18.4bar|<<<<<<<<<<<<<<<<<<<'fluid2"-23 characters
```
{----------------------------------------------------}
```
10949.0				{Q(kW)}
1.100	18.40			{Pg1, Pa2(bar)}
797.0	563.0			{Tg1, Ta2(K)}
586.0	774.0			{Tg2, Ta1(K)}
0.917	0	x.xxxxx		{adjLMTD, mean TD, cryoTD}
0.683	1059.0	0.509e-004	0.07892	{Pr1, Cp1, mu1, k1}
0.683	1059.0	0.509e-004	0.07892	{Pr2, Cp2, mu2, k2}
1.0560	287.07	691.5	0.53197	{P1, R1, Tbulk1, rho1}
18.473	287.07	668.5	9.62605	{P2, R2, Tbulk2, rho2}
49.0	8620.1	{SELECTED}		{m1, dP1 (N/sqm)}
49.0	80000.0	{floating}		{m2,dP2(N/sqm)}

```
{----------------------------------------------------}
```
1	1				{fins1, fins2 1=S, 2=D}
0.200	90.0	433.0	8906.0		{tp(mm), kM, sphtM, rhoM}
0	5.000	1.31266	2.540	0.1016	{geom1, b1, c1, x1, tf1, [ts1]}
9	1.9050	1.05307	2.8222	0.1016	{geom2,b2,c2,x2,tf2,[ts2]}

```
{----------------------------------------------------}
```
The next two lines are STRINGS identifying surfaces
1⇒r.strip(S)mod.19S35| <<<<<<<<<<< 'Surf1'-23 characters
2⇒r.strip(S)24s12 |<<<<<<<<<<< 'Surf2'-23 characters
```
{----------------------------------------------------}
```
{fins =1⇒single-cells(S)}
{fins =2⇒double-cells(D)}
{geom=0⇒variable geometry}
{geom=(1,2,3,···)⇒selected K&L, L&S geometries}
```
{-----+----------------------------------------+---------------------------------}
```

Line 5 of numerical input reads (adjLMTD, mean TD, cryoTD). *The first value* (adjLMTD) *is set at 1.000 during a first design run. From this output, data is produced and used as input for the separate correction program for longitudinal conduction, which produced the value* adjLMTD $= 0.917$ *used in the second run shown here.*

The integer parameter meanTD *is set either to 0 or 1. When mean TD $= 0$, log mean temperature difference is calculated from terminal temperatures. When mean TD $= 1$, a value of* cryoTD *obtained from section-wise energy balances along the exchanger is used. This produces a direct-sizing first approximation to an exchanger with unusual temperature profiles, which can only be properly designed using the stepwise rating method discussed in Chapter 12.*

Notice that there is redundancy in numerical data lines 6 and 7 involving Prandtl number. It was found useful to have the source listing self-check that these values were mutually consistent. Again, the same procedure is applied on lines 8 and 9, but this may only be done for perfect gases.

Line 10 carries the word SELECTED, *which simply indicates that this pressure loss will be 'controlling' in direct sizing, whereas the other pressure loss is chosen large enough so that it will be 'floating'. However, in the source listing which produced the output given below, an overriding requirement was built in to ensure that the sum of the two pressure loss fractions would not exceed 8%. This involved adjusting the first pressure loss progressively and calculating the second from the relationship*

$$\text{sum of pressure ratios } (dPa/Pa2) + (dPg/Pg1) = 0.08$$

The units given below are listed as part of the input data.
UNITS

```
{-----------------------------------------------------------------}
Q                     kW                  {U*S ⇒ Q*1000/LMTD}
Pg1, Pa2              bar
T1, T2                K
MeanTD                K
Pr1, Pr2              Prandtl number, fluid dependent
Cp1, Cp2              J/(kg K)
mu1, mu2              kg/(m s)
k1, k2                J/(m s K)           {gases}
P1, P2                bar
R1, R2                J/(kg K)
Tbulk1, Tbulk2        K
rho1, rho2, rhoM      kg/cm
m1, m2                kg/s
dp1, dp2              N/sqm
```

B.5 INPUT/OUTPUT FOR DIRECT SIZING COMPACT CONTRAFLOW EXCHANGER

```
tp              mm
kM              J/(m s K)        {kM ⇒ kp, kf1, kf2}
sphtM           J/(m s K)
rhoM            kg/cum
fins1, fins2    integers         {1 ⇒ S, 2 ⇒ D, single or double cells}
b1, b2          mm               {plate spacing}
c1, c2          mm               {cell pitch}
x1, x2          mm               {strip length}
tf1, tf2        mm               {fin thickness}
ts1, ts2        mm               {splitter thickness, double cells only}
{-----------------------------------------------------------}
```

Output listing

The reader should be able to identify where **STRINGS** *provided by the input file are printed in the output listing.*

```
{-----+--------------------------------------+------------------------------------}
{Filename BERGOUT – generated by BergFin}
{-----+--------------------------------------+------------------------------------}
volume ··········· BERG:
language ·········· HP-Pascal 3.0
input ············· BERGDAT.TEXT <<<<< Contraflo
program ·········· BergFin.TEXT
output ············ BERGOUT.TEXT

COUNTERFLOW
fluids:         1 ⇒ Gas-1.bar              2 ⇒ Air-18.48bar
geometry:       1 ⇒ r.strip(S) mod.19s35   2 ⇒ r.strip(S) 24s12

OVERALL HEAT TRANSFER
thermal rating .... kW ........................ Q = 10949.0
adjust LMTD for long conduction ..........  adjLMTD = 0.9170

GAS INLET CONDITIONS
gas flow rate ..... kg/s...................... mg = 49.000
gas inlet pressure .. bar .................... Pg1 = 1.1000
gas inlet temperature K ...................... Tg1 = 797.00

AIR INLET CONDITIONS
air flow rate....... kg/s..................... ma = 49.000
air inlet pressure .. bar .................... Pa2 = 18.480
air inlet temperature K ...................... Ta2 = 563.00
```

OUTLET TEMPERATURES AND INITIAL NTUs

gas-side outlet temperature K	T_{g2} = 586.00
air-side outlet temperature K	T_{a1} = 774.00
mean temperature difference K	MEAN = 21.091
gas-side Ntu from terminal temps	tempNTU1 = 9.1379
air-side Ntu from terminal temps	tempNTU2 = 9.1379

The values tempNTU were calculated from terminal temperatures only, and were not used in the direct-sizing calculation. Calculated values of NTU, obtained later, are larger than those obtained from the terminal temperatures. This is due to reduction in LMTD to allow for longitudinal conduction, but it can also be caused by poor assumptions as to mean bulk temperatures of both fluids used in evaluating physical properties. This will eventually affect the value of heat transfer coefficients, and hence NTU values. Thus temporary values tempNTU are to be regarded as providing no more than a useful check on the accuracy of computation.

MEAN GAS FLUID PROPERTIES

mean Prandtl number		Pr_1 = 0.6830
mean specific heat	J/(kg K)	Cp_1 = 1059.0
mean absolute viscosity	kg/(m s)	mu_1 = 0.0000590
mean thermal conductivity	J/(m s K)	k_1 = 0.07892
mean pressure	bar	P_{gm} = 1.0560
gas gas-constant	J/(kg K)	R_g = 287.07
mean bulk temperature	K	T_{gm} = 691.50
mean density	kg/cum	rho_g = 0.5320

MEAN AIR FLUID PROPERTIES

mean Prandtl number		Pr_2 = 0.6830
mean specific heat	J/(kg K)	Cp_2 = 1059.0
mean absolute viscosity	kg/(m s)	mu_2 = 0.00005090
mean thermal conductivity	J/(m s K)	k_2 = 0.07892
mean pressure	bar	P_{ag} = 18.473
air gas-constant	J/(kg K)	R_a = 287.07
mean bulk temperature	K	T_{am} = 668.50
mean density	kg/cm	rho_a = 9.6261

LOCAL GEOMETRY

plate thickness	mm	t_p = 0.2000
metal thermal cond.	J/(m s K)	k_M = 90.000
metal specific heat	J/(kg K)	Cp_m = 433.00
metal density	kg/cum	rho_M = 8906.0

CONSTRAINTS

gas-side max. core press. loss	N/sqm	dPg = 8620.1
air-side max. core press. loss	N/sqm	dPa = 80000.0

The above values of pressure loss are input values.

PLATE THICKNESS AND THERMAL CONDUCTIVITY

plate thickness	mm	tp = 0.2000
plate thermal conductivity	J/(m s K)	kp = 90.000
gas-side fins th. conductivity	J/(m s K)	kf1 = 90.000
air-side fins th. conductivity	J/(m s K)	kf2 = 90.000

GAS-SIDE GEOMETRY

plate spacing	mm	b1 = 5.000
cell pitch	mm	c1 = 1.3127
fin length	mm	x1 = 2.5400
fin thickness	mm	tf1 = 0.1016
4*hydraulic radius K&L geom.	mm	4*rh1 = 1.8750
hydraulic diameter M&B geom.	mm	Mhyd1 = 1.8744

The Manglic and Bergles (M&B) definition is used, and the Kays and London (K&L) definition provides a check on calculated values.

AIR-SIDE GEOMETRY

plate spacing	mm	b2 = 1.9050
cell pitch	mm	c2 = 1.0531
fin length	mm	x2 = 2.8222
fin thickness	mm	tf2 = 0.1016
4*hydraulic radius K&L geom.	mm	4*rh2 = 1.2105
hydraulic diameter M&B geom.	mm	Mhyd2 = 1.2097

COMPUTED GEOMETRICAL FACTORS

(Stotal1/Vexchr)	1/m	alpha1 = 1319.762
(Stotal1/Vtotal1)	1/m	beta1 = 1928.172
(Sfins1/Stotal1)		gamma1 = 0.808606
(Stotal1/Sexchr)		kappa1 = 4.820430
(Sfins1/Sexchr)		lambda1 = 3.897830
(Aflow1/Afront1)		sigma1 = 0.903853
(Sexchr/Vexchr)	1/m	omega1 = 273.7851
(Stotal2/Vexchr)	1/m	alpha2 = 737.0883
(Stotal2/Vtotal2)	1/m	beta2 = 2826.472
(Sfins2/Stotal2)		gamma2 = 0.664395
(Stotal2/Sexchr)		kappa2 − 2.692215
(Sfins2/Sexchr)		lambda2 = 1.788695

(Aflow2/Afront2)   sigma2 = 0.855332
(Sexchr/Vexchr)   1/m   omega2 = 273.7851

When geometrical values are evaluated correctly, the value of Sexchr/Vexchr *should be the same for both sides. This provides a partial check on geometry.*

HEAT TRANSFER COEFFICIENTS

side-1 correlation coeff.	J/(sqm s K)	h1 = 350.569
side-2 correlation coeff.	J/(sqm s K)	h2 = 607.539

FIN AND SURFACE EFFECTIVENESS

gas-side mY	mY1 = 0.692267
gas-side TANH(mY)	tanh1 = 0.599436
gas-side fin performance factor	phi1 = 0.865904
gas-side factor [1-gamma*(1-phi)]	eta1 = 0.891569
air-side mY	mY2 = 0.347215
air-side TANH(mY)	tanh2 = 0.333904
air-side fin performance factor	phi2 = 0.961662
air-side factor [1-gamma*(1-phi)]	eta2 = 0.974528

GAS, WALL, AIR AND OVERALL COEFFICIENTS

gas coefft. referred to plate	J/(sqm s K)	u1 = 1506.655
wall coefficient plate	J/(sqm s K)	uw = 450000.0
air coefft. referred to plate	J/(sqm s K)	u2 = 1593.962
overall coeff. referred to plate	J/(sqm s K)	U = 773.209
gas-side NTU		Ntu1 = 10.0043
air-side NTU		Ntu2 = 10.0043
gas-side Ntu from terminal temperatures		tempNTU1 = 9.1739
air-side Ntu from terminal temperatures		tempNTU2 = 9.1739

All heat transfer coefficients are eventually referred to the plate surface. This means that surface areas are quoted in terms of plate surface only. Other design approaches may refer heat transfer coefficients to the total surface on one side only of the exchanger, and care is then necessary when comparing two designs.

CELL BASIS SOLUTION

length of plate exchanger	m	L = 0.3471
edge-length of single plate	m	E = 1934.1
notional side of block (equal sides)	m	side = 2.6579
aspect ratio (L/notional-side)		aspect = 0.1306
X-section of exchanger	sqm	X-sect = 7.0644
plate surface area	sqm	Sexchr = 671.40
volume of exchanger block	cum	Vol = 2.4523

B.5 INPUT/OUTPUT FOR DIRECT SIZING COMPACT CONTRAFLOW EXCHANGER

mass of exchanger block   kg   Mblock = 3404.3
mean block porosity   Py = 0.8441

The notional side of block and the corresponding aspect ratio both assume that the cross-section of the exchanger is square, in order to provide a 'feel' for the size of the block. The direct-sizing method permits any choice of cross-section, which is constrained only by the number of divisions of the edge length of the exchanger.

ACTUAL CORE PRESSURE LOSS
gas-side core pressure loss   N/sqm   dPg = 8620.1
air-side core pressure loss   N/sqm   dPa = 2760.5
sum of pressure ratios $(dPa/Pa2) + (dPg/Pg1) = 0.0799$

PRESSURE DROP LIMITS
$2505.347 < dPg < 501076.105$
$784.271 < dPa < 204410.184$

HEAT TRANSFER, FREE-FLOW AND FRONTAL AREAS
gas-side surface area   sqm   Stotal1 = 3236.4307
air-side surface area   sqm   Stotal2 = 1807.5498
total surface area   sqm   (S1+S2) = 5043.9806
gas-side flow area   sqm   Aflow1 = 4.5452291
air-side flow area   sqm   Aflow2 = 1.7411726
gas-side frontal area   sqm   Afront1 = 5.0287277
air-side frontal area   sqm   Afront2 = 2.0356677

Total surface areas are provided for those who insist on knowing them.

MASS VELOCITY AND REYNOLDS NUMBER
gas-side mass velocity   kg/(sqm s)   G1 = 10.78054
air-side mass velocity   kg/(sqm s)   G2 = 28.14195
gas-side Reynolds number   Re1 = 397.0041
air-side Reynolds number   Re2 = 668.8159

FRICTION FACTOR, STANTON NUMBER AND HEAT TRANS. COEFF.
gas-side friction factor   f1 = 0.106528
air-side friction factor   f2 = 0.058462
gas-side Stanton number   St1 = 0.030707
air-side Stanton number   St2 = 0.020386
gas-side correlation coeff.   J/(sqm s K)   h1 = 350.569
air-side correlation coeff.   J/(sqm s K)   h2 = 607.539

CHECK ON CALCULATION

design thermal rating	kW	Q = US*LMTD = 10949.0
thermal rating check	kW	U*S*LMTD = 10949.0
specific rating	MW/cum	Qspec = 4.46482

Specific rating is a measure of the 'goodness' of an exchanger.

OUTLET CONDITIONS

gas-side outlet pressure	bar	Pg2 = 1.0138
air-side outlet pressure	bar	Pa1 = 18.452
gas-side outlet temperature	K	Tg2 = 586.00
air-side outlet temperature	K	Ta1 = 774.00

MEAN CONDITIONS

mean gas pressure	bar	Pgm = 1.0569
mean air pressure	bar	Pam = 18.466
mean gas temperature	K	Tgm = 691.50
mean air temperature	K	Tam = 668.50
mean gas density	kg/cum	rhog = 0.5324
mean air density	kg/cum	rhoa = 9.6225

After a first run, improved values of the above mean values become available; they can then be inserted into the input data before the second run.

CORE VELOCITIES

gas-side flow velocity	m/s	Vel1 = 20.265
air-side flow velocity	m/s	Vel2 = 2.9235

Numerical values of flow velocities can sometimes come as a surprise. These are important in the assessment of prospects for fouling and for erosion.

INPUT DATA FOR LMTD CORRECTION

gas-side mass flow rate	kg/s	mg = 49.000
air-side mass flow rate	kg/s	ma = 49.000
gas-side specific heat	J/(kg K)	Cp1 = 1059.0
air-side specific heat	J/(kg K)	Cp2 = 1059.0
gas-side local Ntu		localNtu1 = 19.494
air-side local Ntu		localNtu2 = 20.624
density of metal	kg/cum	rhoM = 8906.0
longitudinal conduction area	sqm	Ac = 0.5833
length of exchanger	m	L = 0.3471
thermal conductivity metal	J/(m s K)	kM = 90.000

The above data is input data for the LMTD correction program.

B.6 ALGORITHM FOR TRANSIENT RESPONSE OF CONTRAFLOW EXCHANGER

INPUT DATA FOR TRANSIENTS

gas-side NTU		Ntu1 = 10.004
air-side NTU		Ntu2 = 10.004
gas-side local Ntu		localNtu1 = 19.494
air-side local Ntu		localNtu2 = 20.624
gas-side mass flow rate	kg/s	mg = 49.000
air-side mass flow rate	kg/s	ma = 49.000
mass of exchanger block	kg	Mblock = 3404.3
specific heat of block	J/(kg K)	Cblock = 433.00
gas-side residence mass	kg	Mgg = 0.8400
gas-side specific heat	J/(kg K)	Cp1 = 1059.0
gas-side ratio of thermal capacities		Rg = 1657.0
gas-side residence or transit time	sec	Tau1 = 0.0171
air-side residence mass	kg	Maa = 5.816
air-side specific heat	J/(kg K)	Cp2 = 1059.0
air-side ratio of thermal capacities		Rf = 239.33
air-side residence or transit time	sec	Tau2V = 0.1187
constant P in transient equations		P = 6.9234

The above data is input for the exchanger transient response program.

Finally, note that the concept of 'effectiveness' is not used in direct sizing. As all terminal temperatures are known before sizing commences, the effectiveness value will be known at that time. Also, the concept of effectiveness becomes invalid for design as soon as LMTD is adjusted to allow for longitudinal conduction.

B.6 ALGORITHM FOR TRANSIENT RESPONSE OF CONTRAFLOW EXCHANGER

The Pascal algorithm used in Chapter 8 for transient response of an exchanger, including wall storage, is written in its simplest explicit forward-difference form. Although the characteristic grid is given in Fig. 8.2, it was found easier to write the algorithm using the physical grid of Fig. 8.1, and to obtain finite-difference increments using Fig. 8.2.

Derivation of the finite-difference increments was explained in Chapter 8, and these expressions can be coded in front of the algorithm, permitting use of the simple expressions ($\Delta\alpha$, $\Delta\beta$, ΔT) in the algorithm itself, which is in HP-Pascal 3.0.

```
                {set up increments}
                dA : = SQRT(1 + P * P)/(m * (1 + P));
                dB : = dA;
                dT : + = SQRT(2) * dA;

                {initialise all temperature values}
                FOR j : = 0 TO m DO                    {over total field}
BEGIN           Th[j,0] : = 0.0;                       {hot fluid}
                Tc[j,0] : = 0.0;                       {cold fluid}
                Tw[j,0] : = 0.0                        {solid wall}
END;

                {Th input disturbance, terminating at z}
                FOR k : = 0 TO z DO
                Th[0,k] : ={chosen disturbance function};

                {transient algorithm}
                FOR k : = 0 TO n DO                    {to end transient}
BEGIN           FOR j : = 0 TO (m - 1) DO              {B characteristic}
BEGIN           slTh : = -Nh * (Th[j,k] - Tw[j,k]);    {slope of Th}
                Th[j + 1,k] : = Th[j,k] + dA * slTh
END;
                FOR j : = 0 TO m DO                    {T characteristic}
BEGIN           slTw : = +(Nh/Rh) * (Th[j,k] - Tw[j,k])
                  -(Nc/(P * Rc)) * (Tw[j,k] - Tc[j,k]);
                Tw[j,k + 1] : = Tw[j,k] + dT * slTw
END;
                FOR j : = 1 TO m DO                    {A characteristic}
BEGIN           slTc : = +Nc * (Tw[j,k] - Tc[j,k]);
                Tc[j - 1,k + 1] : = Tc[j,k] + dB * slTc
END;
END;            {temperature field is now known}
```

Results of this computation may be seen in Figs. 8.3 and 8.4. To obtain Fig. 8.5 it is necessary to return to Fig. 8.1 and allow for the time delay from $T = 0$ to the end of the steady state at each station across the exchanger. This is left as a coding exercise.

B.7 SPLINEFITTING OF DATA

Cubic splinefitting is the preferred method for representing both flow-friction and heat-transfer data. When using interpolating splinefits there is no need to worry whether data is being used outside its range of validity in design, as extrapolation is not built in.

Original data is required in the form of tables of values, which may contain experimental errors. The splinefitting algorithm of Woodward (1970) allows for experimental errors by including an estimated standard deviation of each ordinate. Woodward's method also allows the smoothing spline to be an arbitrary polynomial, but the author found the cubic polynomial to be adequate for most applications. The exception is when the curve being fitted goes through a point of infinite gradient, but this can be fitted by two splinefits with points adjacent to the infinite gradient being fitted by more elementary means.

de Boor (1978) examined a number of splinefitting procedures in his book, and in assessing cubic splines, exponential splines and taut splines, he observed that it is often difficult to improve upon the performance of the simple cubic spline. When oscillations are found, the trick is simply to include additional knot points.

When data are sparse and there are considerable changes in ordinates, with sharp changes in direction, the variable power spline of Soanes (1976) is capable of providing a smooth fit. With variable power splines, a possible technique is to calculate sufficient intermediate points from the variable power spline and then to use this new data to fit the standard interpolating cubic spline.

The author tested both taut splines and variable power splines as alternatives to the cubic splinefit for the representation of data. With sufficient knots, no significant advantage over cubic splines was found in comparison with taut splines. But there are other engineering applications in which taut splines are preferred. It was not found necessary to use variable power splines.

One of the thrusts in looking at different methods of fitting data was to find a twice-differentiable representation that may be useful in certain other applications; see e.g. Young (1988) who calculates the thermodynamic properties of steam from a few fundamental properties.

REFERENCES

Anon (1960–1978) Index by subject to algorithms, *Communications of the Association for Computer Machinery (ACM)*.(Subsequently as loose-leaf *Collected Algorithms from ACM*).

de Boor, C. (1978) *A Practical Guide to Splines*, Berlin: Springer Verlag, *Applied Math. Sciences No. 27*, p. 303.

Press, W. H., Flannery, B. P., Teukolsky, S. A. and Vettering, W. T. (1989, 1992) *Numerical Recipies in Pascal*, Cambridge: CUP, 1989; *Numerical Recipies in Fortran*, Cambridge: CUP, 1992.

Soanes, R. V. (1976) VP-splines, an extension of twice differentiable interpolation, *Proceedings of the 1976 Army Numerical Analysis and Computer Conference*, ARD Report 76-3, US Army Research Office, PO Box 12211, Research Triangle Park, NC, pp. 141–152.

Woodward, C. II. (1970) An algorithm for data smoothing using spline functions, *B. I. T.*, **10**, 501–510.

Young, J. B. (1988) An equation of state for steam for turbomachinery and other flow calculations, *Transactions ASME. Journal of Engineering for Gas Turbines and Power*, **110**, 1–7.

C

OPTIMISATION OF RECTANGULAR OFFSET-STRIP PLATE-FIN SURFACES

C.1 TREND CURVES

The graphs which follow comprise the set from which Figs. 4.12 and 4.13 are taken. The objective is to indicate the most profitable direction in which changes in the local geometry of rectangular offset-strip plate-fin geometries may be made to optimise thermal performance of the exchanger. The figures were generated for the single-cell configuration. An almost identical set was also found for the double-cell configuration.

Each heading specifies the parameter being explored. The left-hand figures are for the high pressure side of the exchanger, and the right-hand figures are for the low pressure side of the exchanger. The relevant constraint is given in the caption.

As these graphs were generated by changing the plate spacing b, cell width c and strip length x, one at a time, the reader should not expect to find that selection of three individually optimised parameters will lead to a fully optimised design. The result will be close, but when all three parameters are changed together there may exist a design better than any indicated by these graphs.

Only one contraflow exchanger was used to generate the graphs. For other exchangers the data should be used only as an indication of the *direction* in which local geometry is to be changed.

Note that there has been no attempt to explore the effect of varying fin thickness on exchanger performance. Although this could have been done, it might have pushed the Manglic and Bergles correlations just a little too far. However, there is no reason why such work should not be completed so that results obtained can be compared with other papers directly concerned with the effect of fin thickness on exchanger performance, for example, Xi, G., Suzuki, K., Hagiwara, Y. and Murata, T. (1989) Basic study on the heat transfer characteristics of offset fin arrays (effect of fin thickness on the middle range of Reynolds number), *Transactions of the Japan Society of Mechanical Engineers – Part B*, **55**(519), 3507–3513.

C.2 BLOCK VOLUME

Trend curves showing how block mass changes as rectangular offset strip-fin parameters (b, c, x) are varied.

LP-side fixed pressure loss

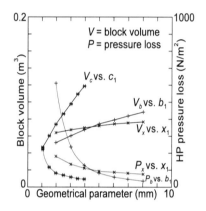

Fig. C.1 High pressure side: LP-side fixed pressure loss

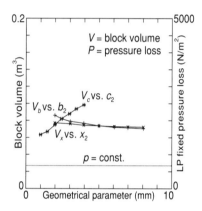

Fig. C.2 Low pressure side: LP-side fixed pressure loss

HP-side fixed pressure loss

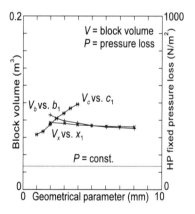

Fig. C.3 High pressure side: HP-side fixed pressure loss

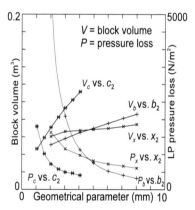

Fig. C.4 Low pressure side: HP-side fixed pressure loss

C.3 BLOCK MASS (EXCLUDING FLUIDS)

Trend curves showing how block mass changes as rectangular offset strip-fin parameters (b, c, x) are varied.

LP-side fixed pressure loss

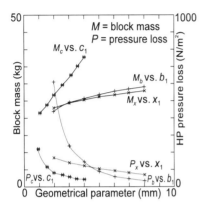

Fig. C.5 High pressure side: LP-side fixed pressure loss

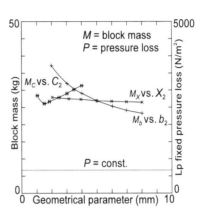

Fig. C.6 Low pressure side: LP-side fixed pressure loss

HP-side fixed pressure loss

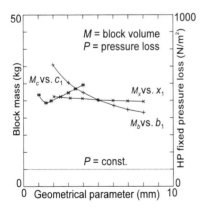

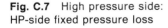

Fig. C.7 High pressure side: HP-side fixed pressure loss

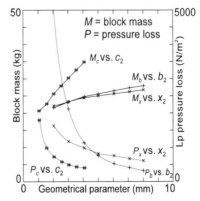

Fig. C.8 Low pressure side: HP-side fixed pressure loss

C.4 BLOCK LENGTH

Trend curves showing how block length changes as rectangular offset strip-fin parameters (b, c, x) are varied.

LP-side fixed pressure loss

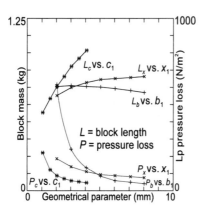

Fig. C.9 High pressure side: LP-side fixed pressure loss

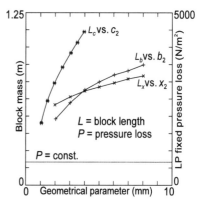

Fig. C.10 Low pressure side: LP-side fixed pressure loss

HP-side fixed pressure loss

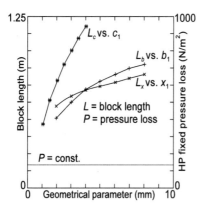

Fig. C.11 High pressure side: HP-side fixed pressure loss

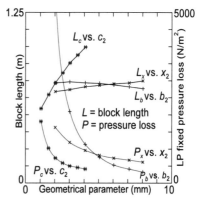

Fig. C.12 Low pressure side: HP-side fixed pressure loss

C.5 FRONTAL AREA

Trend curves showing how frontal area changes as rectangular offset strip-fin parameters (b, c, x) are varied.

LP-side fixed pressure loss

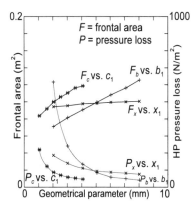

Fig. C.13 High pressure side: LP-side fixed pressure loss

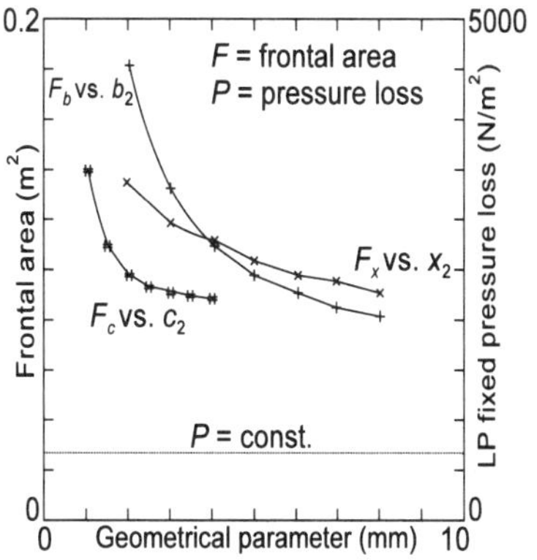

Fig. C.14 Low pressure side: LP-side fixed pressure loss

HP-side fixed pressure loss

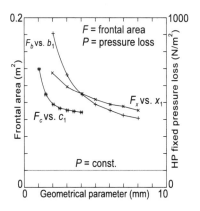

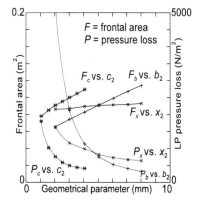

Fig. C.15 High pressure side: HP-side fixed pressure loss

Fig. C.16 Low pressure side: HP-side fixed pressure loss

C.6 TOTAL SURFACE AREA

Trend curves showing how total surface area changes as rectangular offset strip-fin parameters (b, c, x) are varied.

LP-side fixed pressure loss

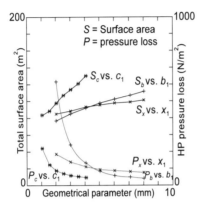

Fig. C.17 High pressure side: LP-side fixed pressure loss

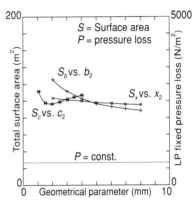

Fig. C.18 Low pressure side: LP-side fixed pressure loss

HP-side fixed pressure loss

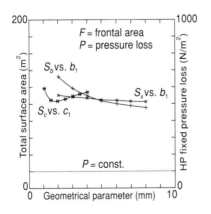

Fig. C.19 High pressure side: HP-side fixed pressure loss

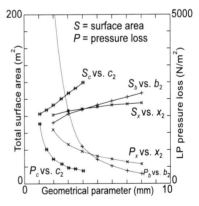

Fig. C.20 Low pressure side: HP-side fixed pressure loss

D

PERFORMANCE DATA FOR RODBAFFLE EXCHANGERS

D.1 FURTHER HEAT TRANSFER AND FLOW FRICTION DATA

Towards completion of this volume, the writer received a number of experimental datasets for the RODbaffle geometries from Dr. C.C. Gentry of the Phillips Petroleum Company, Oklahoma. It seemed useful to plot these for comparison, and to generate a set of smoothed data for the geometry 02WARA, which geometry is not identical to that used in chapter 7.

The RODbaffle codes (e.g. 02WARA) do not refer to dimensions of the RODbaffle geometry. The first two digits (e.g. 02, 03, 04) denote the specific test sequence. The letter symbols (O, W) denote the test fluid, O for oil and W for water. The final three letters identify baffle ring geometry.

Figures D.1 and D.2 correspond to Figs. 7.2 and 7.4. Curves at lower Reynolds numbers with hollow symbols are for oil, whereas curves at higher Reynolds numbers are for water. Although the curves suggest the possibility of a unified correlation for shell-side heat transfer and baffle loss coefficients which might be useful in optimisation (cf. Manglic and Bergles, 1990, Chapter 4), it is evident from the consistency of individual datasets that better designs would always result from using individual correlations, as recommended for plate-fin designs by Kays and London (see Chapter 4).

It is usually known in advance whether the shell-side fluid is to be water or oil, and universal correlations may perhaps be more easily sought when they are generated using the same fluid.

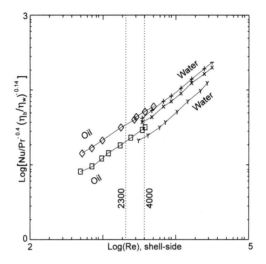

Fig. D.1 Shell-side heat transfer correlations for RODbaffle geometries (experimental data courtesy of C. C. Gentry, Phillips Petroleum Company)

Light oil − 02OARA (◇), 04OARE (□)
Water − 02WARA (+), 03WARB (×), 02WARE (Y)

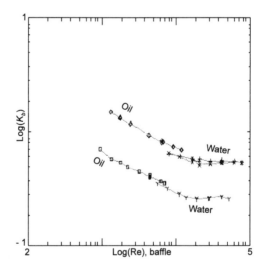

Fig. D.2 Baffle-loss coefficient for RODbaffle geometries (experimental data courtesy of C. C. Gentry, Phillips Petroleum Company)

Light oil − 02OARA (◇), 04OARE (□)
Water − 02WARA (+), 03WARB (×), 02WARE (Y)

Table D.1 provides a comparison of the ARA geometry used in Chapter 7 with the additional five sets of data provided separately by Gentry. The geometries are quite different.

Table D.1 Geometrical data used in the example of Chapter 7

Bundle Geometry	Tube o.d., d (mm)	Tube pitch, p (mm)	Baffle spacing, L_b (mm)	p/d	L_b/d
ARA	38.10	44.45	150	1.1666	3.937
02WARA	12.70	17.4625	124.46	1.375	9.80
02OARA	12.70	17.4625	124.46	1.375	9.80
03WARB	12.70	17.4625	248.92	1.375	19.60
02WARE	15.875	19.050	76.2	1.200	4.80
04OARE	15.875	19.050	76.2	1.200	4.80

Tables D.2 and D.3 which follow are smoothed datasets for configuration 02WARA. Two tables of differing Reynolds numbers are provided because

1. Regular values of shell-side Reynolds number are useful in setting up an interpolation scheme for the group containing Nu, Pr and (η_b/η_w).
2. Regular values of baffle flow Reynolds number are useful in setting up an interpolation scheme for the baffle loss coefficient (k_b).

A relationship between the two Reynolds numbers exists for the test data, but this depends on geometry, mass flow rate and thermodynamic conditions; it was not set up in the tables below.

Table D.2 Shell-side heat transfer for 02WARA (cubic splinefit smoothed data)

Shell-side Re	$\dfrac{\text{Nu}}{\text{Pr}^{0.4}(\eta_b/\eta_w)^{0.14}}$
30 580	232.207
30 000	228.041
25 000	193.808
20 000	161.215
15 000	128.626
12 000	107.946
10 000	93.109
8 000	78.180
6 000	63.275
5 000	55.426
4 000	47.309
3 500	43.153
3 292	41.408

Table D.3 Baffle loss coefficient for 02WARA (cubic splinefit smoothed data)

Baffle flow Re	Baffle loss coefficient (k_b)
77 936	0.547 95
60 000	0.554 79
50 000	0.556 97
40 000	0.555 54
30 000	0.559 04
25 000	0.559 04
20 000	0.571 13
15 000	0.594 93
12 000	0.615 11
10 000	0.629 95
8 391	0.642 24

E

EVALUATION OF SINGLE-BLOW OUTLET RESPONSE

E.1 LAPLACE TRANSFORMS

The required inverse Laplace transforms

$$L^{-1}\left\{\exp\left(\tfrac{n}{s-a}\right)\right\} = \delta(t) + e^{at}\frac{nI_1(2\sqrt{nt})}{\sqrt{nt}}$$

$$L^{-1}\left\{\tfrac{1}{s-a}\exp\left(\tfrac{n}{s-a}\right)\right\} = e^{at}I_0(2\sqrt{nt})$$

may be obtained by series expansion and term-by-term inversion. While deriving these inversions, it was considered that a gap existed in published tables of inverse transforms.[1] Tables E.1, E.2 and E.3 provide a sequence that includes the inversions required. I_0 and I_1 are modified Bessel functions

Table E.1 Laplace transforms—elementary

Transform $\hat{f}(s)$	Inversion $f(t)$
1	$\delta(t)$
$\frac{1}{s}$	1
$\frac{1}{s-a}$	e^{at}
$\frac{1}{s^2}$	t
$\frac{1}{[s(s-a)]}$	$\frac{e^{at}-1}{a}$

---

[1] Dr Jeffrey Lewins, in later private correspondence, reffered the author to some inversions in Carlslaw and Jaeger (1948), which the author had not seen.

Table E.2 Laplace transforms involving exp(n/s)

Transform $\hat{f}(s)$	Inversion $f(t)$
$\exp(\frac{n}{s})$	$\delta(t) + \frac{nI_1(2\sqrt{nt})}{\sqrt{nt}}$
$\frac{1}{s}\exp(\frac{n}{s})$	$I_0(2\sqrt{nt})$
$\frac{1}{s-a}\exp(\frac{n}{s})$	$I_0(2\sqrt{nt}) + a\int_0^t e^{-a\sigma}I_0(2\sqrt{n\sigma})d\sigma$
$\frac{1}{s^2}\exp(\frac{n}{s})$	$\int_0^t I_0(2\sqrt{n\sigma})d\sigma$
$\frac{1}{s(s-a)}\exp(\frac{n}{s})$	$e^{at}\int_0^t e^{-at}I_0(2\sqrt{n\sigma})d\sigma$

Table E.3 Laplace transforms involving exp [n/(s-a)]

Transform $\hat{f}(s)$	Inversion $f(t)$
$\exp(\frac{n}{s-a})$	$\delta(t) + e^{at}\frac{nI_1(2\sqrt{nt})}{\sqrt{nt}}$
$\frac{1}{s}\exp(\frac{n}{s-a})$	$1 + \int_0^t e^{a\sigma}\frac{nI_1(2\sqrt{n\sigma})}{\sqrt{n\sigma}}d\sigma$
$\frac{1}{s-a}\exp(\frac{n}{s-a})$	$e^{at}I_0(2\sqrt{nt})$
$\frac{1}{s^2}\exp(\frac{n}{s-a})$	$t - te^{at} + e^{at}\int_0^t I_0(2\sqrt{n\sigma})d\sigma$
$\frac{1}{s(s-a)}\exp(\frac{n}{s-a})$	$\int_0^t e^{a\sigma}I_0(2\sqrt{n\sigma})d\sigma$

E.2 NUMERICAL EVALUATION OF OUTLET RESPONSE

The following procedure minimises the computational requirement. Assume the inlet disturbance D to be exponential in form, corresponding closely in shape to that obtained from a fast response electrical heater (Fig. E.1).

Then $\qquad D(0, t) = 1 - \exp(\frac{-\tau}{\tau^*})$

$$\text{with non-dimensional time } \tau = \frac{Ntu}{W}\frac{(ut-x)}{L}$$

$$\text{non-dimensional time constant } \tau^* = \frac{Ntu}{W}\frac{ut^*}{L} \text{ at } x = 0$$

The outlet fluid temperature response then becomes

E.2 NUMERICAL EVALUATION OF OUTLET RESPONSE

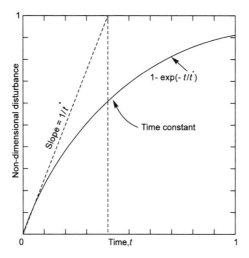

Fig. E.1 Non-dimensional disturbance and time constant

$$G^{\#} = \exp(-n)\left[1 - \exp\left(-\frac{\tau}{\tau^*}\right) + \int_0^\tau \left\{1 - \exp\left(-\frac{\tau - \sigma}{\tau^*}\right)\right\} R(\sigma) d\sigma\right]$$

$$= \exp(-n)\left[1 - \exp\left(-\frac{\tau}{\tau^*}\right) + \int_0^\tau R(\sigma) d\sigma - \exp\left(-\frac{\tau}{\tau^*}\right) \int_0^\tau \exp\left(\frac{\sigma}{\tau^*}\right) R(\sigma) d\sigma\right]$$

The expected response is of the form shown in Fig. E.2.

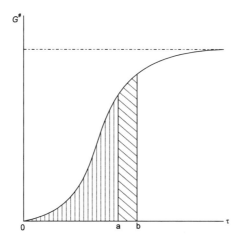

Fig. E.2 Outlet temperature response

Suppose the value of each integral $\int_0^\tau (\ )d\sigma$ is known up to $\tau = a$, then to continue evaluation of the $G^\# - \tau$ curve, the increment (cross-hatched area) is required to continue the summation.

The two integrals to be evaluated are

$$\int_0^\tau R(\sigma)d\sigma \quad \text{and} \quad \int_0^\tau \exp\left(\frac{\sigma}{\tau^*}\right) R(\sigma) d\sigma$$

Let us consider evaluation of the first integral between limits $\tau = a$ and $\tau = b$

$$\int_a^b R(\sigma) d\sigma = \int_a^b e^{-\sigma} \frac{nI_1(2\sqrt{n\sigma})}{(2\sqrt{n\sigma})} d\sigma$$

To avoid difficulties in the denominator when $a = 0$, we change the variable. Putting $\sigma = n\alpha^2$, and $d\sigma = 2n\alpha d\alpha$ the integral becomes

$$\int_{\alpha = \sqrt{a/n}}^{\alpha = \sqrt{b/n}} e^{-n\alpha^2} \frac{nI_1(2n\alpha)}{n\alpha} 2n\alpha d\alpha = 2n \int_{\sqrt{a/n}}^{\sqrt{b/n}} e^{-n\alpha^2} I_1(2n\alpha) d\alpha$$

At a new value of $(\tau = b$, i.e. $\alpha = \sqrt{b/n})$ the new value of the integral is given by

NEW VALUE = OLD VALUE + INCREMENT

Each *increment* of integral may be evaluated using Legendre polynomials in four-point Gaussian quadrature

$$\int_a^b y d\alpha = \frac{1}{2}(b - a) \int_{-1}^{+1} y dA$$

$$= \frac{1}{2}(b - a)[w_1 \cdot y(A_1) + w_2 \cdot y(A_2) + w_3 \cdot y(A_3) + w_4 \cdot y(A_4)]$$

where the A's are abscissae and the w's are weighting values, given in Table E.4 for four-point Gaussaian quadrature described in the paper by Lowan et al. (1954). Values of the modified Bessel function $I_1(2n\alpha) = y(A)$ are computed using an algorithm given by Clenshaw (1962).

In present computations, a top limit of Ntu around 75.0 was obtained before machine overflow occurred within the program. Curves for values of Ntu up to 500.0 have been obtained by Furnas (1930) using graphical methods. It is seldom that Ntu values exceeding 20.0 will be encountered during testing, but values of over 40.0 may occur in real situations.

Table E.4 Four-point quadrature

Position	Abscissae	Weighting
1	−0.861 136 311 594 053	0.347 854 845 137 454
2	−0.339 981 043 584 856	0.652 145 154 862 546
3	+0.339 981 043 584 856	0.652 145 154 862 546
4	+0.861 136 311 594 053	0.347 854 845 137 454

REFERENCES

Carlslaw, H. S. and Jaeger, J. C. (1948) *Operational Methods in Applied Mathematics*, 2nd Edn, Oxford: Oxford University Press.

Clenshaw, C. W. (1962) Chebyshev series for mathematical functions, *National Physical Laboratory, Mathematical Tables* Vol. 5, London: HMSO.

Furnas, C. C. (1930) Heat transfer from a gas stream to a bed of broken solids—II, *Industrial and Engineering Chemistry, Industrial ed.*, **22** (7), 721–731.

Lowan, A. N., Davids, N. and Levinson, A. (1954) Table of the zeros of the Legendre polynomials of order 1–16 and the weight coefficients for Gauss mechanical quadrature formula, *Tables of Functions and Zeros of Functions, NBS Applied Mathematical Series, No. 37*, pp. 185–189.

F

MOST EFFICIENT TEMPERATURE DIFFERENCE IN CONTRA-FLOW

F.1 CALCULUS OF VARIATIONS

A clear exposition of the theory for the Calculus of Variations is give in Hildebrand (1976). Other useful texts are those by Courant and Hilbert (1989), Mathews and Walker (1970), and Rektorys (1969).

The basic problem concerns a function $F(y, \frac{dy}{dx}, x)$ and the finding of a maximum or minimum of the integral of this function

$$J[y] = \int_{x_0}^{x_1} F\left(y, \frac{dy}{dx}, x\right) dx$$

where end values $x_0, x_1, y(x_0), y(x_1)$ are known. Conditions concerning continuity of functions and of their derivatives are covered in the reference texts, and the required solution reduces to solving the Euler equation

$$\boxed{\frac{\partial F}{\partial y} - \frac{d}{dx}\left(\frac{\partial F}{\partial y'}\right) = 0}$$

F.2 GENERALISATION

The problem can be extended to include a constraint in minimisation or maximisation of the integral

$$J[y] = \int_{x_0}^{x_1} F(y, y', x) dx \tag{F.1}$$

where $y(x)$ is to satisfy the prescribed end conditions

$$\left.\begin{array}{l} y(x_0) = C \\ y(x_1) = D \end{array}\right\} \tag{F.2}$$

as before, but a *constraint* condition is also imposed in the form

$$\int_{x_0}^{x_1} G(y, y', x) dx = K \tag{F.3}$$

where K is a prescribed constant. Then the appropriate Euler equation is found to be the result of replacing F in equation (F.1) by the auxiliary function

$$H = F + \lambda G$$

where λ is an unknown constant. This constant, which is in the nature of a Lagrange multiplier, will generally appear in the Euler equation and in its solution, and is to be determined together with the two constants of integration in such a way that all three conditions are satisfied.

F.3 PROBLEM OF GRASSMANN AND KOPP

By inspection, the equations in Section 2.12 are seen to match those given above, viz.

- Integral

$$E_{loss} = T_0 C \int_{T_2}^{T_1} \frac{\Delta\theta}{(T-\Delta\theta)T} dT$$

from which

$$F = \frac{\Delta\theta}{(T-\Delta\theta)T} = \frac{1}{T-\Delta\theta} - \frac{1}{T}$$

- Constraint

$$\frac{\alpha S}{mC} = \int_{T_2}^{T_1} \frac{dT}{\Delta\theta}$$

from which

$$G = \frac{1}{\Delta\theta}$$

F MOST EFFICIENT TEMPERATURE DIFFIERENCE IN CONTRA-FLOW

- End conditions

$$\left.\begin{array}{l}\Delta\theta(T_1) = \Delta\theta_1 \\ \Delta\theta(T_1) = \Delta\theta_1 \end{array}\right\}$$

Then

$$\frac{\partial H}{\partial \Delta\theta} - \frac{d}{dT}\left(\frac{\partial H}{\partial \Delta\theta'}\right) = 0 \qquad \text{where } H = F + \lambda G$$

$$H = \left[\frac{1}{T-\Delta\theta} - \frac{1}{T} + \lambda \frac{1}{\Delta\theta}\right]$$

$$\frac{\partial H}{\partial \Delta\theta} = \left[\frac{1}{(T-\Delta\theta)^2} - \frac{\lambda}{(\Delta\theta)^2}\right]$$

$$\frac{\partial H}{\partial (\Delta\theta)'} = 0$$

and the Euler equation becomes

$$\frac{1}{(T-\Delta\theta)^2} - \frac{\lambda}{(\Delta\theta)^2} = 0$$

with solution

$$\frac{\Delta\theta}{T} = \left(\frac{-\lambda \pm \sqrt{\lambda}}{1-\lambda}\right) = \text{constant}$$

Four possible cases exist

$$\left.\begin{array}{ll}(0 < \lambda < 1), & \text{taking } + \text{ve square root} \\ (0 < \lambda < 1), & \text{taking } - \text{ve square root} \\ (1 < \lambda), & \text{taking } + \text{ve square root} \\ (1 < \lambda), & \text{taking } - \text{ve square root} \end{array}\right\}$$

of which only the first case leads to useful results.
Putting $k = +\sqrt{\lambda}$, and remembering $k \neq 1$

$$\boxed{\frac{\Delta\theta}{T} = \left(\frac{k}{1+k}\right) = \text{const} = b} \qquad (F.4)$$

Equation (F.4) is to be compared with the result given in Section 2.12:

$$E_{loss} \propto \left(\frac{b}{1-b}\right) = \text{const} \qquad (F.5)$$

Using the constraint

$$\frac{\alpha S}{\dot{m}C} = \int_{T_2}^{T_1} \frac{dT}{\Delta\theta} = \left(\frac{1}{b}\right) \int_{T_2}^{T_1} \frac{dT}{T} = \frac{1}{b} \ln\left(\frac{T_1}{T_2}\right)$$

then

$$\frac{\Delta\theta}{T} = \left(\frac{\dot{m}C}{\alpha S}\right) \ln\left(\frac{T_1}{T_2}\right) \qquad (F.6)$$

Values of α and S will not normally be known before the exchanger has been designed. Thus it is not practicable to use equation (F.6) in direct-sizing. Rather it is appropriate to choose a value for b in equation (F.4) appropriate to the minimisation of exergy loss in equation (F.5).

Note that the equations apply to one fluid on one side of the exchanger only.

REFERENCES

Courant, R. and Hilbert, D (1989) *Mathematical Methods of Physics, Volume I*, New York: Wiley, p. 184.

Hildebrand, F. B. (1976) *Advanced Calculus for Applications*, New York: Prentice Hall, 2nd Edn, p. 360.

Mathews, J. and Walker, R. L. (1970) *Mathematical Methods of Physics*, Addison-Wesley, 2nd Edn., p. 322.

Rektorys, K. (Ed) (1969) *Survey of Applicable Mathematics*, Cambridge, MA: MIT Press, p. 1020.

G

PHYSICAL PROPERTIES OF MATERIALS AND FLUIDS

G.1 SOURCES OF DATA

Over the years, the author encountered many delays in attempting to source information on the physical properties of materials of construction. The data is scattered and is often presented in units not generally used by engineers. Some data needs conversion to appropriate engineering SI units, e.g. J/(kg K) for specific heat, J/(m s K) for thermal conductivity and m^2/s for thermal diffusivity. Useful conversion factors are listed in Appendix H. Density in kg/m^3 can be obtained from thermal diffusivity.

G.2 FLUIDS

Particularly near the critical points of fluids, property values tend to change significantly with both temperature and pressure; this behaviour is instanced in later examples of steam tables, e.g the UK Steam Tables in SI Units (1975). For other fluids, the reader may wish to consult Vargaftik (1983), Touloukian (1970) and the IUPAC Series, of which the representative volume on Oxygen (1987) is listed below. Other references can be obtained by consulting the *Journal of Physical and Chemical Reference Data*, a recent issue of the *Chemical Engineers Handbook*, or by seeking information from the manufacturers of working fluids, e.g. the KLEA Refrigerants from ICI Chemicals & Polymers Division.

G.3 SOLIDS

For aluminium, copper and titanium the properties of specific heat, thermal conductivity and thermal diffusivity are presented in Figs. G.1, G.2 and G.3 so that the engineer may see what kind of behaviour exists. These curves are not necessarily typical for other solids, and the series of volumes *Thermophysical Properties of Matter* by Touloukian and others (1970) should be consulted.

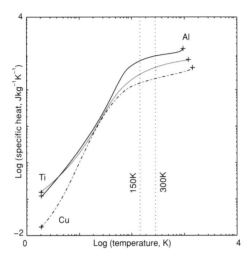

Fig. G.1 Specific heat of aluminium, copper, nickel and titanium J/(kg K)

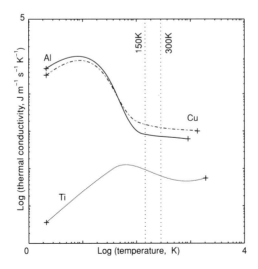

Fig. G.2 Thermal conductivity of aluminium, copper, and titanium J/(m s K)

368 G PHYSICAL PROPERTIES OF MATERIALS AND FLUIDS

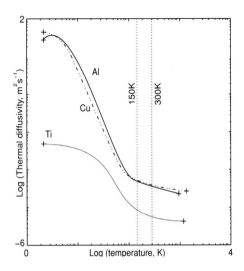

Fig. G.3 Thermal diffusivity of aluminium, copper, and Titanium (m^2/s)

REFERENCES

R. D. McCarty (1977) *Hydrogen Properties*, In: *Hydrogen, Its Technology and Applications* Eds. Cox, K. E. and Williamson, K. D., Boca Raton, FL: CRC Press.

Journal of Physical and Chemical Reference Data (1971 onwards) Washington: American Chemical Society.

Physical Properties Data Service IUPAC Thermodynamic Tables Properties Centre, Department of Chemical Engineering and Chemical Technology, Imperial College of Science, Technology and Medicine, Prince Consort Road, London SW7 2BY.

Touloukian, Y. S. *et al.* (1970 onwards) *Thermophysical Properties of Matter*, Vols. 1–11, Washington: IFI/Plenum Press.

UK Committee on Properties of Steam (1975) *UK Steam Tables in SI Units 1970*, London: Arnold.

Vargaftik, N. B. (1983) *Handbook of Physical Properties of Liquids and Gases*, New York: Hemisphere/Springer.

Wagner, W. and de Reuck, K. M. (1987) *Oxygen, International Thermodynamic Tables of the Fluid State – 9*, Oxford: Blackwell. (See other volumes in the IUPAC Series published by Blackwell, Oxford and Pergamon.)

SOURCEBOOKS ON HEAT EXCHANGERS IN CHRONOLOGICAL ORDER

The following texts should provide excellent sources for tracing other published work on heat exchangers.

REFERENCES

Hausen, H. (1950) *Wärmeübertragung im Gegenstrom, Gleichstrom und Kreuzstrom*, Berlin: Springer.
Wallis, G. B. (1969) *One-Dimensional Two-Phase Flow*, New York: McGraw-Hill.
Hewitt, G. F. and Hall-Taylor, M. S. (1970) *Annular Two-Phase Flow*, Oxford: Pergamon.
Collier, J. G. (1972) *Convective Boiling and Condensation*, New York: McGraw-Hill.
Kern, D. Q. and Kraus, A. D. (1972) *Extended Surface Heat Transfer*, New York: McGraw-Hill.
Gregorig, R. (1973) *Wärmeaustausch und Wärmeaustaucher, Grundlagen der chemishen Technik*, Frankfurt: Sauerländer.
Afgan, N. H. and Schlünder, E. U. (1974) *Heat Exchangers – Design and Theory Sourcebook*, New York: McGraw-Hill.
Hausen, H. (1976) *Heat Transfer in Counterflow, Parallel Flow and Crossflow*, New York: McGraw-Hill. (English edition of 1950 text.)
Shah, R. K. and London, A. L. (1978) *Laminar Forced Flow Convection in Ducts*, New York: Academic Press. Supplement 1 to *Advances in Heat Transfer*.
Shah, R. K., MacDonald, C. F. and Howard, C. P. (1980) *Compact Heat Exchangers – History, Technological Advances and Mechanical Design Problems*, ASME Heat Transfer Division, Vol. 10, New York: ASME.

Walker, G. (1980) *Stirling Engines*, Oxford OUP. (See bibliography.)
Kakac, S., Bergles, A. E. and Mayinger, F. (1981) *Heat Exchangers – Thermal Hydraulic Fundamentals and Design*, Washington: Hemisphere.
Palen, J. (ed) (1981) *Heat Exchanger Sourcebook*, Washington: Hemisphere.
Schmidt, F. W. and Willmott, A. T. (1981) *Thermal Energy Storage and Regeneration*, Washington: Hemisphere.
Hestroni, G (ed) (1982) *Handbook of Multiphase Systems*, Washington: Hemisphere.
Chisholm, D. (1983) *Two-Phase Flow in Pipelines and Heat Exchangers*, London: Longman.
Hausen, H. (1983) *Heat Transfer in Counterflow, Parallel Flow and Cross Flow*, New York: McGraw-Hill.
Kakac, S., Shah, R. K. and Bergles, A. E. (ed) (1983) *Low Reynolds Number Flow Heat Exchangers*, Washington: Hemisphere. 1983.
Taborek, J., Hewitt, G. F. and Afgan, N. H. (ed) (1983) *Heat Exchangers – Theory and Practice*, Washington: Hemisphere.
Kays, W. M. and London, A. L. (1984) *Compact Heat Exchangers*, New York: McGraw-Hill.
Kotas, T. J. (1985) *The Exergy Method of Thermal Plant Analysis*, Boston, MA: Butterworth.
Rohsenow, W. M. and Hartnett, J. P. (1985) *Handbook of Heat Exchanger Applications*, New York: McGraw-Hill.
Rohsenow, W. M., Hartnett, J. P. and Ganic, E. N. (1985) *Handbook of Heat Exchanger Fundamentals*, New York: McGraw-Hill.
Smith, R. A. (1986) *Vaporisers – Selection, Design and Operation*, London: Longman.
Bougard, J., Afgan, N. (1987) *Heat and Mass Transfer in Refrigeration and Cryogenics*, New York: Hemisphere/Springer Verlag.
Kakac et al. (eds) (1987) *Evaporators – Thermal Hydraulic Fundamentals and Design of Two-Phase Flow Heat Exchangers*, Porto, Portuga: NATO Advanced Study Institute.
Kakac, S., Shah, R. K. and Aung, W. (eds) (1987) *Handbook of Single-Phase Heat Transfer*, New York: Wiley.
Vilemas, J., Cesna, B. and Survila, V. (1987) *Heat Transfer in Gas-cooled Annular Channels*, New York: Hemisphere/Springer Verlag.
Wang, B. -X. (ed) (1987) *Heat Transfer Science and Technology*, Washington: Hemisphere.
Bejan, A. (1988) *Advanced Engineering Thermodynamics*, New York: Wiley.
Chisholm, D. (1988) *Heat Exchanger Technology*, Amsterdam: Elsevier.
Kakac, S., Bergles, A. E. and Fernandes, E. O. (eds) (1988) *Two-Phase Flow Heat Exchangers – Themal Hydraulic Fundamentals and Design*, NATO ASI Series E, Vol. 143, Dordrech: Kluwer Academic.
Minkowycz, W. J., Sparrow, E. M., Schneider, G. E. and Pletcher, R. H. (1988) *Handbook of Numerical Heat Transfer*, New York: Wiley.
Saunders, E. A. D. (1988) *Heat Exchangers – Selection, Design and Construction*, London: Longman.
Shah, R. K., Subbarao, E. C. and Mashelkar, R. A. (eds) (1988) *Heat Transfer Equipment Design*, Washington: Hemisphere.
Stasiulevicius, J. and Skrinska, A. (1988) *Heat Transfer of Finned Tubes in Crossflow*, Washington: Hemisphere. (English edition, G. F. Hewitt.)
Zukauskas, A. and Ulinskas, R. (1988) *Heat Transfer in Tube Banks in Crossflow*, New York: Hemisphere/Springer Verlag.

Zukauskas, A. (1989) *High-Performance Single-Phase Heat Exchangers*, Washington: Hemisphere.

Dzyubenko, B. V., Dreitser, G. A. and Ashmantas L. -V. A. (1990) *Unsteady Heat and Mass Transfer in Helical Tube Bundles*, New York: Hemisphere. (The 'helical tubes' are actually 'twisted flattened tubes'.)

Hewitt, G. (ed) (1990) *Hemisphere Heat Exchanger Design Handbook*, Washington: Hemisphere.

Ievlev, V. M., Danilov, Yu. N., Dzyubenko, B. B., Dreitser, G. A. and Ashmantas, L. A. (T. F. Irvine, editor of English edition) (1990) *Analysis and Design of Swirl-Augmented Heat Exchangers*, Washington: Hemisphere.

Shah, R. K., Kraus, A. D. and Metzger, D. (1990) *Compact Heat Exchangers – A Festschrift for A. L. London*, Washington: Hemisphere.

Thome, J. R. (1990) *Enhanced Boiling Heat Transfer*, Washington: Hemisphere.

Foumeny, E. A. and Heggs, P. J. (1991) *Heat Exchange Engineering*, Vols. 1, 2, New York: Ellis Horwood.

Roetzel, W., Heggs, P. J. and Butterworth, D. (eds) (1991) *Design and Operation of Heat Exchangers*, Berlin: Springer Verlag.

Carey, V. P. (1992) *Liquid-Vapor Phase-Change Phenomena*, Washington: Hemisphere.

Martin, H. (1992) *Heat Exchangers*, Washington: Hemisphere.

Organ, A. J. (1992) *Thermodynamics and Gas Dynamics of the Stirling Cycle Machine*, Cambridge: CUP.

Stephan, K. (1992) *Heat Transfer in Condensation and Boiling*, Berlin: Springer Verlag.

Shah, R. K. and Hashem, A (eds) (1993) *Aerospace Heat Exchanger Technology 1993, Proceedings of 1st International Conference on Aerospace Heat Exchanger Technology, Palo Alto, CA, 15–17 February 1993*, New York: Elsevier.

Collier, J. G. and Thome, J. R. (1994) *Convective Boiling and Condensation*, Oxford: OUP.

Hewitt, G. F., Shires, G. L. and Bott, T. R. (1994) *Process Heat Transfer*, Boca Raton, FL: CRC Press.

Lock, G. S. (1994) *Latent Heat Transfer*, Oxford: OUP.

Webb, R. L. (1994) *Principles of Enhanced Heat Transfer*, New York: Wiley.

Sekulic, D. P. and Shah, R. K. (1995) Thermal design of three fluid heat exchangers *Advances in Heat Transfer*, **26**, 219–324. (Substantial paper)

CONVERSION FACTORS

Length (L), m

> 1 inch = 25.4 m
> 1 ft = 0.3048 m
> 1 mile = 5280 ft = 1.609 m

Area (cross-sectional A, surface S), m²

> 1 in² = 645.2 mm²
> 1 ft² = 0.0929 m²

Volume (V), m³

> 1 in³ = 16 387 mm²
> 1 ft³ = 0.028 32 m³
> 1 UK gallon = 0.004 564 m³
> = 4.546 litres
> 1 US gallon = 0.003 785 m³
> = 3.785 litres

I CONVERSION FACTORS

Mass (M), kg

> 1 lbm = 0.4536 kg
> 1 ton = 2240 lbm = 1016 kg
> 1 tonne = 1000 kg

Density (ρ), kg/m³

> 1 lbm/in³ = 2.768 × 10⁴ kg/m³
> 1 lbm/ft³ = 16.0185 kg/m³
> 1 gm/cm3 = 1000 kg/m³

Pressure (p), N/m²

> 1 lbf/in² = 6894.76 N/m²
> 1 kgf/cm² = 98.0665 N/m²
> 1 tonf/in² = 15.44 MN/m²
> 1 Pa = 1 N/m²
> 1 Pa = 1 N/m²
> 1 bar = 14.50 lbf/in² = 10⁵ N/m²
> 1 atm = 1.013 25 bar
> 1 atu = pressure over atmospheric
> (not desirable as a unit)
> 1 mm Hg = 1 torr = 133.322 N/m²
> ($T = 0°C, P = 760$ mm Hg)
> ($g = 9.806\ 65$ m/s²)
> 1 mm H₂O = 9.806 38 N/m²
> ($T = 4°C,\ P = 760$ mm Hg)
> ($g = 9.806\ 65$ m/s²)

Energy (W, D)

> 1 ft pdl = 0.042 14 Nm or J
> 1 ft lbf = 1.356 Nm or J
> 1 Btu = 1055 Nm or J
> 1 therm = 105.5 MNm or MJ
> 1 kWh = 3.600 MNm or MJ

Power, W

> 1 ft lbf/s = 1.356 N m/s or W
> 1 hp = 550 ft lbf/s = 745.6 Nm/s or W

Force, N

> 1 pdl = 0.1383 N
> 1 lbf = 4.448 N
> 1 tonf = 9964 N
> 1 kgf = 9.807 N
> 1 dyne = 10^{-5} N

Torque, N m

> 1 lbf ft = 1.356 Nm
> 1 tonf ft = 3037 Nm

Velocity (u), m/s

> 1 ft/s = 0.3048 m/s
> 1 mile/hr = 0.4470 m/s
> 1 knot = 0.514 444 m/s

I CONVERSION FACTORS

Gas constant (R), J/(kg K)

> 1 ft lbf/(lbm R) = 5.380 95 J/(kgK)

Specific heat capacity (C_p), J/(kg K)

> 1 Btu/(lbm R) = 4187 J/(kg K)
> 1 kcal/kg K = 4186.8 J/(kg K)

Thermal conductivity (λ), J/(m s K)

> 1 Btu/(ft hr R) = 1.730 73 J/(m s K)
> 1 cal/(cm s C) = 418.68 J(m s K)
> 1 kcal/(m hr C) = 1.163 J/(m s K)

Thermal diffusivity $\kappa = \lambda(pC_p)$, m^2/s

> 1 ft^2/hr = 0.000 025 806 m^2/s

Heat transfer coefficient (α, U), J/(m^2 s K)

> 1 Btu/(ft^2 hr R) = 5.678 26 J(m^2 S K)
> 1 kcal/(m^2 hr C) = 1.163/(m^2 s K)

Dynamic (absolute) viscosity (η), kg/(m s)

> 1 lbm/(ft hr) = 0.000 413 kg/(m s)
> 1 poise = 0.1 kg/(m/s)
> 1 centipoise = 0.001 kg/(m s)
> 1 (Ns)/m^2 = 1kg/(m s)

$$1 \text{ lbf}/(\text{ft s}) = 1.488\ 16 \text{ kg}/(\text{m s})$$
$$1 \text{ (kgf s)}/\text{m}^2 = 9.806\ 65 \text{ kg}/(\text{m s})$$
$$1 \text{ slug}/(\text{ft s}) = 47.8802 \text{ kg}/(\text{m s})$$
$$1 \text{ (lbf s)}/\text{ft}^2 = 47.8802 \text{ kg}/(\text{m s})$$
$$1 \text{ gm}/(\text{cm s}) = 0.1 \text{ kg}/(\text{m s})$$
$$1 \text{ (dyne s)}/\text{cm}^2 = 0.1 \text{ kg}/(\text{m s})$$

Kinematic viscosity ($v = \eta/\rho$) – convert to dynamic viscosity (η)

$$1 \text{ stoke} = 10^{-4} \text{ m}^2/\text{s}$$

Surface tension (σ), N/m

$$1 \text{ lbf}/\text{in} = 175.127 \text{ N/m}$$
$$1 \text{ dyne}/\text{cm} = 10^{-3} \text{ N/m}$$

NOTATION

COMMENTARY

The new international standards for notation are followed, with some exceptions. Circumstances always arise where an awkward choice can be avoided and notation simplified, if there is departure from the standard. It was found that the solution of transients deserved such treatment, and the symbol for temperature was changed from T to θ, to allow the use of X, Y, T for dimensionless length and scaled time.

It was relatively easy to accept most of the new symbols, e.g.

- individual heat transfer coefficient (α for h)
- thermal conductivity (λ for k)
- thermal diffusivity (κ for α)
- absolute viscosity (η for μ)

although in the last case the same symbol is now used for absolute viscosity and efficiency, when μ remains available, at least for single species heat transfer.

Systems not arranged by nature are never quite perfect. Although lengthy discussions to arrive at the final preferred list of international symbols must have occurred, this author will plead that the preferred list is for the guidance of the experienced, and for the observance of the novice. Most readers of this volume will fall into the first category, and may understand the dilemmas which arise. Where departure from the preferred convention has arisen, it has been solely to achieve clarity of presentation.

Examples of the important symbols used are

- surface area (S)
- area of cross-section (A)
- fluid mass flow rate $(\dot{m})$
- solid wall mass (M)
- mass velocity of fluid $(G = \dot{m}/A)$
- temperature, steady-state and transient (T)
- temperature difference $(\delta\theta)$
- solution of transient temperatures (θ)
- time, dimensionless time (t, T)

Dimensionless groups are treated at the end of Chapter 2, and will not be further listed in the tables of symbols. One or two of the more uncommon groups are explained where they arise.

CHAPTER 2: FUNDAMENTALS

Symbol	Parameter	Units
A	area of cross-section	m²
C	specific heat at constant pressure	J/(kg K)
E	exergy	J/s
f	friction factor	
G	mass velocity $(\dot{m}/A)$	kg/(m² s)
h	specific enthalpy	J/kg
ℓ	characteristic length	m
$\dot{m}$	mass flow rate	kg/s
N	overall number of transfer units $(US/\dot{m}C)$	
Ntu	larger value of N_h, N_c	
q	heat flow	J/s
Q	exchanger duty	W or J/s
R	ratio $M_w C_w/\dot{m}C$, see Appendix A	
U	overall heat transfer coefficient	J/(m² s K)
p	pressure	bar
S	reference surface area	m²
t	time	s
T	temperature	K
Tspan	temperature span of an exchanger	K
W	ratio of water equivalents $(W < 1)$	
x, y, z	length	m

Greek symbols

α	heat transfer coefficient	J/(m² s K)
γ	ratio of specific heats (C_p/C_v)	
Γ	constant	
δp	core pressure loss	N/m²
Δθ	temperature difference	K
ε	effectiveness	
θ	temperature difference	K
κ	thermal diffusivity ($\lambda/\rho C$)	m²/s
λ	thermal conductivity	J/(m s K)
ξ	normalised length	
ρ	density	kg/m³
τ	residence time	s

Subscripts

fg	latent heat
h, c, w	hot, cold, wall
m	mean
lim	limiting
lmtd	log mean temperature difference
loss	loss
1,2	ends of exchanger

CHAPTER 3: STEADY-STATE TEMPERATURE PROFILES

Symbol	Parameter	Units
A	area of cross-section	m²
C	specific heat at constant pressure	J/(kg K)
f	friction factor	
G	mass velocity ($\dot{m}/A$)	
H, C, W	finite difference temperatures (hot, cold, wall)	K
Kc, Ke	coefficients of contraction, expansion	
L	length of exchanger	m
$\dot{m}$	mass flow rate	kg/s
$\tilde{m}$	residence mass ($\tilde{m} = \dot{m}\tau$)	kg
M	mass of solid wall	kg
n	number of local transfer units ($\alpha S/\dot{m}C$)	
N	number of overall transfer units ($US/\dot{m}C$)	
p	pressure	bar
Q	exchanger duty	W or J/s
R	ratio $M_w C_w/\tilde{m}C$	

Symbol	Parameter	Units
s	curved length of an involute	m
S	surface area	m²
t	angle in radians for an involute	
tp	plate thickness	m
T	temperature	K
U	overall heat transfer coefficient	J/(m² s K)
v	specific volume	m³/kg
V	volume	m³
x, y	length	m
X, Y	normalised length ($X=x/L_x$, $Y=y/L_y$)	

Greek symbols

α	heat transfer coefficient	J/(m² s K)
Δp	core pressure loss	N/m²
ϵ	effectiveness	
η	absolute viscosity	kg/(m s)
κ	thermal diffusivity $\lambda/(\rho C)$	m²/s
λ	thermal conductivity	J/(m s K)
ξ	normalised length (x/L)	
ρ	density	kg/m³
σ	ratio $A_{\text{flow}}/A_{\text{frontal}}$	
τ	residence time	

Subscripts

h, c, w	hot, cold, wall
x, y	directions

Local parameters

A_0, A_1, A_2, A_3	defined in text
p, q, r_2, r_3	defined in text
a_{ij}	matrix coefficients
$\beta_1, \beta_2\ \alpha, \mu$	defined in text

CHAPTER 4: DIRECT SIZING OF PLATE-FIN EXCHANGERS

Symbol	Parameter	Units
a	individual cell flow areas	m²
b	plat spacing	m
c	cell pitch	m
C	specific heat	J/(kg K)

D	cell hydraulic diameter	m
E	edge length	m
f	flow friction coefficient	
G	mass velocity	kg/(m² s)
h, l, s, t	Manglic and Bergles parameters defined in text	
j	Colburn heat transfer coefficient	
L	flow length	m
$\dot{m}$	mass flow rate	kg/s
n	number of local transfer units	$\alpha s/(\dot{m}C)$
N	number of overall transfer units	$US/(\dot{m}C)$
p	pressure	bar
Per	cell pressure to	m
Q	exchanger duty	W or J/s
R	gas constant	J/(kg K)
tf	fin thickness	m
tp	plate thickness	m
ts	splitter thickness	m
U	overall heat transfer coeffieient	$J/(m^2 sK)$
x	strip length	m
z	number of cells	

Greek symbols

α	heat transfer coeffieient	$J/(m^2 sK)$
α, δ, γ	Manglic and Bergles ratios defined in text	
δp	core pressure loss	N/m²
$\delta\theta$	temperature difference	K
θ	temperature	K
κ	thermal diffusivity	m^2/S
λ	thermal conductivity	J/(m s K)
η	absolute viscosity	kg/(m s)
ρ	density	kg/m³

Subscripts

$h, c\ w$	hot, cold, wall
$lmtd$	log mean temperature difference
m	mean
1,2	ends of exchanger

Surface parameters

alpha	Stotal/Vexchr	1/m
beta	Stotal/Vtotal	1/m
gamma	Sfins/Stotal	
kappa	Stotal/Splate	
lambda	Sfins/Splate (kappa × gamma)	
sigma	Aflow/Aplate	

CHAPTER 5: DIRECT SIZING OF HELICAL-TUBE EXCHANGERS

Symbol	Parameter	Units
a	local area	m^2
A	total area	m^2
b	dimensional parameter	
C	specific heat	J/(kg K)
d	tube diameter	m
D	mandrel, wrapper, mean coil diameters	m
f	friction factor	
G	mass velocity	kg/(m^2 s)
ℓ	length of a single tube	m
ℓ_c	length of tubing in one longitudinal tube pitch	
ℓ_p	tubing in projected transverse cross-section	
$K1...K7$	factors defined in text	
L	length of tube-bundle	m
m	integer number of tubes in outermost coil	
$\dot{m}$	mass flow rate	kg/s
n	integer number of tubes in innermost coil	
N	total number of tubes in the exchanger	
p	longitudinal tube pitch	m
p	pressure	bar
P_y	shell-side porosity	
Q	exchanger duty	W or J/s
r	start factor (integer from 1 to 6)	
S	reference surface area	m^2
t	transverse tube pitch	m
T	temperature	K
u	velocity	m/s
U	overall heat transfer coefficient	J(m^2 s K)
V	volume	m^3
y	number of times shell-side fluid crosses a tube-turn	
z	integer number of tubes in intermediate coil	

Greek symbols

α	heat transfer coefficient	J/(m^2 s K)
Δp	core pressure loss	N/m^2
$\Delta\theta_{tmtd}$	log mean temperature difference	K
η	absolute viscosity	kg/(m s)
λ	thermal conductivity	J/(m s K)
ρ	density	kg/m^3
ϕ	helix angle of coiling	

Subscripts

a	annular
i	inside
m	mean
s, t, w	shell-side, tube-side, wall

Note: tube outside diameter (d) has no subscript, as this is the reference surface.

CHAPTER 6: DIRECT SIZING OF BAYONET-TUBE EXCHANGERS

Symbol	Parameter	Units
a, b	constants defined in equation (6.22)	
A, B	constants	
C	specific heat	J/(kg K)
d, D	diameter	m
ℓ	length of tube	m
L	length of exchanger	m
$\dot{m}$	mass flow rate	kg/s
N	number of transfer units ($N = US/\dot{m}C$)	
p	pressure	bar
P	perimeter transfer units ($P = N/L$)	1/m
Q	exchanger duty	W or J/s
s	spacing between two parallel flat plates	m
T	temperature	K
u	velocity	m/s
U	overall heat transfer coefficient	J/(m² s K)
x	distance	m
X	locus of minimum	m
Z	mean tube perimeter	m

Greek symbols

α, β	parameters defined in the text	
Δp	pressure loss	N/m²
ϵ	effectiveness	
η	absolute viscosity	kg/(m s)
θ	temperature for case of condensation	K
ϕ	function	

Subscripts

b, e	bayonet, external
i, o	inner, outer
min	minimum
1,2,3	defined in Figs. 6.1, 6.4, 6.5 and 6.8.

Symbol	Parameter	Units
Embellishments		
^	inner bayonet-tube fluid	
−	mean value	

CHAPTER 7: DIRECT SIZING OF RODBAFFLE EXCHANGERS

Symbol	Parameter	Units
a	flow area per single tube	m^2
A	total flow area	m^2
B	number of RODbaffles	
d	diameter	m
D	shell diameter	m
f	friction factor	
G	mass velocity	kg/(m^2 s)
k	baffle loss coefficient	
L	length	m
L_b	baffle spacing	m
$\dot{m}$	mass flow rate	kg/s
n	number of local transfer units	$\alpha S/(\dot{m}C)$
N	number of overall transfer units	$US/(\dot{m}C)$
p	tube pitch	m
Q	exchanger duty	W or J/s
r	baffle-rod radius	m
T	temperature	K
u	velocity	m/s
U	overall heat transfer coefficient	J(m^2 s K)
Z	number of tubes	
Greek symbols		
α	heat transfer coefficient	J/(m^2 s K)
Δp	core pressure loss	N/m^2
$\Delta \theta_{lmtd}$	log mean temperature difference	K
ϵ	surface roughness	m
η	absolute viscosity	kg/(m s)
λ	thermal conductivity	J/(m s K)
ρ	density	kg/(m s)
Subscripts		
b, p	baffle, plain	
s, t	shell, tube	

Terms from paper by Gentry et al.

C_L	coefficient in correlation $\text{Nu} = C_L(\text{Re}_h)^{0.6}(\text{Pr})^{0.4}(\eta_b/\eta_w)^{0.14}$ where $C_L = \xi_\ell C_\ell$	
C_T	coefficient in correlation $\text{Nu} = C_T(\text{Re}_h)^{0.6}(\text{Pr})^{0.4}(\eta_b/\eta_w)^{0.14}$ where $C_T = \xi_t C_t$	
C_1, C_2	coefficients in correlation $k_b = \phi(C_1 + C_2/\text{Re}_b)$	
D_{bi}	exchanger baffle-ring inner diameter	m
D_{bo}	exchanger baffle-ring outer diameter	m
D_o	exchanger outer tube limit	m
D_s	shell inner diameter	m
ξ_l, ξ_t	expressions defined in papers by Gentry et al.	

CHAPTER 8: TRANSIENTS IN CONTRAFLOW EXCHANGERS

Symbol	Parameter	Units
C	specific heat	J/(kg K)
L	length	m
m	number of space increments in exchanger length	
$\dot{m}$	mass rate of flow	kg/s
$\tilde{m}$	residence mass of fluid	kg
M	mass of exchanger	kg
n	number of local transfer units ($\alpha S/\dot{m}C$)	
N	number of overall transfer units ($US/\dot{m}C$)	
P	ratio of transit times (τ_c/τ_h)	
R	ratio of thermal capacities ($M_w C_w/\tilde{m}C$)	
S	reference surface area	m²
t	time	s
T	scaled time	
u	velocity	m/s
U	overall heat transfer coefficient	J/(m² s K)
x	distance	m
X	normalised distance	
z	number of time intervals in the disturbance	

Greek symbols

α	heat transfer coefficient	J/(m² s K)
α, β, γ	characteristic directions	
Δ	increment	
κ	thermal diffusivity	m²/s
θ	temperature	K
τ	transit or residence time	s

Symbol	Parameter	Units
Subscripts		
h, c, w	hot, cold, wall	
Embellishments		
ˆ	perturbance	
¯	steady-state	
Parameters in Fox's solution for characteristics		
u, v	dependent variables	
x, y	independent variables	
A, B, C, D	parameters, here constants	
α, β, γ, δ	parameters, here constants	

CHAPTER 9: TRANSIENTS IN CROSS-FLOW EXCHANGERS

Symbol	Parameter	Units
A, B, C, D	constants in the transformation equations	
L	length	m
n	number of local transfer units ($\alpha S/\dot{m} C$)	
N	number of overall transfer units ($US/(\dot{m}C)$)	
P	ratio of transit times (τ_c/τ_h)	
R	ratio of thermal capacities, ($M_w C_w/\tilde{m}C$)	
t	time	s
T	scaled time	
u	velocity	m/s
x, y	dimension	m
X, Y	normalised length	
Greek symbols		
α	heat transfer coefficient	J/(m² s K)
α, β, γ	characteristic directions	
Δ	increment	
κ	thermal diffusivity	m²/s
θ	temperature	K
τ	transit or residence time	s
Subscripts		
h, c, w	hot, cold, wall	
Embellishments		
ˆ	perturbance	
¯	steady-state	

CHAPTER 10: SINGLE-BLOW TESTING AND REGENERATORS

Symbol	Parameter	Units
a	arbitrary radius	m
a_0, a_1, b_1	numerical constants	
B	mean solid temperature excess $(\theta_b - \theta_i)$	K
$B^\#$	non-dimensional ratio (B_2/G_1)	
C	specific heat	J/(kg K)
D	non-dimensional inlet disturbance	
G	mean fluid temperature excess $(\theta_g - \theta_i)$	K
$G^\#$	non-dimensional ratio (G_2/G_1)	
α	heat transfer coefficient	J/(m²s K)
k	numerical constant	
L	length of matrix	m
$\dot{m}$	mass flow rate of gas	kg/s
$\tilde{m}_g$	mass of gas in matrix	kg
M_b	mass of matrix	kg
Ntu	number of transfer units (one local value only)	
r	radius	m
R_g	ratio $M_b C_b / \tilde{m}_g C_g$	
s	Laplace transform image of t	
S	surface area	m²
t	time	s
t^*	time constant of inlet exponential temperature disturbance	
u	gas velocity defined as $\dot{m}_g L / \tilde{m}_g$	m/s
V	volume of solid matrix	m³
x	distance into matrix	m

Greek symbols

β	ratio τ/Ntu	
$\delta()$	delta function	
θ	temperature	K
σ	dummy variable	
τ	non-dimensional time	
τ^*	non-dimensional time constant	
ω	rotational speed	1/s
ϖ	non-dimensional rotational speed	

Subscripts

b	bulk solid
h, c	hot, cold
g	gas

Symbol	Parameter	Units
i	initial isothermal reference state	
s	surface	
w	wall	
1, 2	inlet, outlet	

CHAPTER 11: CRYOGENIC HEAT EXCHANGERS AND STEPWISE RATING

Symbol	Parameter	Units
a, b	arbitrary limits	
c	sonic velocity	m/s
C	specific heat at constant pressure	J/(kg K)
h	specific enthalpy	J/kg
k	number of stages of compression	
p	pressure	bar
Q	heat flow	W or J/s
r	compression ratio	
R	gas constant	J/(kg K)
S	entropy	J/(kg K)
T	temperature	K
W	work	W or J/s
x, y	fractions	

Greek symbols

α	blade angle, preferred notation for gas turbines	
γ	isentropic index (C_p/C_v)	
θ	angle	
η	efficiency	

Subscripts

e, n, o, p	equilibrium, normal, *ortho-*, *para-*, (forms of hydrogen)
fg	saturation field
min	minimum
s	isentropic
0	dead state
0, 1, 2, 3	stations in radial turbine analysis

Embellishments

-	mean value

CHAPTER 12: VARIABLE HEAT TRANSFER COEFFICIENTS

Symbol	Parameter	Units
a	numerical constant	
A	area for flow	m^2
B	numerical constant	
c	numerical constant	
C	numerical parameter depending on flow condition	
d	tube diameter	m
E, F, H	parameters in Friedel's correlation	
f	friction factor	
Fl	heat flux	W or J/s
g	acceleration due to gravity	m/s^2
G	mass velocity	$kg/(m^2\ s)$
ℓ	length	m
m	numerical constant	
$\dot{m}$	mass flow rate	kg/s
n	numerical constant	
q	heat flow	W or J/s
T	temperature	
x	dryness fraction	
X^2	ratio defined in text	

Greek symbols

α	heat transfer coefficient	$J/(m^2\ s\ K)$
$\Delta\ell$	length increment	m
Δp	pressure loss	N/m^2
η	absolute viscosity	$kg/(m\ s)$
ρ	density	kg/m^3
σ	surface tension	N/m
ϕ	two-phase flow multiplier	

Subscripts

$crit$	critical
f	liquid
fg	saturation
g	vapour
tp	two-phase

APPENDIX A TRANSIENT EQUATIONS WITH LONGITUDINAL CONDUCTION AND WALL THERMAL STORAGE

Symbol	Parameter	Units
A	wall cross-section for longitudinal conduction	m²
A	flow area	m²
C	specific heat	J/(kg K)
f	friction factor	
L	length	m
$\dot{m}$	mass rate of flow	kg/s
$\tilde{m}$	residence mass of fluid	kg
M	mass of exchanger	kg
n	number of local transfer units ($\alpha S/\dot{m} C$)	
r	radius	m
R	ratio of thermal capacities ($M_w C_w / \tilde{m} C$)	
S	reference surface area	m²
t	time	s
T	temperature	K
u	velocity	m/s
U	overall heat transfer coefficient	J/(m² s K)
V	total volume of the wall	m³
x, y	distance	m

Greek

α	local heat transfer coefficient	J/(m² s K)
κ	thermal diffusivity	m²/s
ρ	density	kg/m³
τ	transit or residence time	s

Subscripts

h, c, w	hot, cold, wall
x, y	directions

AUTHOR INDEX

Abadzic, E.E. 1, 14, 122, 133, 134, 154
Acklin, L. 317
Acrivos, A. 230, 317
Adams, L.H. 45, 50
Adderley, C. 14
Afgan, N. 288, 370
Afgan, N.H. 369, 370
Alberda, G. 260
Alfeev, V.N. 265, 288
Alves, G.E. 49, 50
Ames, W.F. 208, 217, 228, 230, 242, 325
Amundson, N.L. 254, 259
Andrews, J.R.G. 183, 185
Anzelius, A. 244, 259
Arpaci, V.S. 259, 318
Ashmantas, L.-V.A. 15, 205, 231, 371
Aung, W. 370
Austergard, A. 203, 204

Bachman, U. 2, 15
Baclic, B.S. viii, 66, 68, 70, 87, 88, 89, 93, 117, 259
Bailey, R.V. 230
Baines, N.C. 273, 289
Baker, N.S. 231
Bannister, R.L. 12, 15, 229, 230
Barrington, E.A. 206
Barozzi, G.S. 185
Barron, R.F. 117, 263, 276, 288
Bartsch, G. 309
Beckman, L. 230
Bell, K.J. 189, 203, 204, 246, 251, 259
Bejan, A. 49, 50, 88, 89, 134, 154, 257, 259, 370
Bergelin, O.P. 189, 203, 204

Bergles, A.E. 93, 94, 113, 114, 116, 118, 132, 154, 203, 205, 291, 292, 306, 307, 309, 370
Bes, Th. 14, 15
Biancardi, F.R. 317
Boland, D. 89
Bott, T.R. xii, 16, 50, 89, 205, 308, 371
Boucher, D.F. 49, 50
Bougard, J. 288, 370
Bourguet, J.M. 1, 15
Bracha, M. 279, 287, 288
Brewer, R.D. 287, 288
Brinkley, S.R. 255, 259
Brockmeier, U. 94, 117
Brodyansky, V.M. 288
Brown, A. 259
Brum, N.C.L. 185
Burns, D. 260
Bush, J.E. 253, 254, 262
Butterworth, D. 203, 204, 371

Calligeros, J.M. 49, 50
Campbell, J.F. 101, 109, 117, 229, 230
Carey, V.P. 291, 307, 371
Carlslaw, H.S. 361
Carver, M.B. 224, 230
Catchpole, J.P. 49, 50
Cesna, B. 370
Chapman, A.J. 117
Chawala, J.M. 302, 303, 307
Chen, C.C. 305, 307
Chen, H.-T. 241, 242, 322
Chen, J.C. 296, 307
Chen, K.-C. 241, 242, 322
Chen, N.H. 193, 204
Chen, Y.N. 2, 15, 133, 154
Cheruvu, N.S. 15, 230

Chester, M. 213, 230
Chiou, J.P. 74, 76, 89, 118, 255, 259
Chisholm, D. 291, 292, 293, 294, 307, 370
Choi, H.Y. 309
Chu, C.M. 306, 307
Churchill, S.W. 109, 118, 193, 204, 205, 213, 228, 230, 306, 307
Cima, R.M. 317
Clapeyron, B.P.E. xi, xii
Clark, J.A. 255, 259, 317
Clarke, J.M. 118
Clarke, R.H. 305, 307
Clayton, D.G. 23, 50
Clenshaw, C.W. 259, 361
Close, D.J. 246, 259
Cohen, W.C. 317
Collatz, L. 230, 242
Collier, J.G. 291, 292, 293, 294, 306, 307, 369, 371
Collinge, K. 12, 15
Coombs, B.P. 1, 16, 121, 132, 133, 155, 244, 246, 253, 255, 259, 262
Coppage, J.E. 254, 259
Cotta, R.M. 185
Courant, R. 242, 362, 365
Cowans, K.W. 305, 307
Cownie, J. 12, 15, 229, 230
Crisalli, A.J. 12, 14, 15, 229, 230
Crump, K.S. 226, 230

Dabora, E.K. 254, 259
Danilov, Yu. I. 203, 205, 371
Date, A.W. 306, 307
Davids, N. 260, 361
Daugherty, R.L. 201, 205
Das, S.K. 14, 15, 88, 89, 227, 229, 231, 232
de Boor, C. 345
DeGregoria, A.J. 259
Delhaye, J.M. 307
de Reuck, K.M. 368
DeWitt, D.P. 118
Down, W.S. 259
Dreitser, G.A. 15, 205, 231, 371
Dukler, A.E. 292, 303, 309
Duncan, W.J. 201, 205
Duggan, R.C. 258, 262
Dugundji, J. 49, 50
Duhamel, J.M.C. xi, xii
Dusinberre, G.M. 318

Dzyubenko, B.V. 6, 15, 203, 205, 209, 231, 371

Echigo, R. 8, 15
Eilers, J.F. 5, 15, 206
El-Wakil, M.M. 255, 259
ENEA 152, 154
Evans, F. 242, 322
Evans, R. 258, 259

Fernandez, E.O. 370
Fernandez, J. 90
Fiebig, M. 117
Finniemore, E.J. 205
Fischer, K.F. 87, 89
Flannery, B.P. 232, 345
Flower, J.R. 89
Focke, W.W. 14, 5
Foumeny, E.A. 258, 259, 307, 371
Fox, L. 214, 231, 242
Fox, T. 110, 111, 118
Franzini, J.B. 205
Friedel, L. 292, 303, 307
Fujii, M. 309
Fulford, G. 49, 50
Furnas, C.C. 259, 361

Ganic, E.N. 309, 370
Garabedian, P. 242
Gartner, J.R. 322
Gentry, C.C. 5, 15, 187, 188, 189, 195, 197, 200, 205, 206
Gill, G.M. 15, 122, 154
Gilli, P.V. 1, 15, 121, 151, 154
Ginsburg, Ya.L. 181, 185
Gnielinski, V. 132, 141, 154, 201, 205
Golkar Narandji, M.R. 233
Grassman, P. 32, 35, 37, 50, 118, 272, 288
Gregorig, R. 369
Groothuis, H. 298, 308
Guedes, R.O.C. 185
Guentermann, Th. 117
Gungar, K.E. 308
Gurukul, S.M.K.A. 111, 119, 287, 289
Guy, A.R. 89, 204
Gvozdenac, D.D. 87, 89, 231, 242, 317, 322

Hahne, E. 308
Hallgren, L.H. 14
Hall-Taylor, M.S. 291, 308, 369

Hamming, R.W. 260
Hampson, N. 121, 154
Hanamura, K. 15
Handley, D. 260
Harmon, W. 318
Harris, J.A. 228, 232
Harrison, H.L. 322
Harrison, G.S. 15, 154
Hart, J.A. 246, 260
Hartnett, J.P. 309, 370
Hartzog, D.G. 1, 2, 16, 111, 119, 122, 155, 289, 309
Haselden, G.G. 263, 288
Haseler, L.E. 93, 96, 110, 111, 118, 227, 231, 284, 287, 288, 308
Hashem, A. 371
Hausen, H. 1, 3, 15, 89, 121, 154, 211, 228, 231, 242, 244, 258, 260, 290, 308, 369, 370
Heggs, P.J. 66, 68, 89, 93, 117, 259, 260, 307, 371
Hendal, W.P. 296, 308
Herbein, D.S. 88, 89, 101, 118
Hernandez-Guerro, A. 185
Herron, D.H. 224, 231
Hesselgreaves, J. 8, 16, 105, 118, 188, 201, 202, 205
Hestroni, G. 291, 308, 370
Hewitt, G.F. x, xii, 14, 16, 24, 50, 86, 89, 203, 205, 291, 292, 296, 305, 307, 308, 369, 370, 371
Hext, G.R. 262
Hilbert, D. 242, 362, 365
Hildebrand, F.B. 362, 365
Hill, A. 258, 260
Hill, R. xi, xii
Hinchcliffe, C. 262
Hindmarsh, E. 89
Hindsworth, F.R. 262
Hodgkins, W.R. 306, 308
Holger, M. 7, 16
Hovanesian, J.D. 49, 50
Howard, C.P. 251, 254, 260, 261, 369
Huang, Y.M. 231
Huebner, H. 261
Hurd, N.L. 7, 16, 156, 185
Hutchinson, D. 251, 261

Ichikawa, S. 226, 231
Ievlev, V.M. 203, 205, 371
Idlechik, I.E. 181, 185

Incropera, F.P. 118
Ito, H. 132, 142, 154
Ivantsov, A. 288
Izumi, R. 243, 323

Jackson, P. 259
Jacquot, R.G. 226, 231, 242
Jaeger, J.C. 345
Jain, M.K. 242
Jakob, M. viii, xii, 87, 89
Jaswon, M.A. 231, 317
Jeffrey, A. 224, 231
Jegede, F.O. 206
Jendrzejczyk, J.A. 50
Jensen, M.K. 132, 154
Johnson, E.F. 317
Johnson, J.E. 254, 260
Jones, N. 49, 50
Joye, D.D. 185

Kabel, R.L. 306, 308
Kabelac, S. 231
Kahn, A.R. 231
Kakac, S. 370
Kalin, W. 2, 16
Kanevets, G.Ye. 133, 154
Katz, E.F. 246, 251, 259
Kays, W. 4, 16, 89, 93, 94, 97, 113, 114, 115, 117, 118, 205, 370
Kelkar, K.M. 105, 118
Kerschbaumer, H.G. 306, 308
Kern, D.Q. 203, 205, 369
King, J.L. 1, 16, 122, 132, 155, 244, 262
Kishima, A. 226, 231
Klockzien, V.G. 49, 50
Knusden, H. 205
Kohlmayer, G.F. 247, 251, 260
Kondepudi, S.N. 306, 308
Kopp, J. 32, 35, 37, 50, 118, 272, 288
Koppel, L.B. 317
Kotas, T.J. 370
Kowalski, H.C. 49, 50
Kramers, H. 260
Krapp, R. 289
Kraus, A.D. 369, 371
Kreith, F. 118
Kroeger, P.G. 59, 89, 118, 173, 179, 185, 211, 231

L'Air Liquide 1, 16, 121, 154
Laird, A.D.K. 293, 294, 307

Lakshmann, C.C. 228, 231, 232
Lamé, G. xi, xii
Larson, M.A. 318
Laubli, F. 317
Law, V.J. 230
Lee, D. 206, 225, 232
Lee, E.H. xii
Le Feuvre, R.F. 134, 154
Leighton, M.D. 204
Levinson, A. 260, 361
Lessen, M. 49, 50
Liang, C.Y. 246, 260
Liao, N.S. 233
Lindeman, C.F. 232, 242, 322
Linnhoff, B. 86, 89, 90
Little, D.A. 15, 230
Little, W.A. 265, 289
Lock, G.S. 371
Locke, G.L. 251, 260
Lockhart, R.W. 308
Loh, J.U. 307
London, A.L. viii, xii, 4, 16, 89, 93, 94, 97, 113, 114, 115, 117, 118, 205, 254, 259, 317, 369, 370
Lorenz, G. 288
Lowan, A.N. 260, 361
Lou, X. 232
Lu, Y.-D. 206
Luikov, A.V. 41, 50

MacDonald, C.F. 16, 111, 118, 369
Machielson, C.H.M. 306, 308
Macias-Machin, A. 185
Mack, W.M. 119
McAdams, W.H. 242
McCarty, R.D. 368
McNaught, J.M. 306, 307, 308
McQuiggan, G. 15, 230
Makita, T. 280, 289
Manglic, R.M. 93, 94, 113, 114, 116, 118, 203, 205
Manson, J.L. 89
Mansour, A. 286, 289
Marshall, J. 289
Marsland, R.H. 89
Martin, H. 7, 16, 371
Martinelli, R.C. 308
Marumoto, K. 309
Mashelkar, R.A. 370
Massoud, M. 232
Masubuchi, M. 317

Mathews, J. 227, 232, 362, 365
Mayinger, F. 307, 370
Meade, R. 251, 261
Meek, R.M.G. 246, 251, 255, 260, 261
Mehravaran, K. 233
Mesarkishvilli, Z. 309
Metzger, D. 371
Miller, D.S. 181, 185
Minkowycz, W.J. 370
Mitchell, J.W. 232, 242, 243, 317, 322
Mollekopf, N. 287, 289
Mondt, J.R. 246, 261
Mori, H. 15
Mori, Y. 132, 142, 154
Morrison, F.A. 49, 50
Moyle, M.P. 259
Mozley, J.M. 317
Mulready, R.C. 2, 16
Myers, G.E. 232, 240, 241, 242, 243, 322

Næss, E. 203, 204, 205
Nakayama, W. 132, 142, 154
Nelder, J.A. 251, 261
Nesselman, K. 211, 232
Nicholls, J.A. 259
Nikolsky, V.A. 288
Norman, R.F. 242, 322
Nusselt, W. 68, 87, 90, 244, 261

Oakey, J.D. 289
Obot, N.T. 49, 50
Ockendon, J.R. 306, 308
Ogawa, K. 306, 308, 309
O'Neal, D.L. 306, 308
Ontko, J.S. 228, 232
Organ, A.J. 258, 261, 371
Ozisik, M.N. 132, 154

Paffenbarger, J. 287, 289
Pagliaini, G. 185
Pahlevanzadeh, H. 258, 259
Palen, J. 370
Paliwoda, A. 308
Panjeh Shahi, M.H. 206
Parker, M.L. 12, 14, 15, 229, 230
Parkingson, J.M. 251, 261
Patankar, S.V. 105, 119
Pate, M.B. 309
Patzelt, A. 288
Paugh, R.L. 265, 289
Paynter, H.M. 317

Perkeris, C.L. 306, 308
Perrin, A.J. 2, 16
Peschka, W. 289
Petersen, U. 287, 289
Pfeiffer, S. 261
Phadke, P.S. 6, 7, 16, 197, 201, 204, 206
Phillips, R. 259
Picon Nunez, M. 206
Piersall, C.H. 261
Pletcher, R.H. 370
Politykina, A.A. 133, 154
Polley, G.T. 206
Potter, O.E. 228, 231, 232
Prakash, C. 119
Prasad, B.S.V. 93, 111, 119, 287, 289
Press, W.H. 226, 232, 345
Probert, S.D. 258, 259
Profos, O. 2, 16
Pucci, P.F. 246, 261

Razelos, P. 258, 261
Rehme, K. 194, 206
Rektorys, K. 362, 365
Rhee, B.-W. 291, 302, 308
Rhodine, C.N. 231, 242
Ringer, D.U. 287, 289
Rix, D.H. 261
Rizika, J.W. 241, 243
Robertson, J.M. 305, 307, 309
Roetzel, W. 14, 15, 87, 89, 88, 90, 206, 225, 227, 228, 229, 231, 232, 233, 234, 371
Romie, F.E. 233, 241, 243, 258, 261, 323
Rosenberg, D.U. 224, 231
Rohsenow, W.M. 88, 89, 101, 109, 117, 118, 119, 229, 230, 291, 309, 370

San, J.-Y. 243, 258, 261, 323
Sangster, W.A. 206
Saunders, E.A.D. 370
Savkin, N. 309
Schack, A. 290, 309
Schlager, L.M. 309
Schlichter, L.B. 306, 308
Schlichting, H. 47, 50
Schlunder, E.U. 309, 369
Schmidt, F.W. 258, 261, 370
Schneider, G.E. 370
Scholz, H.W. 1, 14
Schumann, T.E.W. 244, 261
Scott, D.M. 262

Scott, R.B. 263, 289
Seban, R.A. viii, xii
Sekulic, D.P. 88, 90, 371
Seshimo, Y. 306, 309
Shah, M.M. 309
Shah, R.K. xiii, xii, 4, 16, 78, 87, 90, 93, 94, 110, 113, 115, 117, 118, 119, 233, 296, 369, 370, 371
Shannon, R.L. 49, 50
Sharifi, F. 209, 227, 228, 233
Shearer, C.J. 251, 261
Shen, C.M. 258, 262
Shires, G.L. xii, 16, 50, 89, 205, 308, 371
Siegla, D.C. 261
Skok, N. 308
Skrinska, A. 370
Small, W.M. 5, 15, 205, 206
Smith, D.M. 87, 90
Smith, E.M. xi, xii, 1, 16, 90, 118, 121, 122, 132, 133, 155, 185, 187, 206, 244, 246, 255, 262, 269, 289
Smith, R.A. 291, 296, 309, 370
Smith, W. 231, 242, 317, 318
Smyth, R. 206
Soanes, R.V. 345
Soland, J.G. 96, 118
Sonju, O.K. 204, 205
Soyars, W.M. viii, xii, 283, 289
Spalding, D.B. 50
Spang, B. 87, 90, 233, 234
Sparrow, E.M. 370
Spendley, W. 251, 262
Spiga, G. 241, 243, 323
Spiga, M. 241, 243, 323
Spindler, K. 308
Spinner, I.H. 217, 233, 317, 318
Stang, J.H. 253, 254, 262
Stansell, J. 287, 289
Stasiulevicius, J. 370
Steadman, J.W. 231, 242
Stephan, K. 371
Stehfest, H. 233
Stermole, F.J. 318
Stevens, R.A. 70, 87, 90
Subbarao, E.C. 370
Suessman, W. 286, 289
Sun, S.-Y. 206
Survila, V. 370
Szomanski, E. 246, 260

Taborek, J. 25, 50, 370

Taitel, Y. 292, 303, 309
Takahashi, Y. 317
Takeshita, M. 308
Tan, K.S. 217, 233, 317, 318
Tanaka, N. 308
Taniuti, T. 224, 231
Taylor, G.I. 213, 233
Taylor, M.A. 111, 119
Teukolsky, S.A. 232, 345
Thom, A.S. 205
Thome, J.R. 307, 371
Thomas, B.E.A. 89
Todo, I. 185, 317, 318
Tong, L.S. 194, 206
Topakoglu, H. 132, 154
Touloukian, Y.S. 280, 289, 366, 368
Townsend, D.W. 89, 90
Treadwell, K.M. 259, 318
Tupper, S.J. xii

Ulken D. 16
Ulinskas, R. 134, 155, 371
Underwood, A.J.V. 87, 90
Usagi, R. 109, 118

Valenti, M. 12, 16, 229, 233
Van den Bulck, E. 258, 262
Vargaftik, N.B. 280, 289, 366, 368
Vettering, W.T. 232, 345
Vilemas, J. 370
von Rosenberg, D.U. 230

Wadekar, V.V. 305, 309
Wagner, W. 368
Walker, G. 370
Walker, M.A. 15, 154
Walker, R.L. 227, 232, 362, 365
Wallis, G.B. 291, 292, 309, 369
Wambsganss, M.W. 50
Wang, B.-X. 370

Wang, C.C. 233
Wang, J.-W. 183, 185
Wanner, M. 288
Ward, P.W. 185
Wardle, A.P. 231
Webb, R.L. 109, 113, 114, 119, 371
Weimer, R.F. 1, 2, 16, 111, 119, 122, 155, 289, 309
Welkey, J.J. 204
Westwater, J.W. 307
White, F.M. 201, 206
Whitfield, A. 273, 289
Williamson, E.D. 45, 50
Willmott, A.J. 233, 258, 260, 261, 262, 370
Wilson, D.G. 12, 16
Winterton, R.H.S. 308
Wolf, J. 233
Woodward, C.H. 345
Woolf, J.R. 90
Worek, W.M. 258, 262
Wursig, G. 289
Wylie, C.R. 79, 90

Xuan, Y. 225, 228, 232, 233, 234

Yagodin, V.M. 288
Yamaguchi, S. 243, 323
Yamashita, H. 241, 243, 323
Yan, C.-Q. 206
Yang, W.J. 246, 260, 318
Yao, L.S. 132, 155
Yoshida, H. 15
Young, A.D. 205, 206
Young, E.H. 308
Young, J.B. 346

Zahn, W.R. 302, 309
Zaleski, T. 185
Zukauskas, A.A. 134, 155, 371

SUBJECT INDEX

Algorithms 326
Annular mist flow 292, 297
Annular no-mist flow 292, 297
Axial conduction 59
Axial conduction for the fluids 211

Bayonet-tube exchangers 156
  bayonet-end pressure loss 181
  condensation 168
  design illustration 169
  direct-sizing 181
  evaporation 157
  explicit non-isothermal solution 173
  isothermal shell-side 156
  non-explicit non-isothermal solutions 179
  non-isothermal shell-side 171
Boiling 296
Buffer zone 111
By-pass control 153

Calculus of variations 362
Carnot efficiency above and below the dead-state 36, 264
Classification of exchangers 1
  bayonet-tube 7
  helical-tube 1
  helically twisted flattened tube 6
  involute curved plate-fin 10
  involute curved tube-panel 3
  plate-fin 4
  RODbaffle 5
  spirally wire-wrapped 6
  wire-woven heat exchangers 8
Compressors 269
Condensation 306
Contra-flow heat exchangers 19
  allowance for longitudinal conduction 59, 65
Conversion factors 372
Core pressure loss 86
Cross-conduction effect 284
Cross-flow heat exchangers 65
  determined and undetermined 83
  longitudinal conduction 74
  mean TD in single-pass 65
  mean TD in two-pass 70
Cryo-expanders 270
  diatomic molecules 266
  effect of pressure ratio on cooling temperature range 272
  monatomic molecules 266
Cryogenic heat exchangers 263
  multi-stream exchangers 281, 283
  three-stream design 111
Cryogenic storage tank 12
  bayonet-tube exchanger 12
  'roll-over' problem 12
Cubic spline fitting 344

Data fitting 344
Dehumidification 306
Differential equations for temperature transients in contra-flow 313
Differential equations for temperature transients in unmixed cross-flow 318
Differential equations for mass and temperature transients in crossflow 323
Dig deeper (to) 38
Dimensionless groups 40
Direct-sizing of heat exchangers viii
  typical input/output 334
Distribution headers 110

Double-tube heat exchanger 298

Effectiveness and Ntu 25
   (ε-Ntu) and (LMTD-Ntu)
      comparison 30
   (ε-Ntu) generalised plot for contra-flow
      and parallel-flow 28
   (ε-Ntu) rating problem 29
   (ε-Ntu) sizing problem 29
Evaporation 291
Exclusions and exceptions 13
Exergy destruction 88
Exponential spline fitting 345

Flow-friction data fitting 345
Flow mal-distribution 305
Friedel's two-phase pressure loss 303
Frosting 306
Fundamentals of heat exchangers 17
   condenser 17, 27, 58
   contra-flow 18, 51, 53
   cross-flow 65, 70
   desuperheating feed-heater 18
   effectiveness 25
   evaporator 17, 28
   log mean temperature difference 19
   parallel flow 18, 58
   rating problem 21, 29
   sizing problem 23, 29
   temperature cross-over 18, 70
   values of Ntu required in cryogenics 36

Gas turbine propulsion systems 12
   compressor inter-cooler 12, 229
   recuperator 101, 229
Gaussian quadrature 360
Grassman & Kopp 32, 363
   most efficient temperature difference in
      contra-flow 32, 363

Heat transfer data fitting 345
Helical-tube exchangers 1, 120
   basic geometry 122
   design for curved tubes 146
   direct-sizing 131
   fine-tuning 141
   simplified geometry 129
   thermal design 131
Hydrogen liquefaction 265
   catalysts 269
   equilibrium hydrogen 266
   normal hydrogen 265

   ortho-hydrogen 265
   para-hydrogen 265
   spins of protons 266

Ice harvesting 306
Icing 306
Intercooler 12, 229
Involute-curved tube panel 3
Involute plate-fin exchangers 10, 73
Inward radial flow turbines 270

Kroeger's method 59
   contraflow longitudinal conduction 59, 211

Labelling of exchanger ends xi
Laplace transforms 357
Leakage buffer 111
Leakage plate 'sandwich' 111
Liquefaction plant 263
   hydrogen 279
   nitrogen 273
LMTD 19
   (LMTD-Ntu) and (ε-Ntu)
      comparison 30
   (LMTD-Ntu) method 19
   (LMTD-Ntu) rating problem 21
   (LMTD-Ntu) sizing problem 23
LMTD reduction factor 65
Lockhart–Martinelli, two-phase flow
   pressure loss 292
Longitudinal conduction 59, 74

Method of characterstics 214, 238
Mist flow 297
Most efficient temperature difference in
   contraflow 32
Multi-stream heat exchangers 281, 283
   cross-conduction effect 284
   three-stream design 111

Nitrogen liquefaction 273
Nucleate boiling 296

Optimisation of rectangular offset-strip
   plate-fin surfaces 104, 347
   block length 350
   block mass 349
   block volume 348
   frontal area 351
   Total surface area 352

Pinch technology 85

Plate-and-frame exchangers 227
Plate-fin heat exchangers 91
  direct-sizing of contra-flow 98
  direct-sizing of unmixed cross-flow 97
  fine-tuning of rectangular strip-fins 104
  flow-friction correlations 93
  heat transfer correlations 93
  involute-curved 10, 73
  one-pass unmixed cross-flow 91, 97
  optimisation of a contra-flow exchanger 106
  plate-fin surface geometry 95
  two-pass unmixed cross-flow 70
Plate-fin and tube exchangers 306
Pressure loss in two-phase flow 292, 303
Properties of materials 366

Regenerators 257
RODbaffle exchangers 187
  characteristic dimensions 189
  configuration 5, 188
  design correlations 191
  direct-sizing 196
  flow areas 190
  further heat transfer and flow friction data 353
  heat transfer 192
  Phadke tube count 197
  pressure-loss shell-side 194
  pressure-loss tube-side 193
  recommendations 203
  tube bundle diameter 197
'Roll-over' 12, 185, 305

Schematic algorithms 326
  direct-sizing in contra-flow 331
  direct-sizing in cross-flow 330
  mean TD in unmixed cross-flow 326
Shell-and-tube exchangers 203
  small tube inclinations 227
Similarity 41
Single-blow testing 244
  analysis of coupled equation 246
  choice of test method 255
  exponential disturbance 249, 253
  harmonic disturbance 249, 254
  Laplace transforms 357
  longitudinal conduction 256
  numerical evaluation of integrals 357
  physical assumptions 245
  step disturbance 249, 250
  outlet response curves 251
Single-pass cross-flow 65, 97
Sizing when Q not specified 31
Small tube inclinations 227
Source books 369
Spline-fitting 344
Steady-state temperature profiles 51
  condensation 58
  contra-flow 53
  evaporation 58
  linear profiles 51
  parallel flow 53
Stepwise rating 283
Stratified flow 297

Taut spline fiting 344
Temperature crossover 18, 70
Thermal storage in wall 214
'Theta' method 24
Time constant 359
To dig deeper 38
Transients in contra-flow 207
  algorithm 343
  axial dispersion 213, 225, 227
  characteristics 218
  designing-out dispersion 225
  direct solution by finite-differences 208, 228, 323
  disturbances in both fluids 223
  engineering applications 229
  fundamental equations 313
  Laplace transforms with numerical inversion 208, 225
  method of characteristics 214, 238
  normalised disturbances 211
  perturbance equations 211
  practical example 220
  scaling and normalisation 210
  temperature calculation scheme 217
  transformation 216
Transients in cross-flow 235
  engineering applications 241
  fundamental equations 318
  method of characteristics 238
  perturbance equations 237
  review of existing solutions 240
  Scaling and normalisation 236
  transformations 239
Transient solutions by Fast Fouricr transform 226

Transient solutions by finite differences 228
Twisted-tube heat exchangers 6, 203
Two-pass unmixed cross-flow 70
Two-phase flow heat transfer 290
Two-phase flow pressure loss 292, 303

Variable heat transfer coefficients 290
Variable power curve fitting 345

Wall, axial/longitudinal conduction 59, 211, 313